Fly Ash and Coal Conversion By-Products: Characterization, Utilization and Disposal III

MATERIALS RESEARCH SOCIETY SYMPOSIA PROCEEDINGS

ISSN 0272 - 9172

MATERIALS RESEARCH SOCIETY SYMPOSIA PROCEEDINGS

MATERIALS RESEARCH SOCIETY SYMPOSIA PROCEEDINGS

MATERIALS RESEARCH SOCIETY SYMPOSIA PROCEEDINGS

MATERIALS RESEARCH SOCIETY CONFERENCE PROCEEDINGS

VLSI-I—Tungsten and Other Refractory Metals for VLSI Applications, R. S. Blewer, 1986; ISSN: 0886-7860; ISBN: 0-931837-32-4

VLSI-II—Tungsten and Other Refractory Metals for VLSI Applications, E.K. Broadbent, 1987; ISSN: 0886-7860; ISBN: 0-931837-66-9

TMC—Ternary and Multinary Compounds, S. Deb, A. Zunger, 1987; ISBN:0-931837-57-x

MATERIALS RESEARCH SOCIETY SYMPOSIA PROCEEDINGS VOLUME 86

Fly Ash and Coal Conversion By-Products: Characterization, Utilization and Disposal III

Symposium held December 1-3, 1986, Boston, Massachusetts, U.S.A.

EDITORS:

Gregory J. McCarthy
North Dakota State University, Fargo, North Dakota, U.S.A.

Fredrik P. Glasser
University of Aberdeen, Aberdeen, Scotland, U.K.

Della M. Roy
The Pennsylvania State University, University Park, Pennsylvania, U.S.A.

Sidney Diamond
Purdue University, West Lafayette, Indiana, U.S.A.

|M|R|S| MATERIALS RESEARCH SOCIETY
Pittsburgh, Pennsylvania

CAMBRIDGE UNIVERSITY PRESS
Cambridge, New York, Melbourne, Madrid, Cape Town,
Singapore, São Paulo, Delhi, Mexico City

Cambridge University Press
32 Avenue of the Americas, New York NY 10013-2473, USA

Published in the United States of America by Cambridge University Press, New York

www.cambridge.org
Information on this title: www.cambridge.org/9781107405622

Materials Research Society
506 Keystone Drive, Warrendale, PA 15086
http://www.mrs.org

First published 1987
First paperback edition 2012

Single article reprints from this publication are available through
University Microfilms Inc., 300 North Zeeb Road, Ann Arbor, MI 48106

CODEN: MRSPDH

ISBN 978-0-931-83751-7 Hardback
ISBN 978-1-107-40562-2 Paperback

Contents

PART III: REACTIONS, MICROSTRUCTURE AND MODELING

PART IV: UTILIZATION

Preface

Vast quantities of inorganic by-products are produced when coal is burned or gasified. In the US, nearly 60 million tons of fly ash are removed from power plant stacks each year and more than 80% of this ash is buried in landfills or stored in holding ponds pending burial. Only 10% has found a commercial market, chiefly in concrete-related uses. Conversion of coal, through gasification and other processes, into other energy sources will lead to additional large quantities of ash in the future. With such vast quantities of ash involved, one needs to be attuned to any environmental consequences of burying ash and to any new possibilities of using this vast resource for industrial or civil engineering applications. Thorough characterization of the ash materials and their reactions provides the scientific basis for safe disposal or effective utilization. It is the purpose of this series to report the materials science and engineering aspects of the characterization, utilization and disposal of coal-derived ash.

Of the papers in this volume, 29 are based on presentations in Symposium N of the 1986 Fall Meeting of the Materials Research Society. This was the the Society's fifth symposium on the subject of coal ash. Proceedings of the first, edited by S. Diamond, were published under the title "Effects of Fly Ash Incorporation in Cement and Concrete," and are available from D.M. Roy of Penn State's Materials Research Laboratory. Relevant papers from the second symposium, edited by G.J. McCarthy, appeared in the July 1984 issue of Cement and Concrete Research. Proceedings of the third and fourth symposia appeared as Volumes I and II of this series (Mat. Res. Soc. Symp. Proc. Vols. 43 and 65). There will be a Volume IV based on a sixth MRS symposium of this title to be held at the 1987 Fall Meeting.

Six additional papers in this volume were presented in a joint session with Symposium M, "Microstructure Development During Hydration of Cement." Although concerned principally with microstructure, the cement and concrete materials discussed in these papers incorporated coal ash. These six papers are also published in **Microstructural Development During the Hydration of Cement**, edited by L. Struble and P. Brown, Materials Research Society Symposium Proceedings Volume 85, 1987.

This volume contains both conventional papers and summaries. The papers were peer-reviewed and handled according to the normal criteria for journal articles and invited review papers. Three of the 35 contributions are short summaries that were "communicated" (in the manner of Cement and Concrete Research and the Materials Research Bulletin) by an editor. These are of two types: summaries of work already published and pointers to that literature, and reports of work in progress that is not yet ready for formal publication or is to be submitted for more extensive publication elsewhere.

D. Rai and coworkers provide a well referenced review of the leaching behavior of electrical utility wastes and deal specifically with the geochemical modeling of calcium as an illustration of the mechanistic approach being employed in their work. To provide data under actual field conditions, fly ash has been placed in "test cells" at a Pennsylvania power plant. L.L. LaBuz et al. describe the design of the experiment and summarize the leachate data obtained over the first year of operation. As part of a study aimed at Al extraction from fly ash, J.S. Watson has done acid leaching tests that also provide insight into the extent and correlation of extractability of various elements in the ash. M.M. Soroczak et al. describe the use of ESCA and SEM to study leached ash. R.I.A. Malek and D.M. Roy report results of a study of electrochemical stability of fly ash-concrete pastes used as beds for pipelines, and provide insights into the redox behavior of several of the important trace elements in that ash. The use of fly ash for stabilization of other wastes is described by C.S. Poon and R. Perry, and by J.G. Laguros and co-workers.

A coal conversion by-product with quite unique characteristics is produced in the emerging atmospheric fluidized bed combustion (AFBC) technology, in which a limestone bed is included in the combustion system to reduce sulfur oxide emissions. In two papers, E.E. Berry and co-workers summarize the types of AFBC units and discuss chemical and physical characteristics, environmental considerations and potential uses of the by-products.

This volume includes significant advances in understanding of fly ash chemistry and mineralogy derived through the application of modern materials characterization methodology. Results of a comprehensive study of low-calcium, high-iron fly ashes are reported by R.T. Hemmings et al. This paper complements a similar study on a high-calcium ash in Volume 65 of this series. The characteristics and extent of variability of high-calcium fly ash from the western U.S. are the subject of papers by R.J. Stevenson and T.P. Huber, G.J. McCarthy et al., R.C. Joshi and B.K. Marsh, and S. Schlorholtz and co-workers. S. Diamond and J. Olek point out the importance of knowing the mineralogy of the calcium in these ashes in addition to the total amount present. W.T. Hester describes strength variations in concrete incorporating fly ash, and notes that this could be due either to inherent variability in the ashes or in curing of the concrete.

The key to many of the uses of coal ash is its hydration behavior. F.P. Glasser, S. Diamond and D.M. Roy provide an extensive review of current knowledge of the hydration reactions of coal ash and similar by-products when incorporated into cement pastes as blending agents. P. Kumarathasan and G.J. McCarthy, and R.C. Joshi and D.T. Lam, describe experiments on the self-hardening hydration reactions in high-calcium ashes. M. Tohidian and J.G. Laguros report studies of chemical additives designed to retard this property.

Microstructures of cement blends incorporating by-products is the subject of a review by M. Regourd, a paper by H.F.W. Taylor and co-workers that includes an SEM/EDX examination of microchemical hydration reactions, and a paper by D.J. Cook et al. The effects of incorporation of by-products on rheology, chloride diffusion, and strength and durability of cement pastes and concrete are described in three papers by D.M. Roy and co-workers.

One of the principal advantages of incorporating blending agents into concrete is the reduction in degradation of properties resulting from the heat of hydration of the portland cement. Papers by S. Kaushal et al. and M.J. Coole and H.J. Harrisson discuss thermal behavior of hydrating blends of concrete and by-products.

The utilization of fly ash in concrete is necessarily quite dependent on the engineering standards in effect in each country. R.M. Majko reviews these standards and provides recommendations for modifications in the ASTM standards in use in the U.S.

Aluminum recovery from fly ash is described in papers by E.E. Berry et al. and M.S. Dobbins and G. Burnet. Berry et al. also describe applications for the residues from the extraction process.

Finally, the reader will note that this volume appears in a uniform typeface. Two-thirds of the papers were submitted on disk and the balance were keyboarded at North Dakota State University. In addition to the editing done during the refereeing process, all papers had additional technical editing by the senior editor. It was the editor's goal that this volume would conform to the highest standards possible in a camera-ready format.

G.J. McCarthy
Fargo, North Dakota
April, 1987

Acknowledgments

Kevin D. Swanson of North Dakota State University (University of Wisconsin after June, 1987) provided dedicated and enthusiastic assistance in the layout, proof-reading and indexing of this book. The editors thank Linda Stoetzer and Kathleen Beeson of NDSU for keyboarding of manuscripts.

The symposium program and publication of this volume were made possible by financial contributions from the following sponsors:

PRINCIPAL SYMPOSIUM SUPPORT

Electric Power Research Institute, Palo Alto, California
American Fly Ash Company, Des Plaines, Illinois

SUPPLEMENTAL SUPPORT

Gas Research Institute, Chicago, Illinois
Western Fly Ash Research Development and Data Center, Grand Forks and Fargo, North Dakota
L.E. Shaw, Ltd., Halifax, Nova Scotia

PART I

Environmental Considerations

LEACHING BEHAVIOR OF FOSSIL FUEL WASTES:
MINERALOGY AND GEOCHEMISTRY OF CALCIUM

DHANPAT RAI, L.E. EARY, S.V. MATTIGOD, C.C. AINSWORTH and J.M. ZACHARA
Battelle, Pacific Northwest Laboratories, P. O. Box 999, Richland, Washington
99352

Received 11 December, 1986; refereed

ABSTRACT

A literature review [1] of the leaching behavior of inorganic constituents contained in fossil fuel wastes indicated that most of the available
information deals primarily with (1) determination of the elemental composition of different wastes, (2) examination of the physical characteristics of
waste solids, and (3) empirical leaching studies involving different solutions
and procedures (e.g., water, acids, extraction procedure, toxicity characteristic leaching procedure). A comprehensive mechanistic approach and data are
needed to predict accurately the composition of pore waters from fossil fuel
wastes. An approach that relates the aqueous concentrations in leachates to
solubility-controlling or adsorption-controlling solid phases in the wastes is
described. The behavior of calcium is used as an example to show the relevance of existing data to predicting leachate composition and to identify the
type of data needed. The application of this approach to pore waters from
flue gas desulfurization sludge shows that Ca concentrations can be accurately
predicted from the nature of the Ca solids present and the thermo-chemical
data descriptive of precipitation/dissolution and complexation reactions.

INTRODUCTION

Fossil fuel wastes contain many chemical constituents, some of which may
become mobile and adversely affect groundwaters and surface waters. Safe and
effective disposal procedures require the ability to predict the leaching
behavior of various waste constituents under different environmental and
disposal conditions. A large volume of literature pertaining to fossil fuel
wastes exists. However, most of this literature deals with (1) the elemental
composition of different wastes, (2) the physical characteristics of waste
solids, and (3) empirical leaching studies involving different solutions and
procedures (e.g., water, acids, extraction procedure, toxicity characteristic
leaching procedure) [1]. Seriously lacking at the present time is a compre-
ehensive mechanistic approach and requisite thermochemical descriptors that
can be used to (1) develop a generic capability for predicting the leaching
behavior of elements contained in fossil fuel wastes, (2) show the relevance
of the existing data to developing such a capability, and (3) identify and
develop the data needed for such a capability.

To develop the ability to predict pore water concentrations of inorganic
elements in fossil fuel wastes, mechanistic data on precipitation/dissolution
and adsorption/desorption reactions in the given system are required. For
redox-sensitive elements, data on the kinetics of redox transformations are
also needed to determine the distribution of redox species and their precip-
itation/dissolution and adsorption/desorption reactions. A schematic of the
mechanisms that affect leachates from fossil fuel wastes is shown in Figure 1.
Many waste characteristics, such as the pH, Eh, solid phase composition,
complexing ligands, competing ions, and gaseous phase, influence both precip-
itation/dissolution and adsorption/desorption reactions. These reactions in
turn influence the pore water concentrations. Of the six waste characteristics mentioned, solid phase composition is one of the most important. The
solid phases control, through their solubility, not only the aqueous concen-

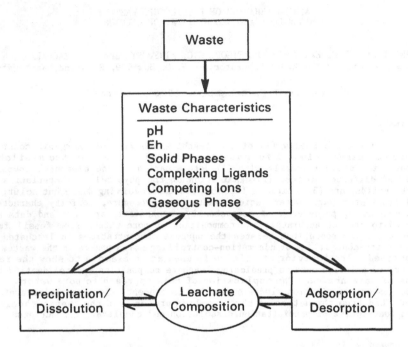

Figure 1. Schematic diagram of the mechanisms that control leachate composition.

trations of inorganic elements, but also the subsequent precipitation of secondary solid phases that may limit pore water concentrations. Solid phases of major elements, such as iron oxides and silicon-aluminum oxides, also provide large reactive surface areas on which the adsorption/desorption of aqueous constituents can occur.

To understand the nature of inorganic compounds in fossil fuel wastes, one must look at the processes generating the compounds and at a detailed chemical characterization of the wastes. The solid phases in fly ash, bottom ash, and oil ash represent the noncombustible portion of the hydrocarbon fuels that are burned to generate electricity. The solid phases in flue gas desulfurization (FGD) sludge represent (1) the solids used as SO_2 scavengers during hydrocarbon combustion (e.g., $CaCO_3$), (2) those produced by the interaction of the scavenger solids with SO_2 and other gaseous phases (e.g., $CaSO_3 \cdot 1/2H_2O$) and their transformation products (e.g., $CaSO_4 \cdot 2H_2O$), and (3) those that result from the admixing or trapping of fine fly ash particles. Depending on the disposal system used, various proportions of ash and sludge may be codisposed, thus increasing the number and types of solid phases that may be encountered in the waste environment. Knowing the nature of an element's solid phases in the fuel source makes it possible in some cases to predict the transformations that will be caused by combustion and thus the nature of the solid phases in the waste. For example, Ca may be present in coal as $CaCO_3$, which is decarbonized to CaO during burning. This type of information, or simply the detailed characterization of the wastes, can be used to predict the aqueous concentrations of an element in the initial leachates derived from the weathering of fresh ash. Thermochemical data can be used to predict the potential nature of the element's solid phases in weathered ash and can also be used, in conjunction with the pH and Eh, the gaseous phase composition, and the nature of the complexing ligands, to establish the types and upper concentration limits of aqueous species. Information on the charge characteristics of aqueous species and the quantity and nature of complexation sites on adsorbent surfaces is needed to quantify the adsorption/desorption reactions.

A comprehensive approach to predicting pore water concentrations is discussed below. The types and availability of mechanistic data needed for this approach are illustrated using Ca as an example.

DISTRIBUTION OF Ca IN WASTES AND LEACHABILITY

Calcium in the solid and aqueous phases exists in the +2 oxidation state. The available literature on (1) the total concentration of Ca in different materials, (2) the distribution and association of Ca with solid phases present in the wastes, and (3) the fraction solubilized in different leaching solutions contacting different wastes is summarized in tables and discussed below. In general, the range of the total Ca concentrations found (~0 to 35%) is similar for the different waste types (Table I). Calcium is not appreciably volatilized during combustion; therefore, the total Ca concentration in fossil fuel wastes (except FGD sludge) is directly proportional to the Ca content of the coal. The large variability in the Ca content of the wastes is caused by the specific enrichment of Ca-containing solids in some wastes (e.g., FGD sludge where Ca solids are used to scrub the flue gases) and/or the variable Ca content of the coals burned.

Calcium in coal is present mainly in the inorganic phase, as carbonates, sulfates, and silicates [3]. A small fraction may also be present in the organic phase [27]. Because Ca is not a volatile element, it is not found to be preferentially enriched in fly ash or associated with the smallest particle size fraction (Table II). Overall, Ca is a major element that is present in fossil fuel wastes in large quantities (Table I), and because it forms relatively stable compounds, it is present primarily as discrete solids. Calcium is also reported to be associated with the glassy matrix [36].

Systematic studies to determine quantities of specific Ca solid phases in different wastes are lacking. Therefore, information from different sources was summarized in Table III to indicate the nature of Ca solids that may be expected in different wastes. A large number of specific compounds of Ca have been reported in both unweathered and weathered fly ash, bottom ash, and FGD sludge, but no data are available for oil ash (Table III). The presence of Ca-bearing solid phases in unweathered fly ash requires further substantiation. Some of the solids frequently reported in the literature (such as CaO, $CaSO_4$, and Ca ferrite) clearly result from high-temperature alteration of the original compounds present in coal, because these solids are not reported to be present or stable in either coal or the natural weathering environment. Some of the other solid phases may either have passed through the burning process unaltered or result from rehydration and recarbonation of the fossil fuel wastes shortly after combustion. The nature of the major Ca compounds in fly ash, bottom ash, and FGD sludge is similar, except for $CaSO_3 \cdot 1/2H_2O$, which is found almost exclusively in FGD sludge. Calcite and $CaSO_4 \cdot 2H_2O$ are commonly reported to be present in both weathered and unweathered wastes.

Although a large number of discrete Ca solids are reported to be present in fossil fuel wastes, no studies are available in which systematic attempts were made to relate the nature of the solid phases or other generic properties of the wastes to leaching characteristics. In general, the data in Table IV and other summarizations of Ca leaching data [5,6,16,36,44,72] indicate that: (1) there is a large variability in the amount of Ca leached, depending on the type of leaching medium used (e.g., more Ca is leached in acids than in water); (2) no relationship can be found between the amount leached and the

Table I

Concentration of Calcium in Utility Wastes, Coal, Soils, and the Lithosphere

Material	Ca Concentration (%) Range	Average	Reference
Fly ash	1.58 - 16.20	7.44	Ainsworth et al. [2] (30 samples)
	0.11 - 12.60	4.40	Valkovic [3] (literature review)
	1.48 - 17.71		Adriano et al. [4] (literature review)
	1.68 - 18.01		Suloway et al. [5] (11 samples)
	0.14 - 22.16	5.86	Roy et al. [6] (58 samples, literature review)
Bottom ash	0.59 - 30.70	6.50	Ainsworth et al. [2] (30 samples)
	0.6 - 13.8		Dahlberg [7], Gladney et al. [8], Holland et al. [9], Kaakinen et al. [10], Klein et al. [11], Kopsick and Angino [12], Liskowitz et al. [13], Small [14]
FGD sludge	12.94 - 31.02		Terman [15] (3 samples)
	0 - 0.25		Summers et al. [16] (wet sludges, literature review)
	10.95 - 34.50	27.39	Ainsworth et al. [2] (10 samples)
Oil ash	0.01 - 33.0		Summers et al. [16] (28 samples, literature review)
	0.14 - 1.46	0.87	Ainsworth et al. [2] (3 samples)
	0.10 - 19.3		Kuryk et al. [17], Stinespring [18], Giavarini [19], Burriesci [20], Henry and Knapp [21] (10 samples total)
Coal	0.05 - 2.67	0.54	Valkovic [3] (literature review)
		0.54	Adriano et al. [4] (literature review)
		0.50	Smith [22] (literature review)
Soils	0.7 - 50.0	1.37	Lindsay [23] (literature review)
		1.00	Jackson [24] (literature review)
Continental lithosphere		4.10	Krauskopf [25] (literature review)
		3.63	Mason [26] (literature review)

Table II

Solid Phase Association of Calcium in Unweathered Fly Ash

Physical Association		Chemical Association with Particles				
Surface Enrichment (% of total)	Size Fraction Dependence	Discrete Solids	Glass	Mullite	Metal Oxides	Other
12 (Hansen and Fisher [28])[a]	No increase in concentration with decreasing particle size (Block and Dams [29]; Campbell et al. [30];[b] Coles et al. [31];[c] Davison et al. [32];[d] Hansen and Fisher [28];[a] Natusch et al. [33]; Smith et al. [34];[e] Smith et al. [35][e])	Warren and Dudas [36]	Warren and Dudas [36]			

[a]Determined from leaching with HF or HCl (using pulverized western coal).

[b]Low-sulfur western coal; samples collected on electrostatic precipitator.

[c]Pulverized western coal; stack fly ash.

[d]Low-sulfur western coal; stack fly ash.

[e]Volatile trace elements increase with decreasing particle size for 1- to 10-μm sizes; submicron particle concentrations become independent of particle size.

Table III

Observed and Predicted Solid Phases of Calcium in Different Fossil Fuel Wastes

Type of Waste	Unweathered Predicted	Unweathered Observed	Weathered Predicted	Weathered Observed	Mineral Name
Fly ash	$CaCO_3$[a]	$CaCO_3$[b]	$CaCO_3$[c,d]	$CaCO_3$[e]	Calcite
	$CaSO_4$[f]	$CaSO_4$[g]	$CaSO_4$[h]		Anhydrite
	CaO[i]	CaO[j]			Lime
		$CaSO_4 \cdot 2H_2O$[k]		$CaSO_4 \cdot 2H_2O$[e]	Gypsum
		$Ca(OH)_2$[l]	$Ca(OH)_2$[m]	$Ca(OH)_2$[n]	Portlandite
		Ca-Ferrite[o]			
		Ca_2SiO_4[p]			Bredigite
		Ca_3SiO_5[p]			
		$Ca_3Al_2O_6$[p]			
		$CaSiO_3$[q]			Wollastonite
		$CaMgSi_2O_6$[q]			Diopside
		$Ca_2MgSi_2O_7$[r]			Melilite
		$Ca_3Mg(SiO_4)_2$[s]			Merwinite
		Ca_3SiO_5[t]			Alite
		$Ca_6Al_4Fe_2O_{15}$[t]			Brownmillerite
		$CaMg(CO_3)_2$[t]	$CaMg(CO_3)_2$[c]		Dolomite
			$CaAl_2Si_4O_{12} \cdot 4H_2O$[v]		Laumontite
			$Ca_6Al_2(SO_4)_3(OH)_{12} \cdot 26H_2O$[v]	$Ca_6Al_2(SO_4)_3(OH)_{12} \cdot 26H_2O$[w]	Ettringite
				Ca-silicate[n]	
				Ca-aluminate[n]	
Bottom ash		$Ca_2MgSi_2O_7$[s]			Melilite
		Pyroxene[s]			
		$CaSO_4 \cdot 2H_2O$[s]			Gypsum
FGD sludge		$CaSO_3 \cdot 1/2H_2O$[x]	$CaSO_3 \cdot 1/2H_2O$[y]	$CaSO_3 \cdot 1/2H_2O$[y]	Hannebachite
		$CaSO_4 \cdot 2H_2O$[s]	$CaSO_4 \cdot 2H_2O$[y]	$CaSO_4 \cdot 2H_2O$[y]	Gypsum
		$CaSO_3$[z]			
		$CaSO_4$[aa]			Anhydrite
		$CaCO_3$[bb]		$CaCO_3$[y]	Calcite
		$Ca(OH)_2$[bb]			Portlandite
		$Ca_6Al_2(SO_4)_3(OH)_{12} \cdot 26H_2O$[z]			Ettringite
Oil ash					

[a]Bauer and Natusch [37]; [b]Luke [38], Brzakovic [39], Harvey [40], Plank [41], Page et al. [42]; [c]Talbot et al. [43]; [d]Talbot et al. [43] reported aragonite, as well as calcite, as the carbonate form; [e]Warren and Dudas [44], Liem et al. [45]; [f]Raask [46]; [g]Simons and Jeffery [47], Brzakovic [39], Manz [48], Styron [49], Natusch et al. [50], Harvey [40], Luke [38], Plank [41], Scheetz and White [51], Groenewold and Manz [52], Humenick et al. [53], Kokubu [54], McCarthy et al. [55], Mattigod and Ervin [56]; [h]Roy and Griffin [57]; [i]Theis et al. [58]; [j]Simons and Jeffery [47], Brzakovic [39], Manz [48], Styron [49], Natusch et al. [50], Harvey [40], Luke [38], Plank [41], Scheetz and White [51], Groenewold and Manz [52], Humenick et al. [53], Kokubu [54], McCarthy et al. [55], Mattigod and Ervin [56], Liem et al. [45], Biggs and Bruns [59]; [k]Simons and Jeffery [47], Rehsi [60], Page et al. [42], Liem et al. [45], Scheetz and White [51]; [l]Harvey [40], Warren and Dudas [36]; [m]Malek and Roy [61]; [o]Simons and Jeffery [47], Mattigod [62], Mattigod and Ervin [56]; [p]Scheetz and White [51]; [q]Humenick et al. [53]; [r]Rehsi [60], McCarthy et al. [55]; [t]Roy et al. [63]; [u]Plank [41]; [v]Mattigod [64]; [w]Simons and Jeffery [47], McCarthy et al. [55], Malek and Roy [61]; [x]Selmeczi and Knight [65], Liem et al. [45], McCarthy et al. [45]; [y]Ainsworth et al. [2]; [z]Selmeczi and Knight [65]; [aa]Selmeczi and Knight [65], Groenewold and Manz [52]; [bb]Liem et al. [45].

type of the waste; (3) rates of Ca leaching decrease with time; and (4) long-term leaching produces steady-state Ca concentrations. Although these results are helpful in defining general behavior patterns, they are not conducive to developing a generic predictive capability for the leaching behavior of Ca in fossil fuel wastes.

MECHANISTIC DATA FOR Ca

Except for a limited number of cases (Table V) for which fragmented data on solubility-controlling solids are either available or hypothesized, nearly all data on the chemistry of Ca in fossil fuel wastes result from descriptive studies of (1) the physical and chemical characteristics of wastes, (2) short-term, site-specific leaching rates, (3) the leaching of wastes in atypical solutions where experimental parameters are not controlled (Table IV), and (4) Ca leaching behavior observed at a selected waste disposal site. Comprehen-

Table IV

Leachability of Calcium from Different Wastes

Type of Waste	Coal Type	Percent Leached in Different Solutions[a]			Solution-to-Solid Ratio (ml/g)	pH$_{H_2O}$	Time (h)	Temp. (°C)	Reference
		H$_2$O	Acid	Chelate					
Fly ash	Lignite[b]	26.0	101,99		1000	11.2	168-336	--	Green and Manahan [66]
	Subbituminous[c]	4.7	42,68	35	5	11.9	3	--	Dreesen et al. [67]
	Subbituminous	2.7-21.2			5-80	12.4	24	--	Elseewi et al. [68]
	Western[d]		64	26,29	100	--	168	20	Harris and Silberman [69]
	Bituminous	<1.3			50	--	48	20	Natusch et al. [50]
	Bituminous	<5.8			50	--	12	--	Linton et al. [70]
	Bituminous[e]	11.4-12.6	45.3		5,16	4.1-11.2	24	--	Roy et al. [71]
	Lignite[e]	2.0	22.0		5,16	12.4	24	--	Roy et al. [71]
	Low sulfur[f]	25.0	69.0		100	4.2	24	21	Hansen and Fisher [28]
	--	18.5			4	7.9	72[g]	20	Kuryk et al. [17]
	--	20.4			4	7.7	24[g]	90	Kuryk et al. [17]
	--[h]	13.0	24		16	4.5	24	--	Small [14]
	Bituminous[i]	22.0	63.6		20	4.5	6,18	105	Ainsworth et al. [2]
	Bituminous[i]	15.8	69.6		20	8.5	6,18	105	Ainsworth et al. [2]
	Bituminous[i]	8.7	85.3		20	11.7	6,18	105	Ainsworth et al. [2]
	Subbituminous[i]	3.6	77.7		20	12.0	6,18	105	Ainsworth et al. [2]
	Lignite[i]	3.4	72.5		20	11.6	6,18	105	Ainsworth et al. [2]
Bottom ash	Bituminous[i]	2.9	50.9		20	8.0	6,18	105	Ainsworth et al. [2]
	Subbituminous[i]	1.6	73.2		20	10.3	6,18	105	Ainsworth et al. [2]
	Lignite[i]	1.4	72.4		20	9.1	6,18	105	Ainsworth et al. [2]
FGD sludge	--	3.2			4	12.3	24[g]	20	Kuryk et al. [17]
	--	1.3			4	9.1	192[g]	90	Kuryk et al. [17]
	Bituminous[i]	3.1	76.8		20	8.7	6,18	105	Ainsworth et al. [2]
	Lignite[i]	4.06	81.6		20	8.9	6,18	105	Ainsworth et al. [2]
Oil ash		20.7			4	3.7	24[g]	20	Kuryk et al. [17]
		23.3			4	2.3	72[g]	90	Kuryk et al. [17]
			3.3		100		24	--	Stinespring et al. [18]
	i	4.8	72.6		20	6.6	6,18	105	Ainsworth et al. [2]

[a]Based on total amount present.

[b]Acid extractants: 1 $\underline{M}$ HCl and 6 $\underline{M}$ HNO$_3$, respectively.

[c]Acid extractants: 1 $\underline{M}$ HCl and 1 $\underline{M}$ HNO$_3$, respectively (0.1, 0.01, and 0.001 $\underline{M}$ HNO$_3$ were also used). Chelating agent: 0.1 $\underline{M}$ citric acid.

[d]Acid extractant: 0.5 $\underline{M}$ HCl. Chelating agents: EDTA and citrate, respectively. Additional extracting solutions reported were histidine, synthetic serum, tris buffer, and 0.5 $\underline{M}$ NH$_4$OH.

[e]EP acid extractant: acetic acid at pH 5; for lignite EP sample, pH was 9.4. Solution-to-solid ratio was 5 for H$_2$O extraction, 16 for acid extraction. Additional H$_2$O extraction data are available for up to 142 d.

[f]Acid extractant: 0.14 $\underline{M}$ HCl (0.029, 0.057, 0.086, 0.21, 0.29, 0.43, and 0.57 $\underline{M}$ HCl and HF were also used). All extractions were carried out on <2.2-μm fraction.

[g]The extraction time is the time needed to reach a maximum solution concentration over an 8-d extraction period.

[h]For H$_2$O extraction, pH was adjusted to 4.5 with H$_2$SO$_4$. For acid extraction, pH was adjusted to 1.0 with H$_2$SO$_4$.

[i]Fly ash analyses are mean values for acid, neutral, and basic leachate from bituminous coal, subbituminous coal, and lignite (n = 5, 6, 8, 6, and 5, respectively). Bottom ash analyses are mean values for bituminous-, subbituminous-, and lignite-derived ashes (n = 19, 6, and 5, respectively). The FGD sludge analyses are mean values for bituminous- and lignite-derived waste (n = 7 and 3, respectively). Oil ash values are an average of values for three samples. Water extraction was for 6 h at 105°C. Acid extraction was for 18 h using 2 $\underline{M}$ HNO$_3$ at 105°C.

sive studies to relate Ca concentrations in solutions to fundamental reactions and waste compositions during all stages of the weathering of fossil fuel wastes under natural conditions have not been undertaken. In the absence of specific mechanistic data for fossil fuel wastes, the available thermochemical data (all from [73], except CaHCO$_3^+$, CaNO$_3^+$, and CaPO$_4^-$ from [74]) and geochemical data were summarized and used to predict the nature of the solubility-controlling solids, dominant aqueous species, and adsorption/desorption reactions of Ca that are important in predicting aqueous Ca concentrations.

Precipitation/Dissolution

Thermochemical data for solid compounds reported to be present in weathered and unweathered fossil fuel wastes (Table III) were used to predict the relative stabilities and solubilities of Ca compounds and the nature of

Table V

Types of Available Calcium Data Obtained from Studies with Utility Wastes

Type of Waste	Precipitation/Dissolution		Adsorption/Desorption	
	Mechanistic	Hypothesized	Mechanistic	Hypothesized
Fly ash		Talbot et al. [43][a]		
		Roy and Griffin [57][b]		
		Warren and Dudas [44], Liem et al. [45][c]		
		Mattigod [64][d]		
		Mattigod [64][e]		
Bottom ash				
FGD sludge	Ainsworth et al. [2][a,f]	Ainsworth et al. [2][f]		
Oil ash				

[a]$CaCO_3$.

[b]$CaSO_4$.

[c]$Ca(OH)_2$.

[d]$Ca_6Al_2(SO_4)_3(OH)_{12} \cdot 26H_2O$.

[e]$CaAl_2Si_4O_{12} \cdot 4H_2O$.

[f]$CaSO_3 \cdot 1/2H_2O$, $CaSO_4 \cdot 2H_2O$.

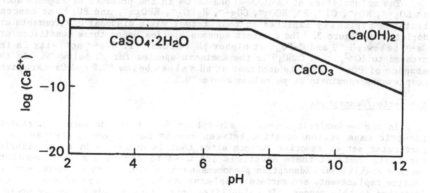

Figure 2. Relative stabilities of calcium solid phases at 25°C for a partial pressure of CO_2 of $10^{-3.5}$ atm and an activity of SO_4^{2-} of $10^{-3.0}$.

the compounds that may control aqueous Ca concentrations as a result of long-term weathering of the wastes. The relative solubilities of selected Ca compounds are shown in Figure 2. The most soluble is CaO, and its high solubility falls outside the graph boundaries. Most of the other Ca compounds fall above the $CaCO_3$ solubility line. Among these compounds, $CaCO_3$ in an alkaline environment and $CaSO_4 \cdot 2H_2O$ in an acidic environment are the least soluble, precipitate rapidly, and are probably the compounds that control aqueous Ca concentrations in weathered fossil fuel wastes over long periods. The results of long-term leaching experiments [44,57] support this conclusion. During the earlier stages of weathering, Ca concentrations in solution may be controlled by compounds that are more soluble than $CaCO_3$ and $CaSO_4 \cdot 2H_2O$. For example, $Ca(OH)_2$, which can form rapidly from the hydrolysis of CaO, may control Ca concentrations in the early stages of weathering under very alka-line conditions and in CO_2-deficient environments. However, with time and as equilibrium with CO_2 levels in air is reached, $Ca(OH)_2$ and other compounds that are metastable in alkaline conditions will convert to $CaCO_3$, followed most likely by buffering of the pH around 8. This pH buffering is expected to result from the conversion of CaO and $Ca(OH)_2$ (which are primarily responsible for the initially alkaline pH of Ca-rich wastes) to $CaCO_3$. Under acidic conditions, where the wastes contain large quantities of $CaSO_4$, Ca concentra-tions will be controlled by $CaSO_4 \cdot 2H_2O$, as it is slightly more stable than $CaSO_4$.

Flue gas desulfurization sludge contains Ca sulfite solids such as $CaSO_3 \cdot 1/2H_2O$, in addition to other Ca compounds, and may exert a transitory control on Ca^{2+} concentrations in oxygen-deficient environments. However, the intrusion of atmospheric oxygen will cause the oxidation of SO_3^{2-} to SO_4^{2-} (although the change may be slow in alkaline conditions), thus promoting the formation of $CaSO_4 \cdot 2H_2O$. Some Ca silicates, such as $CaAl_2Si_4O_{12} \cdot 4H_2O$ [64], and Ca-Al sulfates, such as $Ca_6Al_2(SO_4)_3(OH)_{12} \cdot 26H_2O$ [47,64], may also form in the Si- and Al-rich solution resulting from fly ash dissolution, but direct observations of the formation of such solids during the weathering of utility wastes are available in only a few cases [47].

Aqueous Species

The solubilities of $CaSO_4 \cdot 2H_2O$ and $CaCO_3$ in the presence of ligands such as SO_4^{2-}, NO_3^-, Cl^-, F^-, HCO_3^-, CO_3^{2-}, $H_2PO_4^-$, HPO_4^{2-}, and PO_4^{3-} at concen-trations typically expected in fossil fuel waste disposal environments are depicted in Figure 3. The dominant aqueous species under these conditions are Ca^{2+} below pH ~9 and $CaPO_4^-$ at higher pH values. If SO_4^{2-} activity is in-creased to 10^{-2}, then $CaSO_4^0$ is the dominant species for pH below ~9. In the absence of PO_4^{3-}, Ca^{2+} is dominant at pH values below ~9.2 and Ca-carbonate complexes predominate at pH values above ~9.2.

Adsorption/Desorption

In the mechanistic approach, adsorption data take the form of a stoich-iometric mass action equation between one or more aqueous species and a particular set of reactive surface sites that is described by an equilibrium constant. Sets of these equations are used to describe multiple-solute surface equilibria. Adsorption phenomena such as simple ion exchange, surface lattice replacement, and surface complexation by proton-specific sites are described in this manner. If solubility-controlling solids of Ca are present in fossil fuel wastes or form during weathering, however, the amounts of Ca that can be leached from the wastes will be solubility-limited and independent of adsorption/desorption reactions. Hence, the adsorption/desorption reac-tions of Ca in the presence of solubility-controlling solids can be ignored, even though Ca surface reactions with a large number of solids are well documented.

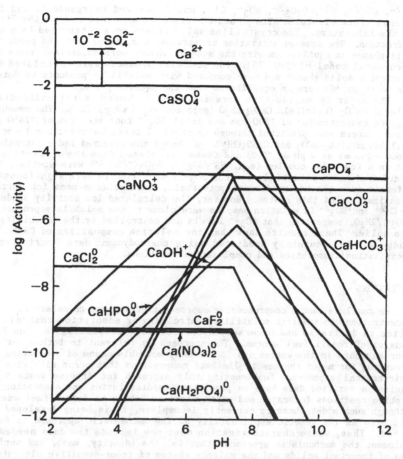

Figure 3. Activities of aqueous Ca species at 25°C where Ca^{2+} activity is controlled by the solubility of $CaSO_4 \cdot 2H_2O$ for acidic pH and by $CaCO_3$ for alkaline pH, for $Cl^- = SO_4^{2-} = 10^{-3}$, $NO_3^- = F^- = 10^{-4}$, $H_2PO_4^- = 10^{-5}$, and p of $CO_2(g) = 10^{-3.5}$ atm. The arrow indicates the increase in $CaSO_4^0$ activity caused by an increase in the SO_4^{2-} activity to 10^{-2}.

Ca BEHAVIOR IN FLUE GAS DESULFURIZATION SLUDGE

The mechanistic approach was applied to determining Ca behavior in pore waters from an FGD sludge site under natural environmental conditions. For this study, special precautions were taken to preserve the chemical integrity of the solid samples and the pore waters collected in the field. As part of this study, the solutions were analyzed by a combination of inductively coupled plasma spectroscope, atomic absorption spectroscope equipped with a graphite furnace, ion chromatograph, colorimeter, lead titrimeter, and ion-selective electrodes for major cations and anions, including Ca, Mg, Na, K,

SO_4^{2-}, SO_3^{2-}, S^{2-}, $S_2O_3^{2-}$, NO_3^-, Cl^-, and dissolved inorganic C, and for several trace elements. The pH and Eh were also measured in both the field and the laboratory. The crystalline solid phases were determined by x-ray diffraction. The aqueous activities of various ions, along with the predicted solid phases in equilibrium with the aqueous phase, were determined using the geochemical model MINTEQ [75]. Ion activity products were calculated for anticipated solid phases and were compared with solubility products to determine which solids were in equilibrium with the aqueous phase.

The major Ca solids that were identified through x-ray diffraction included $CaCO_3$ (calcite), $CaSO_4 \cdot 2H_2O$ (gypsum), and $CaSO_3 \cdot 1/2H_2O$. The measured total Ca concentration of ~600 ppm and total SO_4^{2-} concentration of ~1450 ppm in pore waters were predicted through geochemical model calculations to be in equilibrium with $CaCO_3$ and $CaSO_4 \cdot 2H_2O$. Although the measured redox potentials (about −350 mV at a pH of ~9.0) and those calculated from the activities of ions in a few redox couples (e.g., HS^-/SO_4^{2-}, $S_2O_3^{2-}/SO_4^{2-}$) were similar, the calculated redox potentials for the SO_3^{2-}/SO_4^{2-} couple were significantly different from the measured redox potentials, indicating a need for further investigations of this system. However, the calculated ion activity products for Ca^{2+} and SO_3^{2-} in solution samples were close to the solubility product of $CaSO_3 \cdot 1/2H_2O$, suggesting that SO_3^{2-} levels are controlled by the solubility of this solid. These results show that the solution composition of Ca in FGD sludge may be accurately predicted using thermodynamic data descriptive of precipitation/dissolution and complexation reactions.

CONCLUSIONS

We conclude that a comprehensive approach that strives to relate solution concentrations to specific solubility-controlling or adsorption-controlling solids is desirable and necessary for accurately predicting the leaching behavior of fossil fuel wastes. The approach is relevant to both major and minor elements in the wastes and is the only workable means of predicting and accounting for major changes in leachate composition that occur with time and environmental exposure. Implementing this approach for diverse waste types requires a verified data base on precipitation/dissolution and adsorption/desorption reactions for major and minor elements common to fossil fuel wastes. Although such a data base is currently incomplete, it is being developed and will grow as the generic applicability of the mechanistic approach is documented. Thus, waste characterization must now include the data needed to implement the mechanistic approach, that is, the identity, mass, and surface area of important solids and the valence status of redox-sensitive elements.

ACKNOWLEDGMENTS

This research was funded by the Electric Power Research Institute, Inc. (EPRI) under contract RP2485-08, titled "Leaching Chemistry Studies." We thank Dr. I. P. Murarka, EPRI program manager, for his support, interest, and helpful suggestions. We also thank Jan Baer for editing and Michele Darden for typing the manuscript.

REFERENCES

1. D. Rai, C. C. Ainsworth, L. E. Eary, S. V. Mattigod and D. R. Jackson, Inorganic and Organic Constituents and Their Geochemical Behavior in Fossil Fuel Wastes: A Critical Review, Volume 1 of EPRI EA-(draft report) (Electric Power Research Institute, Palo Alto, California, 1986).
2. C. C. Ainsworth, Dhanpat Rai and T. R. Garland, EPRI EA-(in preparation).
3. V. Valkovic, Trace Elements in Coal (CRC Press, Boca Raton, Florida, 1983).

4. D. C. Adriano, A. L. Page, A. A. Elseewi, A. C. Chang and I. Straughan, J. Environ. Qual. 9, 333–344 (1980).
5. J. J. Suloway, T. M. Skelly, W. R. Roy, D. R. Dickerson, R. M. Schuller and R. A. Griffin, Chemical and Toxicological Properties of Coal Fly Ash, Environmental Geology Notes 105 (Illinois Department of Energy and Natural Resources, Champaign, 1983).
6. W. R. Roy, R. G. Thiery, R. M. Schuller and J. J. Suloway, Coal Fly Ash: A Review of the Literature and Proposed Classification System with Emphasis on Environmental Impacts, Environmental Geology Notes 96 (Illinois State Geological Survey, Champaign, 1981).
7. M. D. Dahlberg, Environ. Sci. Technol. 17, 175–177 (1983).
8. E. S. Gladney, L. E. Wangen, D. B. Curtis and E. T. Jurney, Environ. Sci. Technol. 12, 1084–1085 (1978).
9. W. F. Holland, K. A. Wilde, J. L. Parr, P. S. Lowell and R. F. Pohler, The Environmental Effects of Trace Elements in the Pond Disposal of Ash and Flue Gas Desulfurization Sludge, EPRI 202 (Electric Power Research Institute, Palo Alto, California, 1975).
10. J. W. Kaakinen, R. M. Jorden, M. H. Lawasani and R. E. West, Environ. Sci. Technol. 9, 862–869 (1975).
11. D. H. Klein, A. W. Andren, J. A. Carter, J. F. Emery, C. Feldman, W. Fulkerson, W. S. Lyon, J. C. Ogle, Y. Talmi, R. I. Van Hook and N. Bolton, Environ. Sci. Technol. 9, 973–979 (1975).
12. D. A. Kopsick and E. E. Angino, J. Hydrol. 54, 341–356 (1981).
13. J. W. Liskowitz, J. Grow, M. Sheih, R. Trattner, J. Kohut and M. Zwillenberg, Sorbate Characteristics of Fly Ash (New Jersey Institute of Technology, Newark, 1983).
14. J. A. Small, PhD thesis, University of Maryland, 1976.
15. G. L. Terman, Solid Wastes from Coal-Fired Power Plants: Use or Disposal on Agricultural Lands, Bulletin Y-129 (National Fertilizer Development Center, Tennessee Valley Authority, Muscle Shoals, Alabama, 1978).
16. K. V. Summers, G. L. Rupp and S. A. Gherini, Physical-Chemical Characteristics of Utility Solid Wastes, EPRI EA-3236 (Electric Power Research Institute, Palo Alto, California, 1983).
17. B. A. Kuryk, I. Bodek and C. J. Santhanam, Leaching Studies on Utility Solid Wastes: Feasibility Experiments, EA-4215 (Electric Power Research Institute, Palo Alto, California, 1985).
18. C. D. Stinespring, W. R. Harris, J. M. Cook and K. H. Casleton, Appl. Spectroscopy 39, 853–856 (1985).
19. C. Giavarini, Fuel 61, 549–552 (1982).
20. N. Burriesci, F. Corigliano, P. Primerano, C. Zipelli and M. Petrera, J. Chem. Soc., Faraday Trans. 1 80, 1777–1785 (1984).
21. W. M. Henry and K. T. Knapp, Environ. Sci. Technol. 14, 450–456 (1980).
22. R. D. Smith, Prog. Energy Combust. Sci. 6, 59–119 (1980).
23. W. L. Lindsay, Chemical Equilibria in Soils (Wiley, New York, 1979).
24. M. L. Jackson, in Chemistry of the Soil, edited by F. E. Bear, 2nd ed. (Van Nostrand Reinhold, New York, 1964), pp. 71–141.
25. K. B. Krauskopf, Introduction to Geochemistry (McGraw-Hill, New York, 1967).
26. B. Mason, Principles of Geochemistry (Wiley, New York, 1966).
27. R. B. Finkleman, PhD thesis, University of Maryland, 1980.
28. L. D. Hansen and G. L. Fisher, Environ. Sci. Technol. 14, 1111–1117 (1980).
29. C. Block and R. Dams, Environ. Sci. Technol. 10, 1011–1017 (1976).
30. J. A. Campbell, J. C. Laul, K. K. Nielson and R. D. Smith, Anal. Chem. 50, 1032–1040 (1978).
31. D. G. Coles, R. C. Ragaini, J. M. Ondov, G. L. Fisher, D. Silberman and B. A. Prentice, Environ. Sci. Technol. 13, 455–459 (1979).
32. R. L. Davison, D. F. S. Natusch, J. R. Wallace and C. A. Evans, Jr., Environ. Sci. Technol. 8, 1107–1113 (1974).
33. D. F. S. Natusch, J. R. Wallace and C. A. Evans, Science 183, 202–204 (1974).

34. R. D. Smith, J. A. Campbell and K. K. Nielson, Fuel 59, 661-665 (1980).
35. R. D. Smith, J. A. Campbell and K. K. Nielson, Environ. Sci. Technol. 13, 553-558 (1979).
36. C. J. Warren and M. J. Dudas, J. Environ. Qual. 13, 530-538 (1984).
37. C. F. Bauer and D. F. S. Natusch, Environ. Sci. Technol. 15, 783-788 (1981).
38. W. I. Luke, Nature and Distribution of Particles of Various Sizes in Fly Ash, Technical Report No. 6-583 (U.S. Army Corps of Engineers, Army Engineer Waterways Experiment Station, Vicksburg, Mississippi, 1961).
39. P. N. Brzakovic, in Ash Utilization: Proceedings of the Second Ash Utilization Symposium, Bureau of Mines Information Circular IC 8488 (U.S. Government Printing Office, Washington, D.C., 1970), pp. 205-219.
40. R. D. Harvey, Petrographic and Mineralogical Characteristics of Carbonate Rocks Related to Sulfur Dioxide Sorption in Flue Gases, NTIS Pb-206-487 (Illinois State Geological Survey, Urbana, 1971).
41. C. O. Plank, PhD thesis, Virginia Polytechnic Institute and State University, 1974.
42. A. L. Page, A. A. Elseewi and I. R. Straughan, Residue Reviews 71, 83-120 (1979).
43. R. W. Talbot, M. A. Anderson and A. W. Andren, Environ. Sci. Technol. 12, 1056-1062 (1978).
44. C. J. Warren and M. J. Dudas, J. Environ. Qual. 14, 405-410 (1985).
45. H. Liem, M. Sandstroem, T. Wallin, A. Carne, U. Rydevik, B. Thurenius and P. O. Moberg, in International Conference on Coal Fired Power Plants and the Aquatic Environment, CONF-8208123 (Water Quality Institute, Hoersholm, Denmark, 1982), pp. 338-366.
46. E. Raask, J. Inst. Energy, June 1980, 70-75.
47. H. S. Simons and J. W. Jeffery, J. Appl. Chem. 10, 328-336 (1960).
48. O. E. Manz, in Technology and Use of Lignite, Proceedings of a Bureau of Mines and University of North Dakota Symposium, Bureau of Mines Information Circular IC 8650 (U.S. Government Printing Office, Washington, D.C., 1974), pp. 204-219.
49. R. W. Styron, in Ash Utilization: Proceedings of the Second Ash Utilization Symposium, Bureau of Mines Information Circular IC 8488 (U.S. Government Printing Office, Washington, D.C., 1970), pp. 151-164.
50. D. F. S. Natusch, C. F. Bauer, H. Matusiewicz, C. A. Evans, J. Baker, A. Loh, R. W. Linton and P. K. Hopke, in International Conference on Heavy Metals in the Environment; Symposium Proceedings Volume II, Part 2 (Toronto, 1975), pp. 553-575.
51. B. E. Scheetz and W. B. White, in Fly Ash and Coal Conversion By-Products: Characterization, Utilization and Disposal I, edited by G. J. McCarthy and R. J. Lauf, Mat. Res. Soc. Symp. Proc. Vol. 43 (Materials Research Society, Pittsburgh, 1985), pp. 53-60.
52. G. H. Groenewold and O. E. Manz, Disposal of Fly Ash and Fly Ash Alkali FGD Waste in a Western Decoaled Strip Mine -- Interim Report (Engineering Experiment Station, University of North Dakota, Grand Forks, 1982).
53. M. J. Humenick, M. Lang and K. F. Jackson, J. Water Pollut. Control Fed. 55, 310-316 (1983).
54. M. Kokubu, in Proceedings of the Fifth International Symposium on the Chemistry of Cement, Vol. 4: Admixtures and Special Cements (Cement Association of Japan, Tokyo, 1969), pp. 75-113.
55. G. J. McCarthy, K. D. Swanson, P. J. Schields and G. H. Groenewold, Mineralogical Controls on Toxic Element Contamination of Groundwater from Buried Electrical Utility Solid Wastes. 1. Solid Waste Mineralogy. 2. Literature Review of Fly Ash Mineralogy, PB-83-265116 (North Dakota State University, Fargo, 1983).
56. S. V. Mattigod and J. O. Ervin, Fuel 62, 927-931 (1983).
57. W. R. Roy and R. A. Griffin, Environ. Sci. Technol. 18, 739-742 (1984).
58. T. L. Theis, M. Halvorsen, A. Levine, A. Stankunas and D. Unites, in Report and Technical Studies on the Disposal and Utilization of Fossil

Fuel Combustion By-Products, submitted to the U.S. Environmental Protection Agency, 1982.

59. D. L. Biggs and J. J. Bruns, in Fly Ash and Coal Conversion By-Products: Characterization, Utilization and Disposal I, edited by G. J. McCarthy and R. J. Lauf, Mat. Res. Soc. Symp. Proc. Vol. 43 (Materials Research Society, Pittsburgh, 1985), pp. 21-29.

60. S. S. Rehsi, in Ash Utilization: Third International Symposium, Bureau of Mines Information Circular IC 8640 (U.S. Government Printing Office, Washington, D.C., 1973), pp. 231-245.

61. R. A. I. Malek and D. M. Roy, in Fly Ash and Coal Conversion By-Products: Characterization, Utilization and Disposal I, edited by G. J. McCarthy and R. J. Lauf, Mat. Res. Soc. Symp. Proc. Vol. 43 (Materials Research Society, Pittsburgh, 1985), pp. 41-50.

62. S. V. Mattigod, Scanning Electron Microscopy II, 611-617 (1982).

63. D. M. Roy, K. Luke and S. Diamond, in Fly Ash and Coal Conversion By-Products: Characterization, Utilization and Disposal I, edited by G. J. McCarthy and R. J. Lauf, Mat. Res. Soc. Symp. Proc. Vol. 43 (Materials Research Society, Pittsburgh, 1985), pp. 3-20.

64. S. V. Mattigod, Environ. Technol. Lett. 4, 485-490 (1983).

65. J. G. Selmeczi and R. G. Knight, in Ash Utilization: Third International Symposium, Bureau of Mines Information Circular IC 8640 (U.S. Government Printing Office, Washington, D.C., 1973), pp. 123-138.

66. J. B. Green and S. E. Manahan, Anal. Chem. 50, 1975-1980 (1978).

67. D. R. Dreesen, L. E. Wangen, E. S. Gladney and J. W. Owens, in Environmental Chemistry and Cycling Processes, edited by D. C. Adrioni and I. L. Brisbin, CONF-760429 (National Technical Information Service, Oak Ridge, Tennessee, 1976), pp. 240-252.

68. A. A. Elseewi, A. L. Page and S. R. Grimm, J. Environ. Qual. 9, 424-428 (1980).

69. W. R. Harris and D. Silberman, Environ. Sci. Technol. 17, 139-145 (1983).

70. R. W. Linton, P. Williams, C. A. Evans, Jr. and D. F. S. Natusch, Anal. Chem. 49, 1514-1520 (1977).

71. W. R. Roy, R. A. Griffin, D. R. Dickerson and R. M. Schuller, Environ. Sci. Technol. 18, 734-739 (1984).

72. D. S. Cherkauer, Ground Water 18, 544-550 (1980).

73. D. P. Wagman, W. H. Evans, V. B. Parker, R. H. Schumm, I. Halow, S. M. Bailey, K. L. Churney and R. L. Nuttall, J. Phys. Chem. Ref. Data, Vol. 11, Supplement 2 (American Chemical Society and the American Institute for Physics, New York, 1982).

74. R. M. Smith and A. E. Martell, Critical Stability Constants. Volume 4: Inorganic Complexes (Plenum, New York, 1976).

75. A. R. Felmy, D. C. Girvin and E. A. Jenne, MINTEQ - A Computer Program for Calculating Aqueous Geochemical Equilibria, EPA 600-3-84-032 (U.S. EPA, Office of Research and Development, Athens, Georgia, 1984).

LEACHATE COMPOSITION AT AN EXPERIMENTAL
TEST CELL OF COAL COMBUSTION FLY ASH

L.L. LABUZ, J.F. VTI' LUME and J.W. BELL
Solid Waste and T   Control Group, Environmental Management Division,
Pennsylvania Power & Light Company, Two North Ninth Street, Allentown, PA
18101-1179, USA

Received 15 October, 1986; refereed

ABSTRACT

In an effort to characterize the waste streams at a typical utility fly
ash landfill, a large outdoor test cell of compacted fly ash was constructed
in 1984 at the Pennsylvania Power & Light Company's Montour Steam Electric
Station. The test cell is a scaled-down version of the active dry fly ash
disposal facility at the plant and includes a liner, a leachate collection
system, and surface water runoff perimeter drains. It is 100 feet by 100 feet
at the base and 10 feet high with 2:1 vertical:horizontal side slopes, giving
a top area of 60 feet by 60 feet. The test cell is instrumented to measure
precipitation, infiltration, surface runoff, unsaturated flow, evapotranspiration and leachate drainage. This paper presents a preliminary evaluation
of the chemical composition of leachate generated from the test cell during
the first year of leachate production. Leachate samples are periodically
collected from a concrete sump to which the leachate collection system drains.
Leachate first appeared in the sump in October 1985, approximately one year
after the test cell was constructed. Samples are analyzed for pH, specific
conductance, 22 metals, and 10 ions. The concentrations of parameters present
in the test cell leachate are compared to drinking water standards and to
standard laboratory extraction results. Parameters not detected in the leachate are also identified. The preliminary findings of the leachate monitoring
effort are providing valuable information on the initial quality of leachate
generated under field conditions. This data will be useful in the design of
leachate control systems, wastewater treatment facilities and ground water
monitoring programs for future ash disposal operations.

INTRODUCTION

The electric utility industry currently produces approximately 60 million
tons of fly ash annually. At many coal-fired power plants, fly ash is sluiced
to settling basins for disposal. The resulting supernatant is treated, if
necessary, and discharged to a nearby body of surface water. Increasingly
stringent discharge standards and the need to optimize land use, however, are
causing many of these plants to convert to dry fly ash handling and disposal
systems.

The Pennsylvania Power and Light Company (PP&L) undertook its first dry
fly ash conversion project in 1982 at its Montour Steam Electric Station
(SES), located in central-eastern Pennsylvania. At the time, only limited
information was available on the waste streams generated at dry fly ash
disposal facilities. Design engineers were asked to build leachate collection
and treatment systems without knowing the actual quantity or quality of
leachate to be produced.

In 1984, PP&L launched a major research effort at Montour SES to
investigate leachate production at the Montour dry fly ash landfill. A
scaled-down model of the landfill was constructed and instrumented to monitor
stormwater runoff and leachate collected in the drainage systems. Pore water
quality and flux are also monitored at various depths throughout the test
cell. This arrangement also provided an ideal outdoor laboratory for the
Electric Power Research Institute to test the performance of various

instruments for measuring and sampling moisture in the unsaturated zone in an adjunct study [1].

Fly ash in the test cell has only been subjected to natural leaching processes for just over two years. However, preliminary results from the project have already been put to use. Information on the quality of leachate generated in the test cell was used to develop an environmental impact assessment for a proposed project to demonstrate the use of fly ash as backfill material. The leachate data has also helped tailor the ground water monitoring program at PP&L's next dry fly ash disposal facility, which will result in sizable cost savings by reducing analytical requirements. Most importantly, the utility industry in Pennsylvania is now in a better position to argue against rigid, overly-conservative design standards for fly ash disposal facilities in pending state residual waste management regulations.

Montour Fly Ash Disposal Operations

PP&L's Montour SES consists of two tangential-fired dry-bottom boilers with a combined generating capacity of 1580 MW. It is fueled by pulverized bituminous coal from western Pennsylvania and produces approximately 360,000 tons of fly ash annually. The fly ash is collected in electrostatic precipitators and conveyed pneumatically to storage silos at the disposal area. There it is mixed with water to achieve an optimum moisture content of approximately 15%. The moisture-conditioned fly ash is trucked to the adjacent landfill, end dumped, and spread in six-inch lifts. A smooth-wheeled vibratory roller then compacts the ash to at least 90% of its Proctor Density [2]. The resulting saturated hydraulic conductivity of the compacted fly ash averages 4.0×10^{-5} cm/s. The ash has limited self-hardening properties and is designated a Class F fly ash.

Test Cell Design and Construction

The test cell was designed and constructed to model the active dry fly ash landfill at Montour SES. As shown in Figure 1, the test cell is 100 feet by 100 feet at the base and 10 feet high with 2:1 side slopes. The resulting top surface is 60 feet by 60 feet and is covered with a 3-inch layer of course bottom ash for dust control. This bottom ash cover increases the infiltration rate for the test cell by an order of magnitude over what would occur in its absence. The side slopes are covered with top soil and soil stabilizing grasses. All surface water runoff is collected in perimeter drains. A ramp was constructed to provide access to the top surface for a drilling rig. The ramp is hydraulically isolated from the test cell with a liner and clay berms.

Leachate draining through the ten feet of compacted fly ash enters a drainage blanket and is conveyed through perforated PVC piping to an adjacent sump. The drainage blanket consists of a 1-foot compacted layer of bottom ash over a 20-mil PVC liner and is covered with a fabric filter. The fabric filter prevents fly ash from entering and clogging the drainage blanket.

Samples of fly ash in the test cell were collected both during and after construction to determine whether the ash is representative of that in the full-scale landfill. The amount of calcium and iron present in the ash gives a general indication of the expected acidity or alkalinity of its leachate. PP&L found that the calcium-to-iron ratio for the test cell fly ash is slightly lower than that for earlier samples of fly ash from the plant. Therefore, the test cell would be expected to produce leachate which is slightly more acidic than that produced in the full-scale landfill.

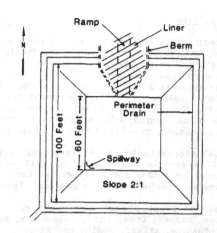

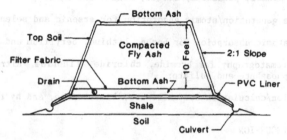

Figure 1. Montour fly ash test cell

Instrumentation

A comprehensive monitoring program was established for the test cell to provide an overall chemical and water balance. The test cell is instrumented with tensiometers (used to measure moisture content as a function of soil "suction"), access tubes for neutron logging probes (also used to measure moisture content), an evaporimeter, a meteorological station, and pore water samplers.

Moisture accounting is routinely performed through direct measurement or calculation of precipitation, infiltration, surface runoff, moisture flux through the fly ash, evapotranspiration and bottom drainage.

Currently, the major emphasis of the monitoring program is on the water quality studies being performed to establish long-term changes in pore water, surface runoff and leachate chemistry. Samples of runoff from the top surface of the test cell are collected automatically after large storm events on a flow-proportional basis. Pore water samples are collected from sixteen porous ceramic pressure/vacuum lysimeters installed at various depths throughout the test cell. Leachate samples are periodically collected from the drainage pipe in the leachate collection sump.

Leachate Sampling and Analysis

The moisture content in the bottom few feet of the test cell increased from 15% in October 1984, when the test cell was constructed, to approximately 30% in October 1985, at which point leachate drainage first occurred. Fourteen leachate samples were collected between October 1985 and April 1986. During this period, approximately 20% of the precipitation falling on the test cell resulted in leachate drainage. Drainage occurred intermittently after April 1986, and only for a short duration, usually within 24 hours of a large storm event. During the summer months, evapotranspiration constitutes a major portion of the moisture balance, resulting in a net loss of moisture in the test cell.

Both filtered (0.45 micrometer) and unfiltered samples were collected, with no significant difference in the results of analyses. Samples were preserved with nitric acid for metals and refrigerated for all other measurements. Leachate analyses were performed by a commercial laboratory using:

- inductively coupled argon plasma for aluminum, barium, calcium, cadmium, chromium, copper, iron, lead, magnesium, manganese, molybdenum, nickel, potassium, silver, sodium and zinc;

- hydride generation/atomic absorption for arsenic and selenium;

- flame/atomic absorption for boron, lithium, beryllium and titanium;

- ion chromatography for bromide, chloride, fluoride, nitrate, nitrite, ortho-phosphate and sulfate;

- titration/calculation for carbonate, bicarbonate, and hydroxide.

RESULTS AND DISCUSSION

The leachate monitoring effort is providing valuable information on the quality of leachate generated under field conditions. Preliminary findings are summarized in Table I along with the Recommended Maximum Contaminant Levels (RMCLs) and Guidance levels that the EPA has proposed for the primary drinking water regulations [3] and the existing RMCLs for the secondary drinking water regulations [4]. These data show that the test cell leachate, in the early stages of the leaching process, exceeds the proposed primary RMCLs for chromium and selenium, the secondary RMCLs for manganese and sulfate and the proposed Guidance Levels for aluminum, sodium and sulfate. The proposed RMCLs for chromium and selenium are exceeded approximately tenfold in each instance. Natural attenuating processes would act to reduce these concentrations in leachate migrating into ground water beneath the disposal area.

The major anions and cations measured in the last leachate sample are shown in Figure 2. Sulfate, calcium, sodium, and potassium account for approximately 90% of the total ionic strength of the leachate. Other major ions include lithium, magnesium, chloride, and molybdate.

The trends in leachate concentrations with time are shown in Figure 3. Concentrations of sodium, potassium, arsenic, boron, chromium, lithium molybdenum and selenium are increasing while concentrations of magnesium, calcium, and manganese are decreasing. No discernible trends for chloride, sulfate, aluminum and barium are apparent. Parameters not detected in the leachate as yet include nitrate, nitrite, bromide, fluoride, beryllium, cadmium, copper, iron, lead, nickel, silver, titanium, and zinc.

PP&L has used this information to modify the ground water monitoring program at the Montour dry fly ash disposal facility. Parameters which are

TABLE I

Montour SES Test Cell Fly Ash Leachate Analysis
(October 1985 – April 1986)

Parameters	Total Elemental Analysis of Dry Fly Ash (mg/kg)	Test Cell Leachate Concentration (mg/l)		EPA Primary & Secondary Drinking Water Regulations**
		Mean	Range	
pH		7.71	7.43-8.11	6.5 to 8.5 (S)
Conductivity, μmhos/cm		3468.	1500.-4190	
Aluminum	140000	0.22	0.1-1.0	0.05 (G)
Iron	109400	<0.05	-	0.3 (S)
Calcium	15900	591.	487.-744.	
Magnesium	4200	52.	18.-94.	
Sodium	1900	165.	73.-238.	
Potassium	22300	182.	71.-315.	20. (G)
Titanium	8500	<0.05	-	
Manganese	290	0.73*	0.13-2.83	0.05 (S)
Sulfate	-	2370.*	2200.-4090.	400. (G), 250. (S)
Nitrate	-	<2.	-	10. (P)
Nitrite	-	<2.	-	1. (P)
Arsenic	286	0.030	0.008-0.038	0.05 (P)
Barium	1003	0.030	0.024-0.034	1.5 (P)
Boron	290	3.44	1.30-4.80	
Cadmium	<0.5	<0.005	-	0.005 (P)
Chromium	218	0.73*	0.06-1.44	0.12 (P)
Copper	185	<0.05	-	1.3 (P)
Lead	114	<0.05	-	0.02 (P)
Lithium	270	13.3	6.1-26.0	
Molybdenum	46	10.79	2.5-23.2	
Nickel	169	<0.05	-	
Selenium	11	0.15*	0.017-0.304	0.045 (P)
Silver	14	<0.05	-	0.15 (G)
Zinc	254	<0.05	-	5.0 (S)
Bromide	-	<10.	-	
Chloride	-	49.	42.-54.	250. (S)
Fluoride	-	<2.	-	1.4 to 2.4 (P)
Sulfur	3200			

*Exceed or Equal EPA Primary and/or Secondary Drinking Water Regulatory Limits.
**(P) Proposed Primary Recommended Maximum Contaminant Level.
  (S) Secondary Recommended Maximum Contaminant Level.
  (G) Proposed Guidance Level.

released in very low or less-than-detectable concentrations have been eliminated from routine monitoring. Conversely, parameters which are released at
high concentrations are used as indicator parameters to identify whether
leachate is migrating from the disposal facility into the surrounding ground-
water.

Field vs. Laboratory Data

Laboratory extractions are frequently used to simulate the leachate
generated from solid waste disposal facilities. In fact, PP&L had to rely
solely on leachate data generated from laboratory extractions when the Montour

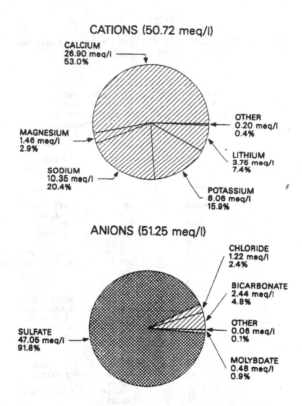

CATIONS (50.72 meq/l)

CALCIUM
26.90 meq/l
53.0%

OTHER
0.20 meq/l
0.4%

LITHIUM
3.75 meq/l
7.4%

MAGNESIUM
1.46 meq/l
2.9%

SODIUM
10.35 meq/l
20.4%

POTASSIUM
8.06 meq/l
15.9%

ANIONS (51.25 meq/l)

CHLORIDE
1.22 meq/l
2.4%

BICARBONATE
2.44 meq/l
4.8%

OTHER
0.06 meq/l
0.1%

SULFATE
47.05 meq/l
91.8%

MOLYBDATE
0.48 meq/l
0.9%

Figure 2.   Ion balance (4/23/86 leachate sample)

landfill was designed. The two tests most often used are EPA's Extraction Procedure (EP Toxicity), which uses an acidified leaching medium at a 1:16 solid-to-liquid ratio, and the ASTM Method A, which uses a leaching medium of distilled water at a 1:4 solid-to-liquid ratio. Although the EP Toxicity Test was designed to simulate leaching conditions in a municipal waste landfill for the purposes of determining whether a waste is hazardous, the results are sometimes regarded as "worst-case" by regulators in nonhazardous waste application.

The results of analyses of EP Toxicity and ASTM A extractions of test cell fly ash are shown in Table II. Both the ASTM A and EP Toxicity Tests yield concentrations of arsenic much higher than those observed under field conditions in the test cell. The EP Toxicity Test also gave higher concentrations of aluminum, iron and zinc. On the other hand, both laboratory tests yield concentrations of chromium, sulfate, molybdenum and lithium much lower than those measured in the test cell. It is interesting to note that for many of the parameters, concentrations are much higher in the ASTM A leachate than the EP Toxicity leachate. It is also apparent that neither test is a reliable measure of leachate quality in the field and should not be used for such purposes without proper consideration of their inherent shortcomings.

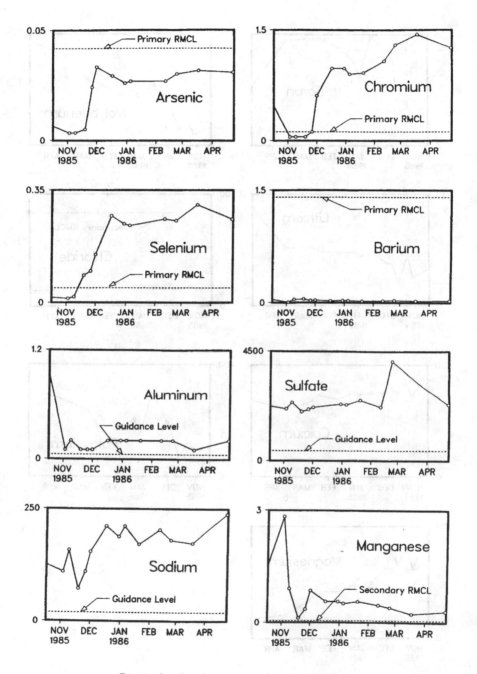

Figure 3. Leachate concentration trends

24

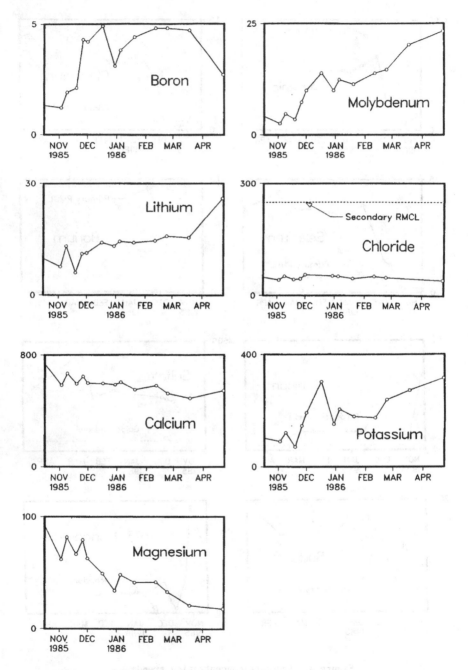

Figure 3. Leachate concentration trends

TABLE II

Montour SES Test Cell Leachate Field vs. Laboratory Extractions

Parameters	Test Cell Leachate Concentration(a)	Fly Ash Leachate Analysis	
		EP Toxicity	ASTM "A"
	Range		
pH	7.43-8.11	4.69	8.16
Conductivity, µmhos/cm	1500.-4190.	1300.	2300.
Analysis in mg/l			
Aluminum	0.1-1.0	2.7	0.4
Iron	<0.05	<0.11	<0.05
Magnesium	18.-94.	6.21	10.0
Manganese	0.13-2.83	0.309	0.014
Calcium	487.-744.	250.	546.
Potassium	71.-315.	11.5	30.6
Sodium	73.-238.	25.1	13.3
Titanium	<0.05	<0.1	<0.05
Sulfate	2200.-4090.	420.	1580.
Arsenic	0.008-0.038	0.126	0.140
Barium	0.024-0.034	<0.1	0.072
Boron	1.30-4.80	0.7	1.9
Cadmium	<0.005	<0.05	<0.005
Chromium	0.06-1.44	<0.05	<0.11
Copper	<0.05	<0.05	<0.05
Lead	<0.05	<0.05	<0.05
Lithium	6.1-26.0	0.43	1.07
Molybdenum	2.5-23.2	0.14	2.52
Nickel	<0.5	0.06	<0.05
Selenium	0.017-0.304	0.050	0.079
Silver	<0.05	<0.05	<0.01
Zinc	<0.05	0.12	<0.01
Chloride	42.-54.	<2.	<2.

(a) Leachate concentrations (range of samples taken from October 1985 through April 1986).

Continuing Studies

Collection of leachate and pore water samples for chemical analysis will continue through 1987, with less emphasis being placed on the water balance, which has now been fairly well characterized. A preliminary review of the chemical data to establish completeness and validity has already begun. Future activities will include performance of regression analyses to determine key relationships among the measured parameters and application of flow and geochemical computer codes to the data. The ultimate goal of the research is to be able to use these models to predict leachate characteristics at other power plant and ash utilization sites where the ash or design/operating features might be different.

REFERENCES

1. B.W. Rehm, et al., <u>Field Evaluation of Instruments for the Measurement of Unsaturated Hydrological Properties of Fly Ash,</u> EPRI Contract 2485-7, Electric Power Research Institute, Palo Alto, CA (in press).
2. R.R. Proctor, "Fundamental Principles of Soil Compaction," <u>Engineering News-Record</u>, August 31, p. 21, and September 7, p. 28, 1933.
3. <u>Federal Register</u>, Vol. 50, p. 46936, Nov. 13, 1985.
4. <u>Federal Register</u>, Vol. 44, p. 42198, July 19, 1979.

COMPARISON OF THE BEHAVIOR OF TRACE ELEMENTS DURING
ACID LEACHING OF ASHES FROM SEVERAL COALS

J. S. WATSON
Chemical Technology Division, Oak Ridge National Laboratory, Oak Ridge, TN
37831-6224

Received 30 October, 1986; refereed

ABSTRACT

The leaching of fly ash from eastern U.S. coals with strong mineral acid
is the initial step in a series of potential processes for producing useful
and marketable materials from the ash. This initial leaching step removes
most, or all, of the more soluble (generally amorphous) materials from the ash
and leaves an inert residue, believed to be mostly mullite and silica. Chemical analyses of the leachate and the residual inert materials indicate the
original distribution of the trace elements in the phases. Significant
differences in crystallography, composition and leaching behavior have been
noted in ash samples from coals from various regions, of various types, and
sometimes even in ash samples from the same or similar coals. Trace element
analyses of fly ash leachates provide a useful means of studying coal characteristics and of determining how the trace elements are incorporated in
various types of fly ashes.

INTRODUCTION

Fly ash from coal combustion power plants is a complex material consisting of several phases containing numerous elements. Although the major phases
of fly ash consist of only a few principal elements, they also contain many
trace elements. Determining the distribution of these trace elements in the
leachable and nonleachable phases of fly ashes is important in many potential
processes for using and in disposal of the ash. To determine the distribution
of both bulk and trace elements among the ash phases, one would prefer to
separate each of the phases from several ashes and determine quantitatively
the composition of each phase. However, quantitative separation of all of the
phases in fly ash is not possible; this brief paper discusses a less ambi-
tious, but simpler, approach in which we determine only the distribution of
the elements between those phases which are easily dissolved in concentrated
mineral acids and those which remain insoluble in acids. The results of this
approach are inherently limited, of course, and there may also be cases where
results are affected by some soluble materials being surrounded by insoluble
phases and thus protected from acid dissolution.
Despite these limitations, this approach provides useful information that
directly addresses the behavior of ashes in dissolution processes. It also
provides information relevant to the leachability of ashes under environmental
disposal conditions, although the aggressive leaching of ash by mineral acids
will certainly dissolve far more material and dissolve it more rapidly than
environmental leaching. These steady acid leaches may even dissolve entire
phases that would hardly be affected under normal environmental disposal
conditions. Nevertheless, these results provide information on the distribution of trace elements in fly ash which should be helpful in considering
environmental issues.
The most soluble phases of fly ashes are believed to be the amorphous
silicates. Ashes from eastern coals often consist principally of silica,
leachable silicates, and mullite [1,2], with lesser amounts of other phases
(such as iron-rich material possibly formed from oxidation of pyrite, and
calcium sulfate-based phases). Alkaline ashes from western coals having a
high calcium content are likely to have little or no mullite but more of the

other phases. Most of the fly ashes used in this study were from eastern coals which contain little calcium. The calcium present is concentrated in the leachable amorphous phases.

Trace elements in fly ashes can be incorporated within a leachable or unleached phase in various ways. Some elements can be incorporated as "substitution" ions within crystalline materials. Others may be "trapped" ions within the crystalline or amorphous materials or may be present as very small particles which are enclosed or otherwise protected from acid dissolution by an insoluble phase(s).

EXPERIMENTAL

The 22 ash samples used in this study were obtained from 19 utility companies participating in an ash utilization study sponsored by the Electric Power Research Institute [3,4]. These samples were composites collected to be representative of the ash from each individual power plant. Most of the samples were from low-calcium eastern bituminous coals, although there was one sample from a high-calcium coal ash and one anthracite coal ash sample.

In our testing procedure, the ash samples were initially leached for 4 h with 6 N hydrochloric acid at reflux (boiling) temperature. Earlier tests [5] indicated that similar results can be obtained by leaching with other strong mineral acids (sulfuric or nitric), but these require somewhat longer leach times. Also, both the current and earlier studies showed that, with hydrochloric acid, longer leach times did not result in significantly greater leaching.

Trace-element analyses of the ashes, leachates, and residues were made with inductively coupled plasma spectrometry (ICP). Several tests were also made using atomic absorption (AA) measurements, but the precision and accuracy of the ICP measurements appeared to be comparable to those of the AA tests. As an additional check of the analyses results, material balances were prepared for each element by analyzing samples of the original ash, the leached residue, and the leachate solution. In most cases, agreement between the ICP analyses and the material balances was within $\pm5\%$ for the major elements and within $\pm20\%$ for the trace elements. When the material balance was poor, the results were either repeated or rejected. Results of the material balances are not reported here, but they may be obtained from an earlier publication [3].

This study examines the portion of several trace metals leached from the different ash samples as a function of either the ash analyses or the fraction of other metals leached. Each data point shown on the figures corresponds to analyses from a separate ash and its leachate (or residue). The study is based upon the reasonable and frequently used assumption that fly ash samples from similar types of coals are likely to have many similarities. Although fly ashes are affected by combustion conditions, modern pulverized-coal combustion systems usually have much in common in terms of temperature and residence time. Because of the possibility of variations in conditions, however, one should expect significant scatter in results from this or any similar study attempting to compare ashes from different coals and power plants. The low concentrations of many elements of interest also make the analyses more difficult and less precise. Typical analyses for several elements are shown in Table I.

RESULTS AND DISCUSSION

Aluminum Leachability

One way to obtain an estimate of the quantity of leachable phases in a coal ash is to look at the aluminum leachability. It is present in most leachable silicate phases, and its leachability usually follows the weight

Table I

Concentration of Several Elements
in the Ashes Used in This Study

Element	Range of Concentrations ppm (as element)
Al	80,000 – 140,000
Ca	6,200 – 150,000
Fe	22,000 – 120,000
Mg	4,000 – 40,000
Sr	200 – 2,600
Na	3,000 – 18,000
K	750 – 37,000
Ni	30 – 200
Co	10 – 90
Pb	120 – 270
Cu	30 – 60

loss from leaching. Previous studies have shown that aluminum leachability depends upon the calcium or total alkali and alkaline earth content of the ash [6]. This is believed to result because the calcium and perhaps other alkali or alkaline earth elements hinder the formation of mullite, an essentially insoluble phase that also contains aluminum. The percentage of aluminum leached as a function of the calcium concentration is shown in Figure 1. In Figure 2, it is shown as a function of the total alkali and alkaline earth concentration. These curves both show a positive trend toward greater leachability as a function of either calcium or total alkali and alkaline earth concentration in the ash. Although there seems to be somewhat better dependence upon the calcium concentration, it would be difficult to defend any claim for differences in the two plots. The data plotted in these two figures are mostly from low-calcium ash samples, and thus the data are concentrated toward the lower left sides of the figures and are not evenly distributed. Nevertheless, the positive dependence is reasonably established by these data and by data from earlier studies [6].

Two data points in these figures differ significantly from the others and should be mentioned separately. The ash sample which showed ~50% aluminum leachability with <5000 ppm elemental calcium was from an anthracite coal (Penn P&L). It is not clear to us how the coal type causes such a large difference in the ash, although it could be related to the combustion methods used. We also have no explanation for the fact that the sample which contained almost 40,000 ppm calcium showed only ~12% aluminum leachability. An examination of partially leached ash by Hulett and co-workers [7] indicated that mullite crystals grow in low-calcium ashes. Heating such ashes for extended times at temperatures of ~900°C can decrease the aluminum leachability considerably, probably because of additional growth of mullite, and it is possible that this ash sample had a longer period than the other ashes at elevated temperatures.

Alkaline Earth Elements

Although other alkaline earth elements generally behave much like calcium, they are unlike calcium in their distribution among the soluble and insoluble phases of fly ashes. Figures 3 and 4, respectively, show the

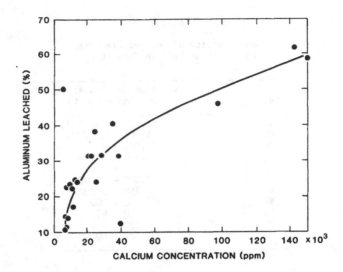

Figure 1. Effects of calcium concentration on aluminum leachability.

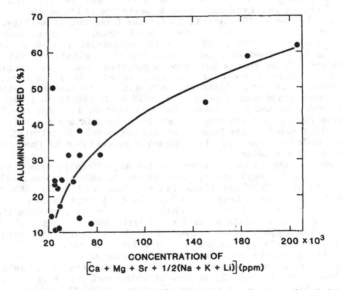

Figure 2. Effects of Groups I and II elements on aluminum leachability.

magnesium and strontium leachabilities plotted as a function of the calcium leachability. The lines shown represent equal leachability for these two elements. Note that leachabilities of both magnesium and strontium are considerably less than those of calcium. Both magnesium and strontium have similar lower leachabilities. The only data point that shows similar leach-

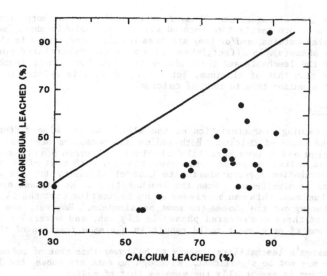

Figure 3. Magnesium leachability vs calcium leachability.

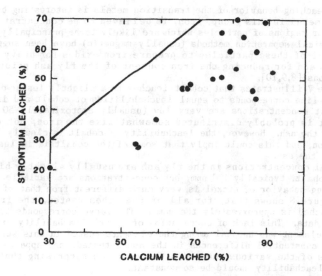

Figure 4. Strontium leachability vs calcium leachability.

ability for calcium and magnesium is from a high-calcium ash (see Figure 3).
These data show that the leachabilities of magnesium and strontium are similar
(within the uncertainty), but they differ from the leachability of calcium.
Even the scatter in the data for strontium and magnesium appears to be similar, although the concentration of strontium is usually an order of magnitude

lower than that of magnesium. These results indicate that both strontium and magnesium are more easily incorporated within an insoluble phase (perhaps the mullite) than calcium, and/or they are less easily incorporated in the soluble amorphous phase(s). In effect, these elements are distributed more evenly throughout the leachable and inert phases than calcium. Their leachabilities are higher than that of aluminum, but their behavior is related more closely to that of aluminum than to that of calcium.

Alkali Metals

The leaching characteristics of the alkali metals also differ significantly from those of calcium. Both sodium and potassium are less leachable than calcium and behave more like aluminum. Figures 5 and 6 compare the leaching of sodium and potassium, respectively, with that of aluminum. The sodium distribution appears similar to that of aluminum, but the scatter in the results is significant. When the leachability of an element approximates that of aluminum, this can be viewed as an indication that the element distributes throughout the phases the same as aluminum. However, one must also bear in mind there are several phases in fly ash, and several alternative distributions of an element could result in the same fraction of the element being leached.

Potassium's leachability appears to be lower than that of sodium, but the difference may not be significant. Although no data are shown, the leachability of lithium is essentially the same as that of sodium.

Transition Metals

The leaching behavior of the transition metals is interesting because the prominent metal in this group, iron, is believed to be concentrated in some particles or regions of particles which are likely to be principally oxides of iron. Physical separation methods (usually magnetic) have been suggested for the removal of these particles to prepare iron oxides for heavy media and other uses and for reducing the iron content of the fly ash prior to other applications [8,9,10].

Figure 7 illustrates that cobalt leaches to a slightly lesser degree than iron. The line corresponds to equal leachabilities of cobalt and iron. Typical cobalt concentrations are very low (usually approximately 50 ppm), and the cobalt is probably distributed somewhat like the iron, not uniformly throughout the ash. However, the leachability of cobalt is clearly lower than that of iron, and this could imply that some of the cobalt is in less soluble regions of the ash.

Nickel concentrations in the fly ash are usually slightly higher than those of cobalt, typically 100 ppm, but concentrations are variable. However, the leaching behavior of nickel is very much different from that of cobalt or iron. Figure 8 shows that, for all of the ashes tested, the fraction of nickel leached is approximately the same. The curve corresponds to a visual fit to the data. This lack of correlation of nickel leachability with that of any other element was unusual among the elements tested in this study. Since there were substantial differences in the ashes tested, and apparently in the quantities of the various phases in these, it is surprising that nickel's degree of leachability should be so constant.

Other Metals

A few other elements appear to have essentially constant or slightly varying leachabilities. Only four ashes showed titanium leachabilities outside the 10 to 20% range. This could mean that a major portion of the titanium is in a separate inert phase, but it suggests that the fraction of titanium in that phase is approximately the same for all coal ashes of the

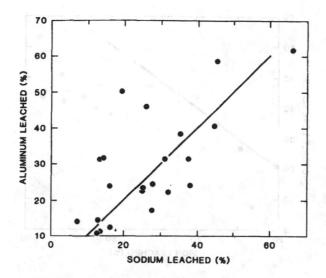

Figure 5. Aluminum leachability vs sodium leachability.

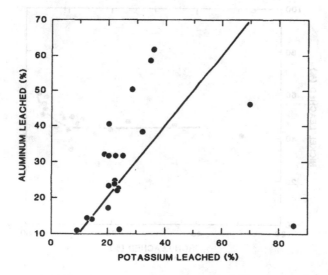

Figure 6. Aluminum leachability vs potassium leachability.

class considered here. Barium also demonstrated very similar leaching behavior in all ashes; practically all of the barium is insoluble in concentrated acid. Only one analysis indicated that more than 10% of the barium was leached, and the average percentage leached was only 3.4%. This can be explained by the presence of sulfates in the ash and the presence of sulfur

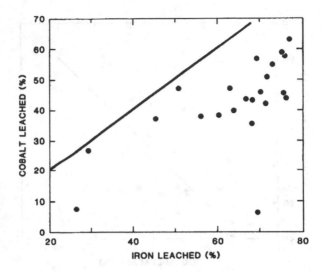

Figure 7. Cobalt leachability vs iron leachability.

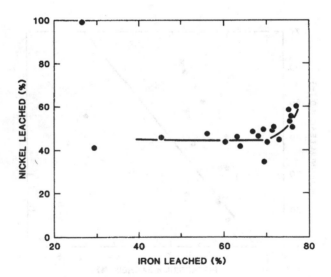

Figure 8. Nickel leachability vs iron leachability.

oxides during the combustion process, since the barium is likely to be associated with anhydrite or some other sulfate phase. Solubility of the barium
sulfate would be extremely low, regardless of the phase in which it is found.
     Other elements showed significant variation in leachability, with no
apparent trends. Phosphorous leachabilities usually ranged from 0 to 60%.

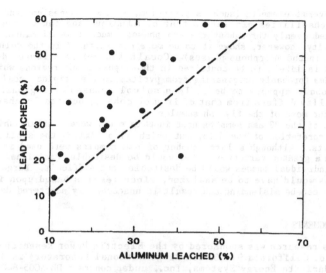

Figure 9. Lead leachability vs aluminum leachability.

One could suspect that significant phosphate would be found in the amorphous phase(s), but the distribution between soluble and insoluble phases would be difficult to predict. Beryllium leachability usually ranged between 20 and 60%, and there was some indication that its leachability increased with aluminum leachability, but the data scatter did not justify claiming such a dependence. Two other elements of potential environmental importance, cadmium and zinc, showed similar leaching behavior. Cadmium leachabilities ranged from 20 to 100%, and zinc leachabilities ranged from 30 to 80%.

In Figure 9, lead leachabilities are plotted against aluminum leachabilities. The lead is shown to be significantly more soluble than aluminum, and this solubility generally increases with the solubility of aluminum. The lead could be incorporated within the soluble silicate phases, but, to behave as shown in Figure 9, the lead would have to be concentrated in the soluble phase to a greater extent than is the aluminum. As noted in Table I, the concentrations of Pb in the ashes were all greater than 100 ppm and appeared to be well within the analytical detection limits. However, the concentrations in the leachate varied more widely and leach solutions from three ashes were too low to be measured quantitatively (a few ppm or below).

SUMMARY AND CONCLUSIONS

By examining the fractions of elements that are leached from fly ash samples produced in several power plants, one can obtain indications of how those elements are distributed in the soluble and insoluble phases of fly ash. Only those fractions of the elements in soluble phases that can contact the acid and are not protected by inert phases are detected by these leach tests. Since aluminum is a major component of the two major phases in the low-calcium ashes from most eastern coals, a comparison of any element's leachability with that of aluminum provides a crude method of estimating whether that element was distributed throughout the mullite and soluble phases of the ash in the same way as aluminum. This approach avoids the difficulties involved in determining the amount of each element in each phase in all of the ash samples. Several elements (such as potassium, sodium, magnesium, and strontium) show leachabilities similar to that of aluminum and can be considered as distributed evenly throughout the ash phases, much like aluminum. Calcium's leachability, however, shows it to be more concentrated in the soluble phases (that is, in the amorphous phases). Cobalt leaches to a lesser degree than iron. It is likely to be contained in those phases and regions which contain iron oxides (probably originating from pyrites in the original coal), but some of the cobalt appears to be in less soluble phases or regions. Nickel's leachability differs from that of iron or cobalt, but its leachability was similar for most of the fly ash samples.

Most of the 22 ash samples used in this study were similar, which simplified interpretation of the data, but which also limited the applicability of the results. Although a large number of ash samples were used in the study, data from a greater variety of ashes would be desirable. Additional microanalyses of individual phases would be desirable, but difficult. Large numbers of particles would have to be analyzed, since results based upon individual particles can be misleading and result in unacceptably scattered data.

ACKNOWLEDGMENTS

This research was sponsored by the Electric Power Research Institute, Palo Alto, California 94303. Oak Ridge National Laboratory is operated by Martin Marietta Energy Systems, Inc., under contract DE-ACO5-84OR21400 with the U.S. Department of Energy.

REFERENCES

1. B.E. Scheetz and W.B. White, in Fly Ash and Coal Conversion By-Products: Characterization, Utilization, and Disposal I, Mat. Res. Soc. Symp. Proc. Vol. 43, edited by G.J. McCarthy and R.J. Lauf (Materials Research Society, Pittsburgh, 1985), pp. 53-60.
2. D.M. Roy, K. Luke, and S. Diamond, in Fly Ash and Coal Conversion By-Products: Characterization, Utilization, and Disposal I, Mat. Res. Soc. Symp. Proc. Vol. 43, edited by G.J. McCarthy and R.J. Lauf (Materials Research Society, Pittsburgh, 1985), pp. 3-20.
3. J.S. Watson, G. Jones, and R.F. Wilder, Recovery of Metal Oxides from Fly Ash Including Ash Beneficiation Products. Vol. 2: Resource Analyses and Evaluation, EPRI CS-4384, Electric Power Research Institute, Palo Alto, California (1986).
4. J.S. Watson, in Fly Ash and Coal Conversion By-Products: Characterization, Utilization, and Disposal II, Mat. Res. Soc. Symp. Proc. Vol. 65, edited by G.J. McCarthy, F.P. Glasser and D.M. Roy (Materials Research Society, Boston, Massachusetts, 1985), pp. 59-67.
5. R.M. Canon, T.M. Gilliam, and J.S. Watson, Evaluation of Potential Processes for Recovery of Metals from Coal Ash, Vol. 1, EPRI CS-1992, Electric Power Research Institute, Palo Alto, California (1981).
6. A.D. Kelmers et al., Res. Conserv. 9, 271-79 (1982).
7. L.D. Hulett, Jr. et al., Science 210, 1356-58 (1980).
8. R.G. Aldrich and W.J. Zacharias, in The Challenge of Change - Sixth International Ash Utilization Symposium Proceedings, Vol. 2, DOE/METC/82-52, edited by J.S. Halow and J.N. Covey (1982), pp. 1-21.
9. D. Mitas, M.J. Murtha, and G.Burnet, in The Challenge of Change - Sixth International Ash Utilization Symposium Proceedings, Vol. 2, DOE/METC/82-52, edited by J.S. Halow and J.N. Covey (1982), pp. 22-45.
10. H.S. Wilson and J.S. Burns, in Proceedings of the Seventh International Ash Utilization Symposium and Exposition, Vol. 1, DOE/METC-85/6018 (1985), pp. 304-16.

AN ESCA AND SEM STUDY OF CHANGES IN THE SURFACE COMPOSITION AND MORPHOLOGY
OF LOW-CALCIUM COAL FLY ASH AS A FUNCTION OF AQUEOUS LEACHING

MYRA M. SOROCZAK*, H. C. EATON** and M. E. TITTLEBAUM
College of Engineering, Louisiana State University, Baton Rouge, LA 70803 USA
*Visiting from: Tennessee Valley Authority, National Fertilizer
Development Center, Muscle Shoals, AL 35660 USA
**Author to whom correspondence should be addressed.

Received 20 October, 1986; refereed

ABSTRACT

The reactivity of coal fly ash is dependent on the chemical composition
of the surface. As reactions occur the ash particle size decreases and new
material is available for reaction. This means that the near-surface chemistry can also be important. In the present study the surface chemistries of
three ashes are determined by x-ray photoelectron spectroscopy both before and
after exposure to a hydrating/leaching environment. Scanning electron microscopy is used to reveal ash morphology. The concentration of sulfur, found at
the ash surfaces as a sulfate, and sodium decreased after leaching while the
amount of iron and aluminum increased. Other elements, including calcium,
increased and decreased with leaching depending on which ash was analyzed.
Changes which occurred in the ash morphology after the removal of leachable
elements are discussed.

INTRODUCTION

The production of steam by conventional coal-fired generators and large
industrial boilers is expected to become increasingly important in the United
States over the next decade. The amount of coal ash produced from conventional
combustion was 47.6 million metric tons in 1983 and is estimated to be 110
million metric tons by 1995. In 1983, about 85% of the ash produced in the
U.S. was sent to disposal sites for ponding, landfilling (including disposal
in surface mines), or interim ponding followed by landfilling [1]. Consequently, disposition of the ash produced as a byproduct of combustion will
become a significant problem for industry.
The mineral matter in coal is composed of micrometer-size crystallites of
clays, carbonates, quartz and pyrite [2]. They are embedded in a carbonaceous
matrix. Fly ash is produced at the combustion front in a coal-burning power
plant. During combustion many of the minerals melt and coalesce producing the
ash. The total amount of ash produced varies with the mineral content but can
range from a few percent of the weight of the unburned coal to as much as 35%.
Bottom ashes are also produced, but standard pulverized coal-fired boilers
typically produce 80 to 90% of the ash as fly ash [1].
Low-calcium (Class F) bituminous coal fly ashes consist principally of
five components: crystalline, substituted magnetite; hematite; amorphous
glass; crystalline, substituted mullite; and quartz [3]. As the particles or
small spheres move through the stack, elements and compounds volatilized in
the combustor condense onto the ash surfaces. Most of the phases in the bulk
portion of the ash particles, e.g. aluminum and silicon oxides, pose no hazard
to the environment, but many of the volatilized elements, e.g. Cd, Co, Cr, Se,
and Pb, can. Many of these metals are present at the fly ash surfaces as
soluble compounds and therefore are available for ground-water contamination
if leaching by rainwater occurs.
In a study of coal ash deposit ponds in eastern Tennessee and northeast-
ern Alabama, Milligan and Ruane [4] found that coal ash leachates were highly
variable solutions characteristically high in dissolved solids, boron,

Figure 1. Scanning electron micrograph of unleached SRM 1633a fly ash.

Figure 2. Scanning electron micrograph of leached SRM 1633a fly ash.

calcium, aluminum, and sulfate. Several constituents, including Cd, Cr, Fe, Mn, and Pb, were found to exceed EPA's criteria for safe drinking water. The acidic ash at one plant produced higher concentrations of metals in the leachate than the alkaline ash at another plant. Milligan and Ruane concluded that further research was needed to determine the toxic species in pond water leachates.

Work reported in the present paper was part of a larger study to determine the chemical composition of several fly ash leachates and to compare these results with surface chemistry of the fly ashes as determined by x-ray photoelectron spectroscopy (XPS). Since XPS data can be taken and interpreted in a few hours as opposed to much longer times needed for leaching test analyses, this technique would represent a novel way to predict disposal characteristics of fly ash.

EXPERIMENTAL

Three low-calcium fly ashes were examined in the present study. They were produced in plants burning eastern bituminous coal; the source of each ash has been discussed previously [5]. Fly ash A is a composite of samples taken from either of two generator units at a power plant in West Virginia. Fly ash B is a composite of samples taken from three generator units at a power plant in Virginia. Both plants used coal-fired turbine-generators which burned pulverized coal in an air suspension with fly ash removed from flue gases by electrostatic precipitators [6]. The National Bureau of Standards (NBS) Standard Reference Material (SRM) 1633a is used primarily as a calibration standard for trace metals in fly ash. The fly ash was supplied to NBS by a coal-fired power plant and is a product of Pennsylvania and West Virginia coals. NBS passed the ash through a No. 170 sieve (90 μm) to produce SRM 1633a [7].

Each fly ash was leached using the American Society for Testing and Materials (ASTM) Method A. The method calls for 350 g of fly ash to be agitated for 48 h in deionized, distilled water. The samples are then filtered and air pulled over the solids until they are dry. Because of previous work which had shown a relationship between particle size and leachate composition, fly ash A was screened into fractions greater than or less than 53 μm before leaching.

Samples were prepared for scanning electron microscopy (SEM) and energy dispersive x-ray analysis (EDXA) by mounting them on aluminum sample holders with double stick tape. Some, but not all, of the prepared samples were coated with gold to enhance conductivity. They were examined in a Cambridge S-10 SEM. Iron pole pieces in the microscope resulted in a small iron peak in the background; all fly ash spectra in this report, however, contained iron peaks above the artifact level.

Samples were prepared for XPS by pressing the fly ash into indium foil. A Perkin Elmer Physical Electronics XPS system was used. Following the method of Campbell, et al. [9], atomic concentration of element A (C_A) was calculated by the relationship:

$$C_A = \frac{I_A/S_A}{Sum(I_x/S_x)} \tag{1}$$

where I_A and S_A are the peak height and sensitivity factor, respectively, of element A. Sensitivity factors were taken from standard tabulations appearing in the literature [10]. Weight percents (wt.%) were calculated by

$$Wt.\% = \frac{C_A * A_A}{Sum(C_x * A_x)} \tag{2}$$

where A_A is the atomic weight of element A.

RESULTS

SEM Characterization of SRM 1633a

Samples of unleached and leached SRM 1633a were examined using SEM/EDX and are shown in Figures 1 and 2, respectively. The particles in the leached samples were more dispersed than in the unleached samples suggesting that either a surface film or irregularities, such as those shown in Figures 3 and 4, caused the unleached particles to agglomerate. Surfaces in the leached sample were generally smoother, indicating that surface material was removed during the leaching process. However, area scans, shown in Figure 5, at low magnifications gave the same EDX spectra for both leached and unleached samples.

Figure 6 gives a general view of unleached SRM 1633a. Particles A, B, and C, marked on the figure with arrows, contained Al, Si, K, Ca, Ti, and Fe in varying amounts. Particle C also contained sulfur. Particle B, shown in

Figure 3. Scanning electron micrograph of an unleached
particle in SRM 1633a fly ash.

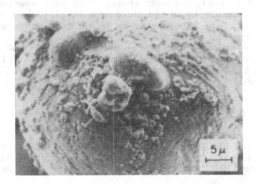

Figure 4. Scanning electron micrograph of an unleached
particle in SRM 1633a fly ash.

Figures 7 and 8 in higher magnifications, was a cloudy glass globule with an
opaque spot. At higher magnification, the opaque spot appeared to be a dendritic surface growth. EDX analysis revealed a significantly higher amount of
iron in the dendritic area.

A general view of SRM 1633a after leaching is shown in Figure 9. The
micrograph shows a lack of agglomeration and little surface detail. The large
smooth particles in Figure 10 are rich in aluminum and silicon, and the particle with dendritic growth patterns was rich in iron.

Figure 11 shows another iron-rich particle with diamond-shaped dendritic
growth patterns. The second large particle is high in aluminum and silicon
and fits the description of an unbroken plerosphere given by Shibaoka and
Paulson [11]. It has the rounded opening suggesting a cenosphere ruptured
from internal gas pressure while the particle was still near the melting point
of the ash. Small spheres, which could have entered through the opening at
any time after solidification, fill the cavity.

SRM 1633a

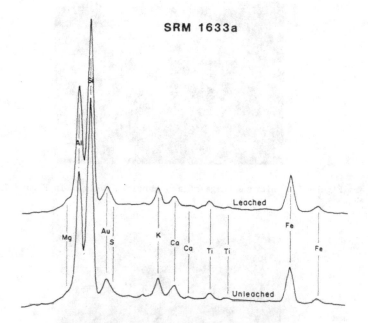

Figure 5. EDX scans from leached and unleached SRM 1633a fly ash.

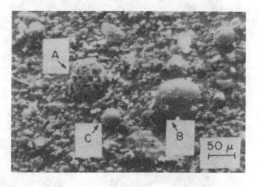

Figure 6. Scanning electron micrograph of unleached SRM 1633a
fly ash. The particles marked A and C have compositions which are discussed in the text.

The large particle shown in Figure 12 had an unusual chemical makeup and
was the only particle examined which contained detectible amounts of barium.
EDX analysis showed that the particle contained major amounts of barium and
sulfur as well as the more common Al, Si, Ca, and Fe.

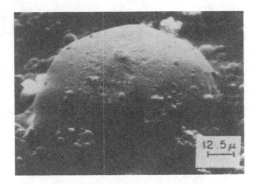

Figure 7. Enlarged image of ash particle B shown in Figure 6.

Figure 8. A further enlargement of particle B.

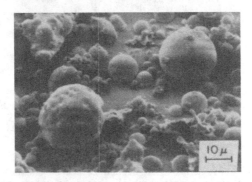

Figure 9. Scanning electron micrograph of leached SRM 1633a fly ash.

Figure 10. Scanning electron micrograph of selected particles in leached SRM 1633a fly ash.

Figure 11. A large iron rich particle (lower portion of micrograph) in leached SRM 1633a fly ash.

Figure 12. A large particle of unusual composition in SRM 1633a fly ash.

Table I

Average Chemical Composition (wt.%) of Fly Ash Surfaces

	SRM 1633a	Ash B	Ash A, <53 μm	Ash A, >53 μm
		Unleached		
C	10.9 +/- 2.8	9.2[a]	8.1 +/- 1.6	13.2 +/- 2.7
O	41.5 +/- 4.1	39.7	40.5 +/- 1.0	34.2 +/- 2.6
Al	12.2 +/- 1.5	14.7	14.2 +/- 0.7	14.5 +/- 1.2
Si	21.8 +/- 2.3	22.2	26.8 +/- 0.4	22.8 +/- 3.2
N	0.39 +/- 0.38	ND[b]	0.46 +/- 0.46	ND
Na	0.24 +/- 0.17	0.52	0.80 +/- 0.16	0.55 +/- 0.34
P	0.11 +/- 0.31	ND	ND	2.79 +/- 2.79
S	6.02 +/- 1.31	1.30	4.93 +/- 0.32	5.07 +/- 1.54
K	0.93 +/- 0.69	1.01	0.88 +/- 0.88	ND
Ca	2.04 +/- 0.65	8.1	0.54 +/- 0.56	1.58 +/- 1.16
Ti	0.40 +/- 0.42	ND	ND	ND
Fe	2.62 +/- 0.75	1.68	0.60 +/- 0.60	5.20 +/- 1.79
Co	0.72 +/- 0.67	0.89	2.26 +/- 0.61	ND
		Leached		
C	12.3 +/- 1.9	10.6 +/- 2.1	11.8 +/- 2.4	10.9 +/- 1.3
O	40.0 +/- 3.7	37.0 +/- 3.7	35.3 +/- 2.7	32.0 +/- 2.4
Al	14.5 +/- 0.5	14.5 +/- 0.6	16.9 +/- 2.5	23.3 +/- 1.8
Si	22.2 +/- 2.0	21.7 +/- 1.5	22.1 +/- 1.5	23.1 +/- 0.9
N	0.57 +/- 0.35	ND	ND	ND
Na	0.06 +/- 0.12	0.52 +/- 0.31	0.45 +/- 0.51	0.29 +/- 0.29
P	2.31 +/- 1.12	ND	ND	ND
S	2.49 +/- 0.74	2.66 +/- 0.82	3.33 +/- 0.24	0.53 +/- 0.92
K	ND	0.92 +/- 0.92	0.38 +/- 0.66	0.25 +/- 0.43
Ca	0.36 +/- 0.36	8.40 +/- 0.92	3.53 +/- 0.84	3.86 +/- 0.86
Ti	0.57 +/- 0.50	ND	ND	ND
Fe	3.25 +/- 0.52	3.71 +/- 0.36	6.20 +/- 1.69	5.81 +/- 0.92
Co	1.34 +/- 0.97	ND	ND	ND

a. Only one sample analyzed.
b. Not detected.

XPS Characterization of Unleached and Leached Fly Ashes

The surfaces of unleached and leached samples of the fly ashes were examined using XPS. The major elements found at the surface of all samples were carbon, oxygen, aluminum, and silicon. Minor elements detected in varying amounts were N, Na, P, S, K, Ca, Ti, Fe, and Co. Table I lists the average surface compositions for each ash. Table II gives the leachate compositions for each ash as determined by Eisenberg [8].

Sodium, potassium, calcium and sulfur were found in higher amounts on the surface of the unleached SRM 1633a than on the leached surfaces. Aluminum and phosphorous were significantly higher and nitrogen, iron, and cobalt were only slightly higher on the leached surfaces.

Table II

Leachate pH and Chemical Analysis [8]

	pH	Ca	SO₄	P
		mg/L		
SRM 1633a	4.15	310	916	0.078
Ash A, <53 μm	7.25	123	284	0.016
Ash A, >53 μm	7.40	69	148	0.053
Ash B	11.1	273	332	<0.01

	Fe	Co	Cd	Cr	Cu	Pb	Mn	Ni	Zn
					mg/L				
SRM 1633a	0.159	0.066	0.047	0.435	0.365	<0.05	0.240	0.730	0.679
Ash A, <53 μm	0.068	<0.05	<0.025	0.741	<0.05	<0.05	<0.05	<0.01	<0.01
Ash A, >53 μm	0.290	<0.05	<0.025	0.662	<0.05	<0.05	<0.05	<0.01	<0.01
Ash B	0.051	<0.05	<0.025	0.743	<0.05	<0.05	<0.05	<0.01	<0.01

Sulfur was also found in higher amounts on the unleached surfaces of Ash B than on the leached surfaces. Carbon was somewhat higher in concentration on the unleached surfaces. Iron and, possibly, oxygen were detected in higher amounts on the leached surfaces.

Sulfur, sodium and cobalt were higher on the unleached than leached surfaces of both size fractions of the Ash A. Calcium was higher on the leached surfaces of both fractions. Nitrogen was detected only on the surface of the unleached <53 μm Ash A, and phosphorous was detected only on the unleached surfaces of the >53 μm fraction. Potassium was greater on the leached surface of the >53 μm fraction, but unchanged on the <53 μm fraction. Iron was much higher after leaching on the <53 μm fraction, but only slightly higher on the larger size fraction.

Of the four major elements, only aluminum was higher after leaching for both size fractions of Ash A. Carbon was higher after leaching the <53 μm fraction, but was higher before leaching the larger size fraction. Silicon and oxygen were higher on the unleached <53 μm fraction, but did not significantly change with leaching of the >53 μm fraction.

The binding energy of the sulfur 2p peak on all the ash surfaces indicated that the sulfur was present as sulfate. The carbon 1s peak on the SRM 1633a and both Ash A fractions was an overlapping double peak with the peak of lower binding energy usually showing up as a shoulder. The carbon 1s peak on Ash B was also a double peak, but the peaks were so nearly the same height that on some scans it looked more like a single peak with a flattened top. The separation in binding energy of the two carbon peaks was approximately 2 eV regardless of ash source or leaching. When an electron gun was used to neutralize charging so that the Si 2s peak at 103.4 eV was used as the energy standard, carbon 1s peak splitting remained at 2 eV. However, if the neutralization was changed so that the carbon 1s peak at 284.6 eV was used as the energy standard, the two carbon peaks converged.

DISCUSSION AND CONCLUSIONS

The chemistry and morphology of coal fly ash are complex functions of the coal mineralogy and the combustion conditions. A variety of ash morphologies have been reported in the literature and have been observed also in the present study. Spherical plerospheres are seen as well as irregularly shaped particles which may be agglomerates or more complex products of incomplete combustion.

Spheres with a complex surface morphology were observed. On certain portions of the surface, an iron-rich growth developed. The structure was similar in appearance to dendritic morphologies commonly observed in slowly cooled alloy systems. Dendrites normally develop because of crystallographic anisotropy in the rate of solidification. This suggests that the observed surface features were crystalline deposits on an amorphous, glass sphere of ash.

Carbon was observed in significant amounts in the XPS spectra. The carbon peaks were observed to be split by approximately 2 eV. Graphitic carbon and carbonate peaks are separated by approximately 4 eV [10]. It is likely, therefore, that the split peaks observed in the present study are due to a combination of residual coal, coal char (graphitic) or adventitious carbon. These results suggest that XPS may be useful for studies of incomplete combustion.

XPS was determined to be an unreliable method for quantitatively predicting the composition of leachate. For example, Cr, Cu, Zn and Ni were all detected in the leachate but were not observed at the surface of the ash particles prior to leaching. This means that either the concentrations of these elements were below the limits of detection of XPS or that the surface concentration was low.

Some elements were observed at the surface of the ash particles and not in the bulk. For example, Na, K, Ca and S were all found in higher concentrations at the surface of the unleached SRM 1633a ash. This was not true for the other ashes examined suggesting that this is not a characteristic of all ashes. Here we are assuming that observation of an element at the surface of the unleached ash and not at the surface of the leached ash implies that prior to leaching the element resided only at the surface. It might be argued that the leachant might allow the transport of interior elements to the surface. We consider this to be possible but not a major mechanism since most of the ash particles observed were of the glassy type into which water penetration would be difficult.

Some particle size dependent compositions were also observed. This is consistent with the observations of others. Nitrogen was detected on the surface of the unleached <53 μm Ash A material. Phosphorus, on the other hand, was found only on the unleached >53 μm Ash A. Clearly, further work is warranted to determine the reasons for these observations.

In summary, XPS and SEM were used to characterize the surface chemistry of selected fly ashes produced by the combustion of Eastern coals. The ashes were examined both before and after leaching by distilled, deionized water. XPS confirmed the surface enrichment of certain elements but was determined to be inappropriate for the quantitative prediction of leachate chemistry. Some elements were observed only to be present at the surface of certain size fractions. Differences in surface chemistry along with differences in the XPS carbon peak shapes among the fly ashes suggest that studies of ash chemistry may be useful in diagnosing the efficiency and characteristics of a particular combustor.

ACKNOWLEDGEMENTS

The authors wish to thank the U.S. Department of Energy for its financial support through the LSU Center for Energy Studies, and the Tennessee Valley Authority for graduate student support.

REFERENCES

1. C.J. Santhananam, C.B. Cooper, and A.A. Balasco, Chemical Engineering Progress, 82, 42 (1986).
2. M.O. Amdur, A.F. Sarofim, M. Neville, R.J. Quann, J.F. McCarthy, J.F. Elliot, H.F. Lam, A.E. Rogers, and M.W. Conner, Environ. Sci. Technol., 20, 138 (1986).
3. R.M. Canon, T.M. Gilliam and J.S. Watson, Evaluation of Potential Processes for Recovery of Metals from Coal Ash, EPRI CS-1992, Volume 1 (Electric Power Research Institute, Palo Alto, CA, 1981).
4. J.D. Milligan and R.J. Rurane, Effects of Coal Ash Leachate on Ground-water Quality, TVA-EPA Program Report, EPA-600/7-80-066 (1980).
5. M.D. Patil, H.C. Eaton, and M.E. Tittlebaum, Fuel, 63, 788 (1984).
6. G.T. Seese, Jr., Master's Thesis, West Virginia University (1980).
7. National Bureau of Standards, Analysis Sheet on SRM 1633a (1980).
8. S.H. Eisenberg, Master's Thesis, Louisiana State University (1982).
9. J.A. Campbell, R.D. Smith, and L.E. Davis, Applied Spectroscopy, 32, 316 (1978).
10. C.D. Wagner, W.M. Riggs, L.E. Davis, J.F. Moulder, and E. Muilenberg, Handbook of X-Ray Photoelectron Spectroscopy (Perkin Elmer Corporation, Physical Electronics Division, Eden Park, Minnesota, 1978).
11. M. Shibaoka and C.A.J. Paulson, Fuel, 65, 1020 (1986).
12. J.R. Brown, B.I. Kronberg, and W.S. Fyfe, Fuel, 60, 439 (1981).
13. G.E. Cabaniss and R.W. Linton, Environ. Sci. Technol., 18, 271 (1984).
14. S.J. Rothenberg, P. Denee, and P. Holloway, Appl. Spec. 34, 549 (1980).

PROPERTIES AND ENVIRONMENTAL CONSIDERATIONS RELATED TO AFBC SOLID RESIDUES

E.E. BERRY* and E.J. ANTHONY**
*PO Box 7261, Oakville, Ontario, Canada
**E.J. Anthony, CANMET, Energy Mines and Resources Canada, Ottawa, Ontario, Canada

Received 2 February, 1987; communicated by G.J. McCarthy

ABSTRACT

Atmospheric-pressure fluidized bed combustion (AFBC) produces solid residues that are different from the familiar pulverized coal ashes. When limestone beds are used to adsorb SO_x, high-Ca residues, comprised largely of CaO and SO_4, are produced. Leachates from high-Ca AFBC residues are strongly alkaline (pH >11) and contain high levels of dissolved solids (TDS >3000 mg/L). If water is added during handling, hydration of CaO may cause a temperature rise and hydration of $CaSO_4$ may result in premature hardening of the residues. Trace elements and organic components may leach from disposal sites. This paper presents an overview of the nature of AFBC residues and the factors influencing their disposal.

INTRODUCTION

Atmospheric-pressure fluidized bed combustion (AFBC) is rapidly gaining acceptance in North America as a means to use "problem" fuels, such as low-grade or high-sulphur coals for thermal power generation. In AFBC, coarsely crushed fuel is suspended in a combustion chamber with a bed-material, such as sand or limestone. The suspension is sustained by fluidizing the mass with air injected from below the bed.

There are two general classifications of the AFBC process, illustrated schematically in Figure 1, namely:

Bubbling or dense-phase AFBC in which coal is combusted in a mass of limestone or an inert materials that is fluidized at high solids density in such a way that it remains largely in place as a well defined bed.

Circulating AFBC in which both the fuel and a bed of low solids density are suspended in a circulating gas stream.

Although, in principal a number of solids could be used as the bed material, in practice two general types can be distinguished. These in turn produce two distinct classes of AFBC residues:

Limestone or dolomite beds are used to absorb acidic combustion gases (sulfur oxides) when higher sulfur fuels are burned. The solid wastes from limestone-bed systems, termed high-calcium residues in this paper, are usually produced in two forms: a coarse bed-drain comprising spent limestone granules; and a fine baghouse or carry-over ash.

When low-grade coals or coaly wastes are burned, a bed of sand or spent shale, is frequently used, in which little or no absorption of acidic gases occurs. The solid wastes from such systems are termed low-calcium residues in this paper. They are also produced in two forms, of differing particle sizes, both comprise largely calcined clay or shale.

Recently considerable attention has been given to the nature, disposal and possible use of the solid residues from AFBC boilers [1-18]. As has been noted by previous authors [5], the properties of fluidized bed wastes is very

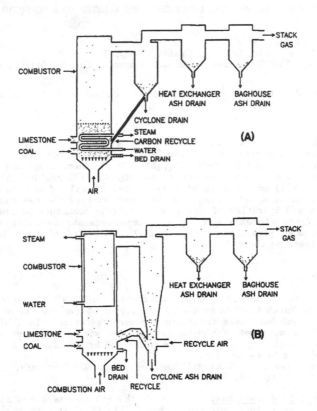

Figure 1. Schematic of material flow in: (A) a bubbling AFBC system;
(B) a circulating AFBC system.

dependent upon a large number of interacting variables. Consequently a wide
variety of materials with differing properties can result from different AFBC
installations or from the same installation when different fuels are burned or
when operating conditions are changed.

This paper provides an overview of the nature of AFBC solid wastes and
the issues of concern regarding their disposal, with particular emphasis on
high–Ca residues. In a companion paper [19], closer attention is directed to
the utilization potential of high–Ca AFBC residues.

CHEMICAL PROPERTIES OF AFBC SOLID WASTES

A limestone fluidized bed combustor is a chemical reactor in which fuel
is burned, limestone is calcined and much of the sulphur oxides produced are
adsorbed. Three chemical processes predominate and in turn influence the
nature of the combustion residues:

$$CaCO_3 \quad \text{-----}> \quad CaO + CO_2 \qquad\qquad (1)$$

$$\text{Sulphur compounds} \quad \text{--oxidation-->} \quad SO_2 \qquad\qquad (2)$$

Table I

Chemical Analyses (wt.%) of AFBC Residues

Major Oxides	High-Ca Residue[a]		Low-Ca Residue[b]	
	Bed Drain	Baghouse	Coarse ash	Fine ash
SiO_2	6.6	11.7	50 – 55	44 – 48
Al_2O_3	2.7	5.8	22 – 26	21 – 25
Fe_2O_3	2.4	10.9	5 – 7	6 – 7
CaO	62.7	41.8	4 – 6	6 – 8
MgO	0.1	1.5	1.5 – 2	1.5 – 2
K_2O	0.3	0.7	4 – 5	3.5 – 4
Na_2O	0.2	0.3	1	1
SO_3	28.6	17.2	3 – 5	5 – 6.5

a. From Berry [7]
b. From Lotze and Wargalla [8]

$$2 \; CaO \; + \; 2 \; SO_2 \; + \; O_2 \quad \text{-----} \rightarrow \quad 2 \; CaSO_4 \qquad (3)$$

Thus, as indicated by previously reported x-ray diffraction data [1,2,4], the residues may contain, $CaSO_4$, CaO, $CaCO_3$ and coal ash components such as cal-cined clays, quartz and iron compounds. In some cases, residues may contain minor concentrations of CaS. In general, residues from bubbling beds contain <0.2% CaS. Circulating bed solids may contain somewhat larger quantities (up to approx. 2%), depending upon from where in the circulating mass the solid is discharged [12].

Low-Ca AFBC residues are largely calcined clay or shale minerals, very little calcium sulphate is formed in the absence of substantial calcium car-bonate in the original shales [8].

For optimal sulphur capture, fluidized bed combustion is conducted at temperatures in the range 815 to 870°C. Because these are much below the temperatures reached during combustion of pulverized coal, melting of the inorganic constituents does not occur. Without the opportunity for ion trans-port through a molten phase, mixing of components is greatly restricted and the formation of glasses is retarded. AFBC residues are thus very different in properties from the fly ashes and bottom ashes formed during pulverized coal combustion.

To illustrate the properties of AFBC residues, data on materials from two sources are discussed in this paper. Solid residues from an AFBC installation at Summerside, Prince Edward Island have been the subject of a number of investigations in Canada [7,14,16,17]. This installation uses limestone and burns a high-sulphur coal; the residues are typical of the high-Ca type. Lotze and Wargalla [8,18] have reported studies on residues from a circulating AFBC system at Lunen, in the Federal Republic of Germany. The fuel used is a high-ash waste coal and a small, but unspecified quantity of pulverized lime-stone is added to the circulating bed. The resulting residues (CaO = 4 to 6%) are typical of the low-Ca class of materials. Because of the site-specific nature of AFBC residues, it is expected that considerable differences will be observed from source-to-source. However, the properties of the two selected wastes are illustrative of the extreme differences between the low- and high-Ca types.

Chemical analyses of wastes from both sources are given in Table I. The expected differences in their overall chemical compositions are clearly seen as is the extent to which separation of constituent elements occurs between the bed-drain and baghouse discharges.

In high-Ca wastes, the bed drain material is largely derived from the calcined and partially sulphated limestone. Coal-ash components, such as

Table II

Estimated Phase Composition (wt.%) Of High–Ca AFBC Residue[a]

	Bed Drain	Baghouse
$CaSO_4$	48.7	29.3
CaO	42.6	29.8
Other[b]	8.7	41.0

a. From Berry [7]
b. Includes carbon and coal ash components

Table III

Trace Element Concentration (mg/L) Of High–Ca AFBC Residue[a]

Element	Bed drain	Baghouse
As	310	220
Ba	154	213
Be	0.1	2.1
B	25	63
Cd	<1	15
Cr	10	23
Co	<5	<5
Cu	8	29
Pb	45	140
Li	8	30
Mn	849	635
Mo	<20	<20
Ni	27	47
P	290	650
Ag	<5	<5
Sr	123	60
Ti	574	1090
V	9	38
Zn	209	343
Zr	20	24

a. From Berry [7]

SiO_2, Al_2O_3 and Fe_2O_3 are more concentrated in the baghouse residue. For the Summerside residues, an estimate was made of the CaO and $CaSO_4$ content of the wastes by calculation from the total Ca and S content. The results are given in Table II and confirm that much of the sulphated limestone remains in the bed material.

Trace elements in high–Ca residues (Table III) are distributed between the bed-drain and baghouse residues, with increase in concentration of some metallic elements in the baghouse discharge.

Because high–Ca residues contain CaO and $CaSO_4$, they react with water:

$$CaO + H_2O \longrightarrow Ca(OH)_2 \qquad (4)$$

$$CaSO_4 + 2H_2O \longrightarrow CaSO_4 \cdot 2H_2O \qquad (5)$$

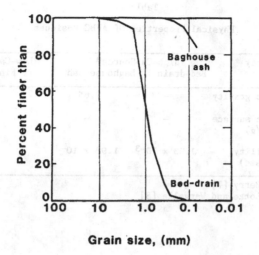

Figure 2. Grain-size distributions of AFBC residues from Summerside PEI.

Both reactions are exothermic ($\Delta H = 1140$ kJ/kg and $\Delta H = 126$ kJ/kg, for reactions 4 and 5, respectively), and the hydration of calcium sulphate results in the formation of a cemented mass. Thus, if high-Ca residues from AFBC are permitted to contact water during handling or disposal, temperature rise and the possible solidification of the material is to be expected. Most of the temperature rise will result from slaking of lime (Reaction 4). The reactivity of lime varies considerably with the manner in which it is formed; so-called "dead-burned" CaO is relatively slow to react and thus the temperature reached during slaking may be substantially lower than for a reactive "quick lime." Reactions of AFBC residues with water are also impeded by the presence of a layer of $CaSO_4$ at the surface of the lime particles (see below).

Physical Properties

From the point of view of their disposal or utilization, the physical properties that are of interest are those having a direct influence on their behavior as engineering materials or in compacted masses. The physical properties that have been reported in the literature for materials from different sources include:

- Particle size and gradation, typical of which are the data shown in Figure 2;
- Specific gravity;
- Particle morphology and internal structure.

Stone [6] has presented data on other physical properties of spent high-Ca bed residue from the Alexandria installation (Table IV).
Low-Ca residues are strongly pozzolanic [18]. In this regard they are markedly different from high-Ca residues which are generally unreactive towards lime or Portland cement [16].

Table IV

Physical Properties Of AFBC Residues

| Property | High-Ca Source[a] | | Low-Ca Source[b] |
	Bed-drain	Baghouse ash	Fine ash
Specific gravity	-	2.65	2.60
Specific surface (cm^2/g)	-	-	9565
Permeability (cm/sec)	1.93×10^{-3}	1.93×10^{-3}	-

a. From Berry [7]
b. From Lotze and Wargalla [8]

Microstructure

It has been reported that the bed-drain material from limestone systems is heterogeneous at the level of individual particles. Early studies present-ed some evidence that bed drain particles are only partially sulphated, with the formation of a coating of $CaSO_4$ over a core of unreacted lime [9,10].

Kalmanovich et al. [11] have examined the microstructure of the wastes from the Summerside source in some detail. This work has confirmed that the particles of residue comprise an inner core of CaO surrounded by a layer of anhydrite ($CaSO_4$), as shown in Figure 3. In some instances a thin outermost layer, rich in Si and Fe, is also present (Figure 4).

Although thermodynamically favorable, sulphation of calcined limestone is impeded by $CaSO_4$ that blocks most of the pores after about 35% of the lime-stone has been used. The effects of this are evident in Figure 5, in which the formation of relatively non-porous regions in the otherwise porous lime-stone is clearly seen.

Environmental Aspects of the Nature of AFBC Residues

Disposal of high-Ca AFBC residues presents a number of potential prob-lems, some of which are similar to those found with other sulphated residues such as flue gas desulphurization (FGD) sludges. Of particular concern are the following:

• Leachates from AFBC residues are of high pH (>13) due to the presence of free-CaO and contain substantial quantities of dissolved solids.
• When moisture is present, or if water is added during handling, hydration of CaO may cause a substantial temperature rise of the material.
• Trace element and organic components may leach from disposal sites.

Leaching Characteristics

The chemical analyses of high-Ca residues indicate that several slightly water-soluble constituents are present. In particular, CaO and $CaSO_4$ are both potential sources of dissolved solids. The solubility of calcium sulphate in the presence of lime has been widely studied. At 25°C the invariant solution in contact with both lime and gypsum contains 1.04 gL^{-1} of CaO and 1.70 gL^{-1} of $CaSO_4$ [20]. Hence a pH over 11 and a total dissolved solids concentration

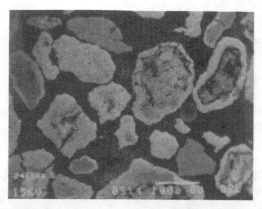

Figure 3. Polished section of high-Ca AFBC bed-drain residue from Summerside PEI. Note outer layer of calcium sulphate surrounding inner core of CaO.

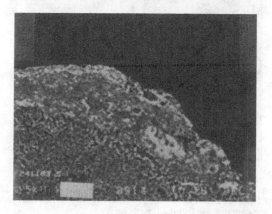

Figure 4. Polished section of high-Ca AFBC bed-drain residue from Summerside PEI. Note detail of ash-rich outer region of particles.

(TDS) in excess of 3000 mg/L is to be expected for water that is saturated with lime and gypsum. It is clear that high-Ca AFBC residues have a substantial potential to pollute surface and ground water systems if adequate control of their disposal is not maintained.

In addition to calcium oxide and calcium sulphate, AFBC residues contain trace constituents derived from the original limestone and coal. Some of these may be subject to mobilization by leaching.

To understand the potential of a waste as a source of leachate, it is necessary to conduct experimental leaching under laboratory conditions in an effort to simulate the consequences of field exposure to water or other solutions. Several different protocols for leachability evaluation have been proposed. For the purposes of illustrating the potential leachability of AFBC residues, data obtained using a protocol developed for recent Canadian studies [13] has been selected. The procedure requires multi-cycle leaching of the

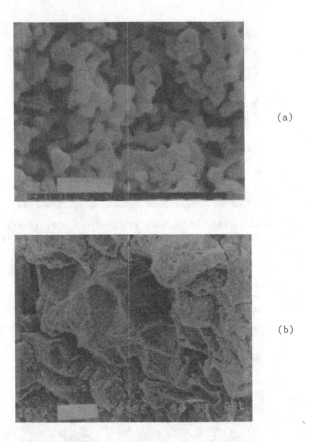

(a)

(b)

Figure 5. Scanning-electron micrographs of high-Ca AFBC bed-drain residue
from Summerside PEI: (a) porous region of un-sulphated limestone;
(b) non-porous sulphated region.

residue with either distilled water or an acidic medium (pH 4.5) at two
liquid-to-solid ratios: 4:1 and 20:1. For the purposes of this discussion,
data from the fourth cycle of water leaching at a 4:1 ratio of liquid-to-solid
are shown in Table V [13]. The substantial quantities of Ca and sulphate in
the leachate are clearly seen; other elements are present only in trace quantities. It can be concluded that from the perspective of water quality the
principal concern regarding the disposal of high-Ca AFBC residues must be with
the potentially high levels of TDS and the high pH of leachate solutions.

Waste Management and Disposal

A major concern of published studies of AFBC residues has been directed
to determining whether they are "hazardous" for the purposes of regulatory
compliance. A study by Grimshaw et al. [14] concluded that bubbling bed AFBC
residues were non-hazardous by U.S. RCRA standards. Similarly a study by Sun

Table V

Elemental Concentration (mg/L) Of Leachate
From Two High-Ca AFBC Residues[a]

Element	Minto ash	Devco ash
Al	<.01	<.01
As	<.005	<.005
Ba	<.005	0.17
Be	<.0005	0.0005
B	0.03	0.05
Cd	0.03	0.02
Ca	1600	838
Cr	<.01	0.13
Co	<.05	<.05
Cu	<.008	<.008
F	0.69	0.87
Fe	0.08	0.08
Pb	<.05	<.05
Mg	<.01	1.02
Mn	<.01	<.01
Mo	<.3	0.4
Ni	0.11	0.07
P	0.9	0.8
K	<1	27
Si	0.05	12.1
Na	<1	11
Sr	0.79	5.86
S (as SO_4)	1410	1440
Ti	<.005	<.005
V	<.005	0.06
Zn	0.12	0.11
Zr	<.05	<.05

a. From [13]

and Peterson [15] concluded that the potential environmental impact of AFBC wastes is less than that of FGD wastes. It should be noted that FGD wastes are also substantially composed of calcium oxide (usually hydrated) and calcium sulphates.

The management of FBC wastes encompasses storage, handling, conditioning, placement and monitoring. Their chemical and physical properties bring specific problems to each of these areas.

To avoid exothermic reactions and premature solidification of the residues, they must be stored and conveyed either in a dry state or in very dilute suspensions, where large volumes of water will be required. Unlike most pulverized fuel ashes, conditioning of AFBC wastes with small amounts of water, to reduce dusting and aid compaction, will likely result in temperature rise and setting of the materials soon after addition of water.

In the dry state AFBC residues are difficult to compact and are relatively permeable, with hydraulic conductivity of 10^{-2} to 10^{-4} cm/sec [14]. Grimshaw et al. [14] reported that, for high-Ca waste, permeability decreases over a period of one year to approximately 10^{-3} to 10^{-5} cm/sec. It is probable that if the residues are moistened and compacted the cementing action of calcium sulphate will lead to development of much lower permeability shortly after placement. This remains to be demonstrated under field conditions.

58

In summary, the properties of high-Ca AFBC wastes will require that novel approaches are taken to their handling and disposal with particular regard to the following potential concerns:

- safe handling of an alkaline dust containing free calcium oxide;
- control of temperature rise and solidification induced by contact with water;
- prevention of impairment of ground and surface waters by relatively soluble components in the ash capable of producing leachate of high pH and TDS.

REFERENCES

1. J.L. Minnick, Development of Potential uses for the Residue from Fluidized Bed Combustion Processes, Reports Under Contract No. EF-77-C-01-2549.
2. J.L. Minnick, Development of Potential Uses for the Residue from Fluidized Bed Combustion Processes. Reports Under Contract No. DE-AC21-77-ET10415.
3. R.H. Miller, Proc. 5th Int. Conf. Fluidized Bed Combustion, Vol. II (1977) pp. 800-816.
4. R.J. Collins, J. Testing and Evaluation 8, 259-264 (1980).
5. C.C. Sun, et al., Experimental/Engineering Support for EPA's FBC Program: Final Report Vol. III. Solid Residues Study. EPA-600/7-80-015C (1980).
6. R. Stone and R.L. Kahle, Environmental Assessment of Solid Residues from Fluidized Bed Combustion Processes, Report PB-282-940 (1978).
7. E.E. Berry, An Evaluation of Uses for AFBC Solid Wastes: Final Report, DSS Contract OSQ83-00077 (1984).
8. J. Lotze, and G. Wargalla, Zem.-Kalk-Gips, 38, 239 (1985).
9. I. Johnson, Sorbent Utilization, Enhancement and Regeneration, Proc. DOE/WVU Conf. Fluidized-bed Combustion System Design and Operation (Morganstown, West Virginia, October 27-29, 1980) pp. 333-360.
10. J. McLaren and D.F. Williams, J. Institute of Fuel, pp. 303-308 (1969).
11. D.P Kalmanovitch, V.V. Razbin, E.J. Anthony, D.L. Desai and F.D. Friederich, The Microstructural Characteristics of AFBC Limestone Sorbent Particles, Proc. 8th Int. Conf. Fluidized Bed Combustion, Vol. I, (1985) pp. 53-64.
12. E.J. Anthony, The Presence of Calcium Sulphide in Solid Wastes from Circulating Fluidized Bed Combustors, CANMET Report, ERL Division Report 86-53(TR) (1986).
13. Dearborn Environmental Consulting Services, Preliminary Evaluation of AFBC Solid Waste Properties, Environment Canada Report, EPS 3/PG/2 (1985).
14. T.W. Grimshaw, R.A. Minear, A.G. Eklund, W.M. Little and H.J. Williamson, Assessment of Fluidized Bed Combustion Solid Wastes for Land Disposal, U.S. EPA Report, EPA 600/7-85/007a (1985).
15. C.C. Sun, and C.H. Peterson, Fluidized Bed Combustion Residue Disposal Environmental Impact and Utilization, Proc. 6th Int. Conf. on Fluidized Bed Combustion, (1980).
16. E.E. Berry, An Evaluation of Uses for AFBC Solid Wastes - Phase II, Examination of Waste from CFB Summerside as a Potential Soil-cement Component: Final Report, DSS Contract OSQ84-00037 (1984).
17. INTEG Ltd., Investigation of the Utilization and Disposal of Boiler Ash from C.F.B. Summerside, P.E.I., CANMET CONTRACT 78-9037-1, Interim Report (1979).
18. J. Lotze and G. Wargalla, Zem. -Kalk -Gips., 38 374 (1985).
19. E.E. Berry and E.J. Anthony, Evaluation of Potential Uses of AFBC Solid Wastes. This volume.
20. F.E. Jones, Trans. Faraday Soc. 35, 1484 (1939).

ELECTROCHEMICAL STABILITY OF EMBEDDED STEEL AND TOXIC ELEMENTS IN FLY ASH/CEMENT BEDS

R.I.A. MALEK and D.M. ROY*
Materials Research Laboratory, The Pennsylvania State University, University Park, PA 16802 USA
*Also affiliated with the Department of Materials Science and Engineering.

Received 2 December, 1986; refereed

ABSTRACT

The electrochemical stability in fly ash/cement beds is of major concern to the durability of construction metals (iron or steel) embedded in the matrix as well as the stabilization (fixation) of toxic elements. The electrochemical stabilities were evaluated by measuring the redox potential as a function of both time and leach solution. For simulating the field conditions, the measurements were made on leachates of a prepared solution simulating rain composition in the area of application and results were contrasted to those obtained on leachates of standard deionized water. Two leaching techniques were used: the standard EPA-EP test; a test developed at MRL/PSU for simulating field conditions in which leaching fluids are pumped up a fly ash/cement column. The redox potentials (based on hydrogen scale), Eh's, were plotted vs. pH of the leachates and the regions of stability of various construction materials and toxic elements were predicted. Tafel plots were also constructed for iron in contact with different leachates, and its corrosion rate was estimated.

INTRODUCTION

Cement stabilized fly ash beds provide potential means for utilizing fly ash in many practical applications. The Materials Research Laboratory (MRL) of The Pennsylvania State University (PSU) has been studying the utilization of cement stabilized fly ash beds as a pipeline bedding material [1]. The basic requirements for such bedding material are:

a. It should be compatible with the material being embedded as well as the surrounding ground/soil composition;
b. It should comply with federal, state and other pertinent regulatory requirements for the amounts of toxic elements released through leaching. The leaching behavior of cement/fly ash bed was discussed in detail in a previous publication [1]; these results will be summarized.

Summary of Leaching Results

The leaching portion of this study has been reported previously [1]. The results will be summarized here because they figure in the discussion of electrochemical stability. Of major concern to the study is the possible toxicity level caused by percolating rain/runoff water through the bedding material. Because the toxic elements are from the fly ash (constituting major part of the solid matrix), it was important to study the leachability of both the fly ash alone and the cement stabilized fly ash. Comparison of the compositions of leachates obtained from both sources will enable estimating the effectiveness of the stabilizing material (cement) as well as determining the mechanism of release in order to predict long-term behavior.

Leachants. Two leachants were used, deionized water (with acetate buffer in the EPA-EP test), and solution simulating the average rain composition in the area of application. In contrast to DI water, which is relatively neutral and ion free, the rain water had as its major influencing characteristics a low pH (4.2) and ~5 mg/L SO_4^{2-} and NO_3^-.

Leaching Techniques. Two leaching techniques were used. The first was the standard EPA-EP test in which an acetic acid buffer solution (pH 5 or less) was stirred with the solid (20:1 water to solid ratio) for 24 h. The second technique was developed at MRL/PSU for simulating field conditions. In the later technique, a 500 g sample of fly ash is incrementally compacted in a cylindrical sample holder by taping and vibration. This procedure is followed to provide uniform packing throughout the length of the sample. Several permeable barriers are inserted along the length of the sample to reduce the possibility of fluid channeling through the sample. The sample column is connected to a positive displacement pump that allows the fluid to pass from the bottom of the bed to the top. A constant pump pressure is maintained at about 11.5 psi. The top of the column is vented to atmospheric pressure. Four litres of leachants are contained in a reservoir connected to the column's bottom via the pump and to its top for circulating the leaching fluids. A pressure gauge is connected between the pump and column inlet (at the bottom) to monitor continuously the driving pressure. The outlet (at the top) is connected to a three-way tap for periodic collection of leachate sample and for determining the flow rate.

The driving pressure was found to drop initially due to increased permeability, and it was necessary to increase the pumping speed to attain constant pressure. The experiment was continued up to 18 days (d), which was found to be a sufficient period for attaining steady state conditions with respect to most species leached. It was found that this period of experiment is equivalent to the rain water percolation through a bed of equal thickness for ~30 years [based on an average rain accumulation in the area of application of ~51.5 inches/year].

The same system was used also to study the leaching behavior of a cement stabilized fly ash bed. In this experiment a slurry of 93% fly ash plus 7% Type I cement (compositions and mineralogies are given in Tables II and III), having a water/solid ratio = 0.5, was poured inside the cylindrical sample column and held by a support disc provided with a rubber O-ring. The disc had holes that were plugged with screws until curing was complete. Every precaution was taken to keep the stabilized bed saturated with water. The stabilized sample was allowed to cure for 28 d at 38°C in the sample holder prior to initiating the tests. After 28 d, the holes of the supporting disc were unplugged, the tube was connected to the circulating assembly and the leaching experiment was started with periodic collection of samples from top of the column. The solutions were sampled in 30 mL aliquots and analyzed to determine cations by DC plasma emission spectrometry (DCP), and anions by selective adsorption ion chromatography using a Dionex automated ion chromatograph.

The significance of the MRL/PSU flow-through leaching procedure is that it permits subjecting the bedding material to a more severe condition in which the leachant-leachate equilibrium drifts continuously by allowing the solution to move at a certain velocity. This is more advantageous than the batch method in which a certain volume of leachate is allowed to equilibrate with the solid. At equilibrium, the rate of release of material into solution will reach a steady state, and, for many elements, rate of dissolution is equal to rate of deposition of dissolved matter on the surface of fly ash.

The leachate compositions from the two different leaching procedures as well as the leaching mechanisms and long-term behavior were discussed in detail in a previous publication [1]. Table I summarizes the concentrations of toxic elements in final leachates from both leaching procedures. It can be seen that all concentrations are well below the RCRA limits for the EPA-EP test [1].

Table I

Compositions (in mg/L) of EPA-EP and Flow-Through Procedure Leachates
for Fly Ash and Cement Stabilized Fly Ash
(starting fluid was simulated eastern PA rain)[a]

	RCRA Limit for EP Test	Fly Ash Bed		Fly Ash/Cement Bed	
		EPA-EP	Flow Through	EPA-EP	Flow Through
As	5.0	0.20	<0.10	0.1	0.3
Ba	10.0	0.16	0.62	0.75	0.2
Cd	1.0	<0.02	<0.02	<0.03	<0.03
Cr	5.0	0.01	0.31	0.51	<0.03
Pb	5.0	0.35	0.33	1.00	0.08
Hg	0.2	0.03	<0.1	<0.1	<0.10
Se.	1.0	0.35	0.35	<0.10	0.10
Ag	5.0	<0.02	<0.03	<0.05	<0.03

*See ref. 1 for additional details.

The present study is an attempt to evaluate the stabilities of embedded
material (iron or steel) as well as the toxic elements by correlating the Eh's
(redox potential) with pH values of the different leachates.

EXPERIMENTAL

Materials Characterization

Table II shows the bulk chemical (oxide) composition of the as-received
fly ash sample and of the Type I cement used in stabilizing fly ash beds. The
low percentage calcium and relatively higher iron contents classifies this
type fly ash low-calcium (ASTM Class F). X-ray diffraction was used to
characterize the mineralogy of as-received fly ash. A summary of the data is
shown in Table III. Other measured physical properties included: B.E.T.
surface area = 0.89 m^2/g; Blaine surface area = 3251 cm^2/g; kerosene density =
2.35 g/cm^3.

The stabilized fly ash was prepared with 7.0% Type I cement and 93.0% fly
ash and a water:solid ratio = 0.5.

Eh-Measurements

Portions of the leachates from both leaching procedures were collected
and stored in sealed containers without any contact with air. Eh was measured
using a combination of platinum electrode and reference electrode in one body.
For matching the potential characteristics of a conventional Ag/AgCl reference
electrode, the filling electrolyte chamber (containing the reference element)
was filled with 4 M KCl saturated with silver chloride. The Ag/AgCl reference
electrode was selected because it works successfully in high ionic strength
solutions (a Calomel electrode is more successful in solutions having total
ionic strength <0.2 M). The potential developed by the reference electrode
portion relative to the normal hydrogen scale at 25°C is -0.199 volts. The
electrode combination was connected to an ORION model 611 pH/millivoltmeter.
According to the experimental setup, the following cell is established:

+ Pt/solution // Sat. KCl, AgCl/Ag -

Table II

Chemical Analyses (wt.%) of Fly Ash and Cement

	Fly Ash	Type I Cement
SiO_2	48.9	20.84
Al_2O_3	27.4	4.10
TiO_2	1.30	0.27
Fe_2O_3	11.0	2.90
MgO	0.93	4.05
CaO	2.70	63.84
MnO	0.037	0.21
SrO	0.09	0.05
BaO	0.12	0.02
Na_2O	0.23	0.07
K_2O	2.45	0.74
P_2O_5	0.38	0.21
SO_3	0.4	2.67
LOI	3.6	1.32
Rb_2O	----	<0.01
Cs_2O	----	<0.01
TOTAL	99.5	101.27

Table III

Mineralogy of the Low-Calcium Fly Ash

Total
Quartz, mullite, hematite and Al and Mg substituted spinel or magnetite
with dominant silica-rich aluminosilicate glass

Non-magnetic Fraction
Quartz, mullite and aluminosilicate glass

Magnetic Fraction
Hematite and magnetite with minor Al and Mg substituted ferrite spinel

Therefore,

$$E_{measured} = E_{solution/Pt} + E_{Ag/AgCl, KCl_{sat.}} \qquad (1)$$

$$Eh\ (Solution) = E_{measured} - E_{Ag/AgCl, KCl_{sat.}} \qquad (2)$$

i.e.,

$$Eh\ (Solution)\ in\ millivolts = E_{measured} - (-199) \qquad (3)$$

The liquid junction potential effects were avoided by keeping a proper flow of the concentrated filling electrolyte into the sample. This is assured with an open electrolyte filling hole and a constant height of the filling electrolyte above the level of the solution in which the electrode is immersed during calibration and/or measurements.

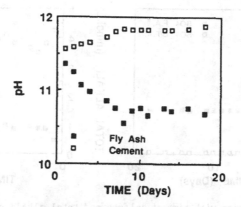

Figure 1.   Variation of pH with time for leachates from
the PSU flow through procedure.

The electrode combination was checked [2] by measuring the potential of
Zobell's solution (solution consisting of 3.33 x 10^{-3}M $K_4Fe(CN)_6$,
3.33 x 10^{-3}M $K_3Fe(CN)_6$ and 0.10M KCl) then used to measure the Eh values of
different solutions. All measurements were done at 25°C and all the potential
values are based on standard hydrogen scale at 25°C and one atmosphere. In
addition, the pH values were measured at 25°C using a glass electrode.

Rates of Corrosion

The rates of corrosion of pipeline material in different leachates were
measured by a potentiostat and the Tafel plots were constructed.

RESULTS AND DISCUSSION

pH and Eh

The variations in pH of the leachates as a function of time (PSU flow
through procedure) for the fly ash and cement stabilized fly ash beds are
shown in Figure 1. The pH values of leachates resulting from the EPA leaching procedure were 4.5–5.0. The variations in concentrations of sulfate and
total alkalis with the PSU procedure (Fig. 2) suggest that the pH of the
leachates (and the equilib-bria in pore fluids) are controlled by concentrations of SO_4^{2-} released from the fly ash beds, whereas it is controlled by
alkali in the cement stabilized fly ash beds [3].

The variations in Eh of the leachates (PSU procedure) as a function of
time and pH are shown in Figures 3 and 4. These variations are typical for
either fly ash or cement stabilized fly ash beds. The Eh values of leachates
of the EPA procedure are 338 mv and 445 mv for fly ash and fly ash + cement
respectively. The implications of these values for stabilities of iron and
toxic elements (as predicted from Eh-pH diagrams) can be summarized in the
following:

Iron will acquire different degrees of passivation. It will be more
passive to corrosion in leachates from cement stabilized fly ash beds
than from the fly ash. The existence of high concentrations of alkali
coupled with high pH in the first leachates, and higher concentrations of

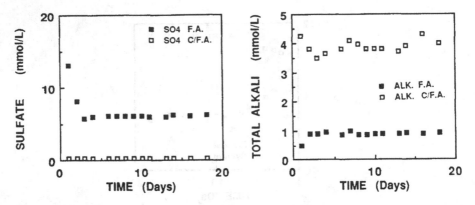

Figure 2. Variation with time of sulfate and total alkali concentration in different leachates from the PSU flow-through procedure.

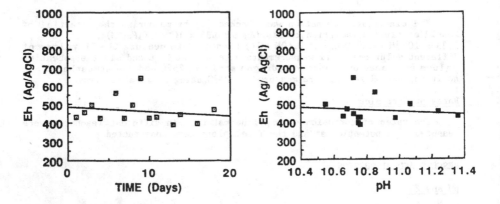

Figure 3. Variation of Eh with time for leachates from the PSU flow through procedure.

Figure 4. Eh vs. pH for leachates from the PSU flow through procedure.

sulfate with relatively lower pH values in the second leachates might explain this variability.

The toxic elements can, under the environment of pore fluids described above, exist as one or more of the following chemical species: Hg, Ag, HgO, PbO, Ba(OH)$_2$, BaSO$_4$, Cd(OH)$_2$, AsO$_4^{3-}$ or HAsO$_4^{2-}$, CrO$_4^{4-}$, and SeO$_4^{2-}$ [4-6]. The very low solubility of these ionic species is consistent with long-term, low-contamination, behavior for Ag, Ba, Cd, Hg, and Pb. Of significance is the fact that the presence of very low concentrations of Se, As and Cr, although they might exist as soluble species, could result from their scarcity in the fly ash used.

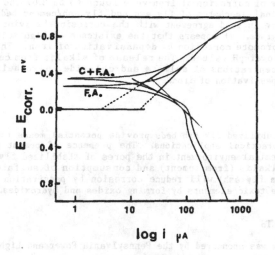

Figure 5. The Tafel plot.

The Tafel Plots

Tafel plots were constructed from tests in which a specimen of iron (similar to the pipeline material) was polarized while in contact with an electrolyte solution (leachate from PSU flow-through leaching experiment). The specimens were polarized at about 300 mv anodically (positive-going potential) and cathodically (negative-going potential) from the corrosion potential, E_{corr}. The resulting Tafel plot showing current, plotted on a logarithmic scale, vs. potential is shown in Figure 5. Anodic or cathodic Tafel plots are described by the Tafel equation:

$$E - E_{corr.} = \beta \log \frac{i}{i_{corr.}} \qquad (4)$$

where $E - E_{corr.}$ = polarization or overpotential (i.e., the difference between the specimen and the corrosion potential); = Tafel constant; i = measurement current density in A/cm^2; $i_{corr.}$ = corrosion current density in A/cm^2. It is evident from equation (4) that $i = i_{corr.}$ at zero polarization (i.e., at zero applied potential). Therefore, the $i_{corr.}$ could be obtained by extrapolating the cathodically polarized and the anodically polarized linear parts of the Tafel plot to zero potential. Ideally, the two Tafel lines should intersect at zero potential ($E_{corr.}$) and the current density at this intersect is $i_{corr.}$. Experimentally (as in the present study), it can happen that the extrapolation of the anodic and cathodic linear Tafel regions do not intersect at $E_{corr.}$. The deviation is probably in the anodic part of the plot, since the metal (iron) specimen is corroding and its surface is continuously changing with application of voltage. The anodically polarized part (which represents dissolution of the metal) could then reflect the combination of several different Tafel slopes. In this case, the $i_{corr.}$ could be accurately measured at the point where the cathodic Tafel extrapolation intersects $E_{corr.}$. The corrosion rate could then be calculated from Faraday's law relating weight of metal dissolved to the quantity of current passed.

The rates of corrosion of iron were found to be 8.67 and 1.43 mills per year in the final leachates of fly ash and fly ash/cement beds respectively. The results are in good agreement with the corrosion behavior predicted from the Eh-pH diagrams. It appears that the existence of high sulfate concentrations will promote corrosion by depassivation of iron. This effect will increase at lower pH values. The release of alkalis from cement combined, with low concentrations of sulfate and chloride, also helps to minimize corrosion by passivation of iron.

CONCLUSIONS

Cement-stabilized fly ash beds provide potential means for utilizing fly ash in many practical applications. The presence of cement will favorably modify the chemical environment in the pores of stabilized fly ash beds. The release of alkalis (from cement) and consumption of sulfate and chloride (released from fly ash) will reduce corrosion by passivation iron, and help stabilize some toxic elements by forming oxides and hydroxides.

ACKNOWLEDGEMENTS

This work was sponsored by the Pennsylvania Power and Light Company.

REFERENCES

1. R.I.A. Malek, P.H. Licastro, and D.M. Roy, in Fly Ash and Coal Conversion By-products: Characterization, Utilization and Disposal II, Mat. Res. Soc. Symp. Proc. Vol. 65, edited by G.J. McCarthy, F.P. Glasser and D.M. Roy, (Materials Research Society, Pittsburgh, 1986) pp. 269-284.
2. D.K. Norstrom, "Thermochemical Redox Equilibria of Zobell's Solution," Geochim. Cosmochim. Acta 41, 1835-1841 (1977).
3. D.M. Roy, R.I.A. Malek, M. Rattanussorn, and M. Grutzeck, in Fly Ash and Coal Conversion By-products: Characterization, Utilization and Disposal II, Mat. Res. Soc. Symp. Proc. Vol. 65, edited by G.J. McCarthy, F.P. Glasser and D.M. Roy, (Materials Research Society, Pittsburgh, 1986) pp. 219-226.
4. M. Pourbaix, Atlas of Electrochemical Equilibria in Aqueous Solutions, Pergamon Press (1966).
5. M. Pourbaix, Thermodynamics of Solutions, London, E. Arnold (1949).
6. R.M. Garrels and C.L. Christ, Solution, Minerals and Equilibria, Freeman, Cooper and Co. (1965).

STUDIES OF ZINC, CADMIUM AND MERCURY STABILIZATION IN OPC/PFA MIXTURES

C.S. POON* and R. PERRY**
*Department of Metallurgy and Science of Materials, Oxford University, Parks Road, Oxford OX1 3PH, U.K.
**Department of Civil Engineering, Imperial College, London, U.K.

Received 5 October, 1986; refereed

ABSTRACT

The utilization of pulverized fuel ash (PFA) (fly ash) for the stabilization of heavy metal waste is described. Solutions of the group IIB elements (zinc, cadmium and mercury) are used as model materials because of their significance as industrial wastes. The study included aqueous chemistry determinations, a leaching test, the use of SEM to examine microstructure, and compressive strength measurements. The use of PFA in a cementitious matrix lowers the alkalinity of the overall system and thus improves the immobilization of the amphoteric metal such as zinc. The interaction between mercuric solution and PFA plays an important role in improving the retention of the blended system for mercury. SEM results show that the microstructure of the ordinary Portland cement (OPC)/PFA blended system is significantly modified by the incorporation of the waste material. The advantages of using the blended system over a pure OPC system are described in physical and chemical terms.

INTRODUCTION

Ordinary Portland cement (OPC), pulverized fuel ash (PFA), slag, silicates and other cementitious materials have long been utilized for the disposal of low level radioactive waste [1]. The growing environmental pressure has recently required the generally regarded less hazardous materials, such as heavy metal wastes, to be stabilized to assure proper disposal. Various commercial processes have been developed during the past two decades for such purposes. A lime-pozzolana process is widely used in the U.S. [2]. In the U.K., however, an OPC-PFA process has been used for a number of years [3].

The objective of the present paper is to increase the basic understanding of the chemistry and mechanism involved between the waste and the stabilizing agents in the OPC/PFA process using solutions of Zn, Cd and Hg as models of heavy metal wastes.

MATERIALS AND METHODS

Materials

Zinc nitrate, cadmium nitrate and mercury chloride and nitrate were "AR" grade and were obtained in the U.K. from BDH. The chemical compositions of OPC and PFA used in this study are presented in Table I. The soluble silicate was a 40% sodium metasilicate (Na_2SiO_3) solution. The heavy metal content in these matrix materials is insignificant when compared to the concentration of the simulated waste solutions.

Aqueous Chemistry

The aqueous chemistry of OPC, PFA and OPC/PFA in water and in solutions containing the heavy metals was studied in a N_2 atmosphere. A volume of 200 mL of water or of solutions containing 2000 ppm (0.2 wt.%) of Zn, Cd or Hg were mixed with 40 g of solid (OPC, PFA, and OPC/PFA in a 3:1 ratio) and

TABLE I

Composition (as wt.% oxides) of OPC and PFA

	SiO_2	Al_2O_3	Fe_2O_3	CaO	MgO	SO_3	Na_2O	K_2O
OPC	20.1	5.3	3.5	64.6	1.3	2.8	0.13	0.77
PFA	47.1	29.8	11.6	2.3	1.6	0.7	1.3	3.6

agitated with a magnetic stirrer. Samples (10 mL) were extracted from the slurry at 1, 3, 5, 10, 15, 20 and 50 min after the initial mixing and filtered through a 0.45 μm membrane. After pH measurement, the filtrates were acidified with dilute HNO_3 and analyzed by atomic absorption spectroscopy (AAS).

Leaching Tests

The 2000 ppm metal solutions were solidified through additions of OPC/sodium silicate and OPC/PFA with liquid/solid ratios of 5 and 0.5, respectively. The specific formulations were:

OPC/silicate: 40 g OPC + 9.6 g sodium silicate in solution + 200 mL of metal solution;

OPC/PFA: 25 g OPC + 75 g PFA + 50 mL of 2000 ppm metal solution.

The mixes were cured at room temperature and humidity for 28 days. After curing the samples were crushed to pass a 2 mm sieve. The crushed products were sealed in containers and treated for one day with 0.15 M dilute acetic acid solution (initial pH = 4.6) at a liquid/solid ratio of 2.5. The leachate was filtered through a 0.45 μm membrane filter. After pH measurement, the filtrates were acidified and analyzed for metals by AAS. The leaching was repeated through four additional one day treatments with acetic acid.

Scanning Electron Microscopy

Scanning electron microscopy (SEM) was performed with a JEOL 35CF system. The samples studied by SEM were prepared with more concentrated 2 wt.% (20,000 ppm) heavy metal solutions and had liquid/solid ratio of 0.5. After curing at room temperature and humidity for one day, hydration was stopped with acetone and fracture surfaces were prepared for SEM examination.

Strength

Samples having a liquid to solid ratio of 0.5 were prepared for compressive strength measurements with water and with 2 wt.% metal solutions of Zn, Cd and Hg nitrates and Hg chloride. The test specimens were moulded into 40 mm^3 cubes and tested at 1, 7 and 28 d.

RESULTS

Aqueous Chemistry

The results of the aqueous phase studies are presented in Figures 1 to 3. Each result on a figure is an average of three repetitions.

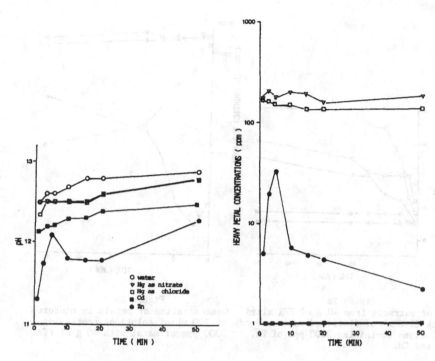

Figure 1a

pH of extracts from 40 g of OPC mixed with 200 mL of distilled water and of solutions containing 2000 ppm of Zn, Hg and Cd.

Figure 1b

Concentration of metals in mixtures of 200 mL of solutions containing 2000 ppm of metal with 40 g of OPC.

The mixture of 40 g OPC with 200 mL distilled water (Fig. 1) produced a highly alkaline solution with pH from 12.5 to 12.9. The initial pH values for the 0.2 wt.% metal solutions were:

Zn(NO$_3$)$_2$ 5.0   HgCl$_2$ 4.1
Cd(NO$_3$)$_2$ 4.4   Hg(NO$_3$)$_2$ 2.4

The effect of adding the 0.2 wt.% metal solutions to the OPC was to lower pH compared to the distilled water mix. Zinc gave the lowest value, followed by Cd and the two Hg solutions. The concentrations of Zn and Cd decreased immediately after mixing (Zn from 2000 ppm to 4 ppm and Cd to an undetectable level). After an initial drop, the Zn concentration increased to 23 ppm after 5 minutes but subsequently decreased gradually to 2 ppm after 50 minutes of mixing. The Cd concentration, however, remained at an undetectable level throughout the experiment.

The PFA/distilled water mixture had a low initial pH of 5.8 which then increased gradually to pH = 9 over the 50 min period (Fig. 2a). The result from using heavy metals with PFA was similar to that found in the OPC system. All lowered the pH of the solutions, with the greatest lowering being due to Zn(NO$_3$)$_2$. Both the Zn and Cd nitrate solutions remained acidic during the course of the experiment. The Hg nitrate and chloride solutions, however,

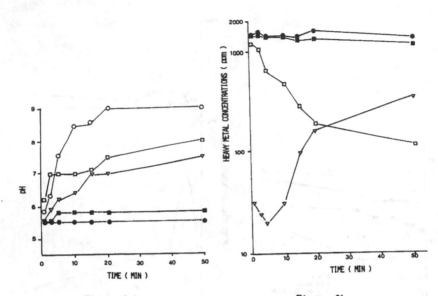

Figure 2a
pH of extracts from 40 g of PFA mixed
with 2000 mL of distilled water and of
solutions containing 2000 ppm of Zn,
Hg and Cd.

Figure 2b
Concentration of metals in mixtures
of 200 mL of solutions containing
2000 ppm of metal with 40 g of PFA.

were initially acidic but increased gradually to a slightly alkaline pH after
50 minutes.

Zinc and Cd concentrations in solutions remained generally high throughout
the experiment although a slight decrease was observed after 50 min (Fig. 2b).
Mercury from the HgCl$_2$ solution decreased gradually from the 1183 ppm measured
after 1 min to 115 ppm over the same period. However, the Hg concentration in
the Hg(NO$_3$)$_2$ solution stayed around 30 ppm for the first 20 min but increased
to 335 ppm by the end of the mixing period.

The corresponding results for the OPC/PFA system are shown in Figure 3.
The alkalinity of OPC was reflected in the relatively high pH's measured for
all of the mixes. Only the Zn(NO$_3$)$_2$ solution lowered the pH by a significant
amount.

The first Zn concentration measured had decreased from 2000 ppm to only
866 ppm, but it decreased steadily to 2.7 ppm after 50 minutes. Concentrations of both Hg solutions showed moderate decreases to 260 and 200 ppm for
the chloride and nitrate solutions, respectively, in 50 min. The initial Cd
concentration was high (225 ppm) but rapidly decreased to an undetectable
level after 15 minutes.

Leaching Test Results

The results of the leaching tests are presented in Table II. The data
indicate that Cd is the least leachable metal in either matrix, with Zn only
slightly more leachable. Mercury was the most leachable metal, with the
chloride salt being less well fixed than the nitrate salt. Generally, the
leachate concentrations in the OPC/PFA matrices were less concentrated than
those from the OPC/silicate formulations, especially for the chloride salt.

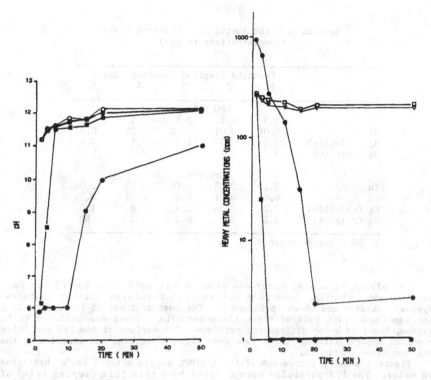

Figure 3a
pH of 40 g of OPC/PFA mixed with
200 mL of distilled water and of
solutions containing 2000 ppm of
Zn, Hg and Cd.

Figure 3b
Concentration of metals in mixtures
of 200 mL of solutions containing
2000 ppm of metal and 40 g of
OPC/PFA.

Zinc and particularly Cd, remained at a constant low concentration during the initial stages of leaching.

The use of two Hg salts in this study was due to the insolubility of $Hg(NO_3)_2$, which was unstable in an aqueous environment. This Hg salt tends to hydrolyse to form a red precipitate of HgO. Although $HgCl_2$ is largely covalent in nature it readily dissolved in water to give the desired 2000 ppm Hg solution. A comparison of the leachate concentration of these two Hg-dosed samples indicates that the nitrate was retained better than the chloride in the OPC/silicate matrix, but had a similar retention in the OPC/PFA mix.

Scanning Electron Microscopy

An electron micrograph of the as-received PFA particles is shown in Figure 4. Particle sizes vary between 0.5 and 40 μm. The particles are mainly spherical in shape and generally have a relatively smooth surface, although a small amount of material is occasionally seen extending from some of the particle surfaces.

TABLE II

Results from the Acetic Acid Leaching Tests
(concentrations in ppm)

| | Leachate Sampling Duration (days) | | | | |
	1	2	3	4	5
	OPC/Silicate				
Zn	1.3	0.1	0.1	0.1	0.2
Cd	0.01	0.01	0.01	0.01	0.01
Hg (chloride)	23.5	20.0	11.5	7.5	5.4
Hg (nitrate)	3.7	1.4	1.7	0.7	0.7
	OPC/PFA				
Zn	0.13	0.05	0.05	0.05	0.04
Cd	0.01	0.01	0.01	0.01	0.01
Hg (chloride)	3.3	3.3	2.7	0.8	1.2
Hg (nitrate)	3.2	1.6	0.7	1.4	ND

ND = Not Determined

The effect of adding water and other metal solutions to PFA has been examined under the SEM. Generally all samples displayed similar features. Typical features are shown in Figure 5. The most dominant of these were long and sometimes thick rods of crystalline deposits. These are believed to be gypsum, based on x-ray diffraction evidence. The surface of the PFA particles remained smooth, although there were some signs of etching and reaction on the smaller sized particles.

Figure 6 shows a micrograph of the PFA/OPC sample after 1 day's hydration in water. The PFA particles were covered by a thin film (duplex film) of hydration products, although it cannot be concluded whether these products were formed directly from a reaction between the alkaline pore solution and PFA particles or were merely deposited from the pore solution during OPC hydration. From looking at the OPC particles, hydration appears to have proceeded normally, producing calcium silicate hydrates (C-S-H). The types of C-S-H produced were fibrous (Type I) and reticular (Type II), as is typical of hydrated OPC.

For the Zn-dosed OPC/PFA sample (Fig. 7), the dominant feature was long rods of crystals, which could be identified as ettringite by their hexagonal cross section. The PFA surface was smooth with few hydration products. On examining the OPC particles, no identifiable C-S-H was observed, but the grains were covered by an amorphous layer of precipitate.

Figure 8 shows an SEM micrograph of the $HgCl_2$-dosed OPC/PFA sample. On the PFA surface, duplex films similar to those observed in the distilled water mix were observed. OPC hydrated normally to produce fibrous Type I and reticular Type II C-S-H. The effect of the $Hg(NO_3)_2$ on the microstructure of the hydration products of PFA/OPC blended system (Fig. 9) was very similar to that of $HgCl_2$, with C-S-H formed on both PFA and OPC surfaces.

The dominant feature under the SEM in the Cd-dosed sample was hexagonal rods of ettringite crystals (Fig. 10), which were associated with both the PFA and OPC particles. A layer of amorphous precipitate was seen on some OPC particles.

Compressive Strength

The results of the strength measurement (Table III) show that addition of metals in the OPC/PFA system did appear to have an effect on the strength

73

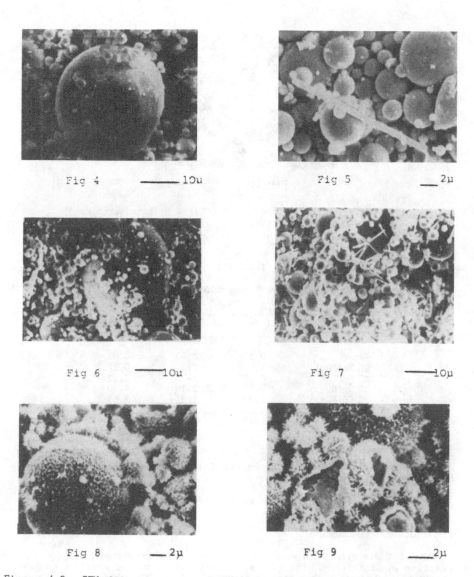

Fig 4   ——— 10u

Fig 5   _ 2μ

Fig 6   ——— 10u

Fig 7   ——— 10μ

Fig 8   _ 2μ

Fig 9   _ 2μ

Figures 4-9. SEM photomicrographs of PFA before (Fig. 4) and after (Fig. 5) 1 d of hydration with water, and of OPC/PFA after 1 d of hydration with water (Fig. 6), with the $Zn(NO_3)_2$ solution (Fig. 7), with the $HgCl_2$ solution (Fig. 8) and with the $Hg(NO_3)_2$ solution (Fig. 9). All solution concentrations were 2% (20,000 ppm) of metal.

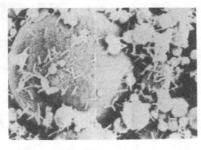

Fig 10   ——2µ

Figure 10. SEM photomicrograph of OPC/PFA hydrated with the
Cd(NO₃)₂ solution.

TABLE III

Compressive Strength Data (MPa)

HYDRATING SOLUTION	1 day	7 day	28 day
Water	0.57	2.35	4.03
2% $Zn(NO_3)_2$	N.D.	N.D.	0.22
2% $HgCl_2$	0.57	–	5.42
2% $Hg(NO_3)_2$	0.48	2.16	4.00
2% $Cd(NO_3)_2$	0.20	1.21	2.68

N.D. = Not Determined

development. The order of 28 d compressive strength as a function of the
hydrating solution was:

$$Zn(NO_3) < Cd(NO_3) < Hg(NO_3)_2 = water < HgCl_2.$$

DISCUSSION AND CONCLUSIONS

When OPC was mixed with water in the aqueous chemistry study, a hydrolysis
reaction was initiated. The high pH values that resulted were due to dissolution and hydration reactions of constituents of OPC, particularly of free
lime and C_3S. The lower pH values encountered in the PFA/water mixtures were
due to the low free lime content of the PFA.

The addition of heavy metals lowered the pH of all systems investigated.
This is due to the consumption of lime to neutralize the acidic nature of the
metal salt solutions. Zinc and Cd solutions gave substantial lowering of the
pH, while $HgCl_2$ and $Hg(NO_3)_2$ solutions had only a slight lowering effect.
The concentration of Zn, Hg and Cd in the OPC and OPC/PFA systems decreased
with respect to time. The two Hg solutions behaved similarly despite their
containing different anions species. The rate of decrease was in the order:

$$Cd > Zn > Hg.$$

This difference in the rate of decrease could be accounted for partly by the difference in initial precipitation behavior of Zn, Cd and Hg in the alkaline environment. Zinc and Cd might be precipitated as complex hydroxides while Hg might only be precipitated as a partially soluble oxide. Cadmium exhibited a much lower solution concentration than Zn due to the lower solubility product of $Cd(OH)_2$ ($K = 10^{-14}$) in comparison with $Zn(OH)_2$ ($K = 10^{-11}$). Furthermore, $Zn(OH)_2$ is more soluble than would be expected from the solubility product constant due to the higher solubility of the neutral species:

$$Zn(OH)_2(s) = Zn(OH)_2(aq) \qquad K = 10^{-6} \qquad (1)$$

In the pure PFA system the concentration of Zn and Cd remained high and it is therefore probable that PFA alone did not stabilize both metals. This was due to the low CaO content of the PFA. For the two Hg solutions, however, different mechanisms were demonstrated to be in operation. The Hg concentration of $HgCl_2$ solution decreased gradually from 1183 ppm to 115 ppm. The Hg concentration in the $Hg(NO_3)_2$ solution stayed low (ca. 30 ppm) after 20 min and then increased. Two mechanisms are thought to be operative for the retention of Hg, namely: (i) a simple precipitation of HgO due to reactions involving $Ca(OH)_2$ and Hg compounds; and (ii) a sorption of Hg by the PFA particles. In the case of $HgCl_2$, the progressive decrease in concentration was mainly due to the precipitation of oxide. There could also be some sorption by PFA. In the case of $Hg(NO_3)_2$, however, the main species in solution were $Hg(NO_3)_2$, $HgNO_3^+$ and Hg^{2+} which are capable of reacting with the hydroxyl ion as well as being chemisorbed onto the PFA surface.

The results of the leaching tests carried out on the OPC/silicate and OPC/PFA systems revealed that the retention of metals was of the order:

$$Cd > Zn > Hg.$$

This followed the trend of the aqueous chemistry study.

The use of two Hg salts in the study indicated different mechanisms by which the stabilizing agent reacts with a metal carrying different ligands. In the OPC/silicate sample, the $Hg(NO_3)_2$-dosed sample gave a much lower leachate concentration than the $HgCl_2$-dosed sample. This may be due to the fact that mercuric nitrate is largely ionic in nature and the retention of Hg in an alkaline environment results from the formation of HgO. Mercuric chloride, however, is largely covalent in nature and thus the extent of its reaction with the hydroxyl ion in solution was significantly less. This contrast in the retention of $HgCl_2$ and $Hg(NO_3)_2$ in OPC/silicate was not observed in OPC/PFA, indicating that Hg was retained by different mechanisms in the OPC/PFA blended system as discussed above.

Comparing the concentration of all three metals in OPC/silicate and OPC/PFA systems, both produced low Cd concentrations. The OPC/PFA system produced a lower Zn leachate concentration than the OPC/silicate. This was probably because while the alkaline environment of OPC immobilized the precipitated Zn in the form of a complex hydroxide, the pH in the PFA/OPC leachate was not high enough to resolubilize the amphoteric Zn due to the replacement of part of the OPC by non-alkaline PFA.

The addition of metal solutions affected the hydration and the microstructure of the cementitious matrices. Zinc greatly retarded the hydration of OPC and PFA by forming a layer of amorphous gel on the particles. The formation of ettringite, however, was enhanced by Zn with long and large rods of crystalline nature observed by SEM. This ettringite was randomly distributed in the pore spaces of the cementitious matrices and is believed to be formed from a through solution mechanism. Previous study has shown that these crystals expanded the pore structure and resulted in increased pore volume and large proportion of pore space with large radii [6]. X-ray diffraction revealed that no $Ca(OH)_2$ was present in the Zn-dosed samples.

The addition of Hg in the form of $HgCl_2$ and $Hg(NO_3)_2$ did not result in significant changes in the cementitious matrices. Porosity measurements [6] however, indicated that the $HgCl_2$-dosed samples have a lower total pore volume than a the control (water hydrated). This is probably due to the chloride promoting the hydration of C_3S and resulting in more hydration products to fill the pore space. Mercuric nitrate, however, gave a similar pore structure to the control. Although $Hg(NO_3)_2$ did slightly retard the reaction, the hydration of OPC and PFA/OPC blended systems proceeded relatively normally to produce the C-S-H revealed by SEM.

Cadmium induced moderate changes in the microstructure of the solidified systems. The morphology of the hydration product was modified. This can be shown by the partially hydrated OPC grains producing the C-S-H and other hydration products. XRD examination indicated the presence of the $Ca(OH)_2$ phase. The production of calcium aluminate sulphate hydrate phases typified by monosulphate and ettringite shifted the pore size distribution to larger pore sizes [6].

The effects of Zn and Hg on the hydration and microstructure of the OPC/silicate matrix has been studied previously [4]. The present study confirms the previous findings that Zn greatly retards the hydration of OPC and produces a significant change in the pore structure through ettringite and monosulphate formation. The present study has also demonstrated that adding PFA to the OPC system does not alter the reactions between the metal and OPC.

Previous studies [7] by extended x-ray absorption fine structure analysis on Zn and Hg indicated that the chemical environments of Zn and Hg in the OPC matrix were not altered by including PFA in the system. This is also in accord with the present SEM study.

The changes in microstructure were reflected in the mechanical strength of the solidified products. Compressive strength measurement revealed that the addition of Zn and Cd greatly reduced the strength of the solidified products. The two mercury solutions, however, provide similar strength to that of the pure water control. This is thought to be related to the of the different metal ions on the hydration of OPC and PFA and the resulting changes in microstructure.

Considering the aqueous chemistry, leaching and strength tests together, the results reported here show that Zn and Cd were well retained by the OPC/PFA matrix, even if the product did show lower strength. Mercury solutions, however, show less metal retention, but produce stronger solidified product. The differences in retention are due mainly to different chemical reactions between the metals and CaO and to differential sorption on the PFA particles. The differences in physical strength are due to the different degree of modification of the microstructure, especially in sulphoaluminate hydrate production, resulting in different pore volume and structure.

REFERENCES

1. C.J. Jantzen, J.A. Stone, and R.C. Ewing (ed.), Scientific Basis for Nuclear Waste Management VIII, Mat. Res. Soc. Symp. Proc. Vol. 44, (Materials Research Society, Pittsburgh, 1985).
2. I.U. Conversion System Inc., Environ. Sci. & Technol., 11, 436-437 (1977).
3. C.L. Chappell, British Patent No. 1485625 (1 June 1973).
4. C.S. Poon, C.J. Peters, R. Perry, P. Barnes and A.P. Barker, Sci. Total Environ., 41, 55-71 (1985).
5. C.S. Poon, A.I. Clarke and R. Perry, Public Health Engineer, 13, 108-111 (1985).
6. C.S. Poon, A.I. Clarke, R. Perry, P. Barnes, and A.P. Barker, Cem. and Concr. Res., 16, 161-172 (1986).
7. C.S. Poon, A.I. Clarke and R. Perry, Environ. Technol. Letters, 7, 461 (1986).

STABILIZATION OF DRILLING FLUID WASTE WITH FLY ASH

GEORGE M. DEELEY*, LARRY W. CANTER** and JOAKIM G. LAGUROS**
*Shell Development Center, Houston, Texas
**School of Civil Engineering and Environmental Science, The University of Oklahoma, Norman, Oklahoma 73019

Communicated by G.J. McCarthy

SUMMARY

Water based drilling muds typically contain clays, barite, lime, caustic soda and other chemicals, such as polymers. Land disposal of these wastes raises the possibility of groundwater pollution which can be abated if the waste is stabilized either by chemical reaction or by solidification through some form of cementation. Many ASTM high-calcium (Class C) fly ashes are cementitious and thus may be useful in stabilization of drilling mud. The basic idea is to stabilize the clay-containing muds using the model of soil and roadbed stabilization with high-calcium fly ash [1]. Fly ash that is not utilized is considered to be a solid waste, so this application would would actually constitute codisposal of two wastes.

A laboratory study of three different drilling fluids mixed with a Class C fly ash is underway. Mixtures of drilling fluid with 10 and 30% fly ash are allowed to cure for 1 to 5 weeks before applying a modified EP toxicity test [2] and analyzing the resulting leachates for arsenic, barium, chromium, lead and zinc. Several preliminary results are summarized below.

The fly ash as well as the drilling fluids were subjected to the same EP toxicity test, and thus it was possible to compare the mixtures with the fly ash and fluids. Arsenic was slightly released by one drilling fluid fly ash mixture, as was barium. Lead was slightly released by two mixtures but strongly taken up by another. Chromium and zinc behaved as if the combination of drilling fluid and fly ash were a simple physical mixture with no chemical fixation effect. The concentrations of the elements released did not approach the total concentrations present in the waste drilling fluids or fly ash, and in no case were the EP toxicity limits exceeded. The utilization of Class C fly ash appears to be a promising means of stabilization of drilling muds.

A full report of these results will be the subject of a paper at the next Symposium.

REFERENCES

1. J.G. Laguros and M.S. Keshawarz, in Fly Ash and Coal Conversion By-Products: Characterization, Utilization and Disposal II, edited by G.J. McCarthy, F.P. Glasser and D.M. Roy, Mat. Res. Soc. Symp. Proc. Vol. 65, (Materials Research Society, Pittsburgh, 1986) pp. 37-46.
2. U.S. Environmental Protection Agency, Test Methods for Evaluating Solid Wastes - Physical/Chemical Methods, SW-846, 2nd Edition (EPA, Washington, DC, 1982).

STABILIZATION OF DRILLING FLUID WASTE WITH FLY ASH

CHARLES, ISAAC, CARTER, CARTER and FRANCIS E. HAMILTON
School of Civil Engineering and Environmental Science, The University of
Oklahoma, Norman, Oklahoma, USA

REFERENCES

PART II

Characterization

SPECIATION IN SIZE AND DENSITY FRACTIONATED FLY ASH
II. CHARACTERIZATION OF A LOW-CALCIUM, HIGH-IRON FLY ASH

R.T. HEMMINGS*, E.E. BERRY*, B.J. CORNELIUS* and B.E. SCHEETZ**
*Ontario Research Foundation, Sheridan Park, Mississauga, Ontario, Canada, L5K 1B3.
**Materials Research Laboratory, The Pennsylvania State University, University Park, PA 16802

Received 25 February, 1987; Communicated by G.J. McCarthy

ABSTRACT

Morphological, chemical and mineralogical speciation of a low-Ca, high-Fe fly ash from a bituminous coal has been investigated by examination of size, density and magnetic fractions. Fractionation by size revealed little information as to speciation among particle types. However, separation of the ash into eight density fractions and into magnetic and non-magnetic components showed major differences in particle properties. It was found that glass-containing particles can be divided into three general types: Type 1, being low-Fe content, low-density hollow spheres comprising aluminosilicate/mullite glass ceramics; Type 2, of intermediate density, being ferroaluminosilicate/mullite glass ceramics; and Type 3, high density composite particles of spinel/hematite crystals embedded in an iron-substituted glass. It is proposed that Type 1 and Type 2 particles are derived from thermal decomposition of clay minerals with a range of Fe contents. Type 3 particles are considered to arise from thermal decomposition of pyrite in the presence of small quantities of aluminosilicate minerals. Two general types of glass were distinguished: Glass I(f), being largely a low-iron aluminosilicate; and Glass II(f), being a ferroaluminosilicate of high Fe-content. XRD and vibrational spectroscopic evidence suggest that, in both glass types, Fe is substituted for Al in an aluminosilicate-type structure.

INTRODUCTION

In early studies of the chemical and mineralogical properties of fly ashes [1-3], it was recognized that Fe was present in various component phases. Crystalline iron-containing phases were identified by x-ray diffraction (XRD) as magnetite (Fe_3O_4), hematite (Fe_2O_3) and glass [1]. Watt and Thorne [3] examined density fractions of a number of ashes from bituminous coals and found both Fe and Ca to be more concentrated in the heavier fractions (SG 2.5-2.6). These authors concluded that most of the Fe found analytically in the fly ashes occurred in separate, Fe-rich particles and not in glassy siliceous particles.

In related work, Lauf [4] identified Fe-rich spheres in reflected light microscopy by their high reflectivity and optical isotropy. The particles were seen to comprise a "light" phase and an internal pattern comprising a dark phase. The light phase was attributed to a spinel or ferrite [5]; the dark phase was optically anisotropic and contained Al and Si. Lauf [4] further suggested that the dark phase was mullite or other clay decomposition products, and that the high-Fe particles may be pseudomorphs of pyrite framboids [6] found in the feed coal.

Studies of ashes from North American bituminous coals [7-9] have identified Fe^{2+} and Fe^{3+} glasses, hematite and $(Fe,Al)_2O_3$, magnetite spinels of the type $(Fe,M)_3O_4$ (where M = Mg or Ca), Fe-substituted mullite, and goethite (alpha-FeOOH). Ashes formed under reducing conditions have been shown to contain hercynite ($FeAl_2O_4$), wustite($Fe_{1-x}O$), fayalite (Fe_2SiO_4) and iron sulphides. By Mossbauer spectroscopy, Aikin et al. [10] identified Fe-containing phases in brown coal ash (CaO content of 16-17%), and showed it to contain iron largely in the form of calcium alumino-ferrite, $Ca_2Fe_{2-x}Al_xO_5$.

Biggs and Bruns [11] have concluded from XRD and chemical analysis that magnetic ash fractions from two midwestern bituminous coals were intimately intergrown or fused mixtures of ferrite spinel, hematite, glass and quartz. Most particles were observed under the reflected light microscope as individual spheres consisting of three phases: ferrite spinel, hematite and glass in varying proportions. Two crystal habits were reported to be common among the ferrite spinels, namely, octahedra and dendritic masses. Dendrites were the more abundant form and were found to be intergrown with a dark matrix considered to be silicate glass. Octahedral cross-sections observed frequently were equilateral triangular faces, considered to be (111) planes of magnetite. This plane was also thought to host lamellae of hematite. As Diamond has shown [12], the glass phase, interstitially dispersed between the crystalline constituents of magnetic particles, can be etched with 1% HF in the same manner as for the non-magnetic glassy spheres [5]. If HF treatment is extended, spheres are found with regular crystalline structures of hexagonal or triangular form that could represent the spinel plates described by Biggs and Bruns [11]. Diamond [12] concluded that the spaces between the crystals are occupied by small amounts of Fe-rich glass.

The intra-particle crystallization of Fe-rich fly ash fractions was also studied by Norton et al. [13] using scanning electron microscopy of polished sections of both magnetic and non-magnetic ash fractions. Clear regions of dendritic growth within otherwise glassy matrix structures are evident from this work.

In summary, it is clear that Fe in fly ash is distributed between crystalline and glassy phases and between "magnetic" and "non-magnetic" particles. A number of distinct crystalline Fe-rich phases have been identified and some evidence for Fe-rich glassy constituents has been found. It is the purpose of this paper to present the results of some work in progress that is directed toward understanding the role of glassy constituents in determining the nature and properties of ash. In particular, this work has been conducted to increase understanding of the role of iron in fly ash glasses.

In a previous paper [14], results of research on a high-Ca, low-Fe ash were presented. It was shown that: (a) glasses of various compositional characteristics can co-exist in one ash sample; (b) Ca can modify the properties of aluminosilicate glasses to a marked degree, and (c) glasses of different composition appear to be associated with particles of differing true particle density. This paper presents data for an ash with low-Ca content but with exceptionally high levels of Fe.

EXPERIMENTAL

Materials

Representative samples of the fly ash for the study were obtained from an eastern Canadian generating station burning a high-Fe, bituminous coal. The raw ash was separated into size fractions by air-classification to produce a coarse (+45 μm) classifier reject and a fine (-45 μm) product. The -45 μm material was further classified to produce an ultra-fine (-10 μm) fraction.

A second sample of the raw fly ash was separated into density fractions using the sink-float method described previously [14]. Eight fractions were produced, as shown schematically in Figure 1. A further sample of raw ash was separated into a "magnetic" and a "non-magnetic" fraction using a Davis tube magnetic separator.

Analytical Methods

Chemical analyses of the raw ash and all fractionated products (Table I) were carried out by ICP spectrophotometry. A LECO analyzer was used for carbon and sulphur determinations. Volume particle-size distributions were

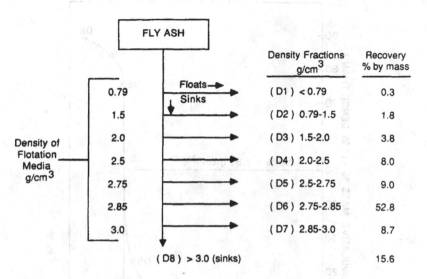

Density Fractions g/cm^3	Recovery % by mass
(D1) < 0.79	0.3
(D2) 0.79-1.5	1.8
(D3) 1.5-2.0	3.8
(D4) 2.0-2.5	8.0
(D5) 2.5-2.75	9.0
(D6) 2.75-2.85	52.8
(D7) 2.85-3.0	8.7
(D8) > 3.0 (sinks)	15.6

Figure 1. Schematic representation of density fractionation of fly ash.

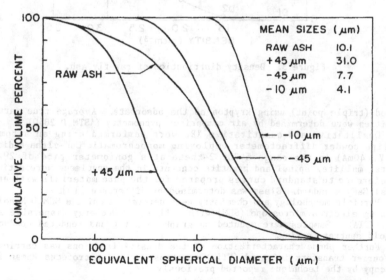

MEAN SIZES (μm)	
RAW ASH	10.1
+45 μm	31.0
−45 μm	7.7
−10 μm	4.1

Figure 2. Particle size analysis of raw and size-fractionated fly ash.

determined by the electrozone technique (Coulter) in aqueous media with Isoton dispersant (Fig. 2). Surface area measurements were performed by the BET

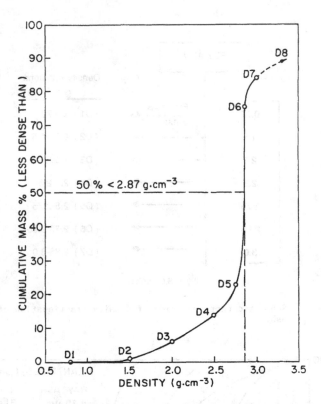

Figure 3. Density distribution of raw fly ash.

method (triple point) using krypton as the adsorbate. Average true particle densities were determined by air comparison pycnometry (ASTM D 2840).

Qualitative and quantitative XRD were performed using an automated Philips powder diffractometer employing monochromatic CuK-alpha radiation (40kV, 40mA) in the range 6-70° 2-theta at a goniometer rate of 2°/min. Quartz, mullite, spinel and hematite contents in the specimens were estimated by reference to standard curves prepared for the pure materials with an internal CaF_2 standard. Glass was determined by difference [15].

Particle morphology and chemistry were determined using a SEMCO Nanolab 7 scanning electron microscope (SEM) with a KEVEX 7000 energy dispersive analyzer (EDXA). Samples were mounted on graphite paint and rendered conductive by gold sputter coating.

Further phase characterization of the density fractions was carried out by Fourier transform infrared (FTIR) spectroscopy and microfocus Raman spectroscopy by the technique reported previously [16].

RESULTS AND DISCUSSION

Density Distributions

The distribution of the mass of each density fraction as a function of density is given in Figure 1 and plotted in cumulative form in Figure 3.

Three features are noted:

    a. As was found to be the case for other fly ashes [14], the distribution of density appears to be continuous up to a true particle density of approximately 2.8 g/cm^3.

    b. The major part (53%) of the ash was separated into a fraction of narrow density range 2.75 to 2.85 (D6).

    c. At higher density some discontinuity was apparent, suggesting the presence of a separate group of high density particles perhaps of different origin from the main body of the ash.

As is discussed below, chemical, mineralogical and morphological differences were observed between the density fractions. However, for the fractions D1 through D7 (i.e. for particles of true density less than 3.0 g/cm^3), it is considered that the major density differences and the continuous form of the distribution of density should be attributed to the influence of distributions in shell wall thickness, void volume and particle diameter. For particles of density >3.0 g/cm^3 it is evident that the major factor determining their density is the presence of iron-rich phases (including spinels and hematite) in substantial quantities (see below).

It is interesting to note that the proportion of lower density particles in this high-Fe ash is considerably less than that found [14] in the high-Ca ash, i.e. 14% and 61% <2.5 g/cm^3, respectively.

Morphological Speciation

The raw fly ash, as collected from the electrostatic precipitator, is typically heterogeneous, being a mixture of particles of different size, shape, density and chemical composition. As is clear from the SEM data (Fig. 4), most of the particle types common to fly ash were present in this material. Size classification favoured the concentration of large carbon particles and detrital quartz into the coarse (+45 μm) fraction.

Maximum and mean particle sizes were reduced as size-classification progressed, with some indication that proportionally more spherical particles were present in the finest fractions (Fig. 4).

Morphological speciation was more distinct among the density fractions (Fig. 5) and marked segregation of particle types was apparent. Fraction D1 (density <0.79 g/cm^3) comprised cenospheres or floaters. The particles were predominantly well-formed, smooth-surfaced spheres ranging in size from 15 to 60 μm in diameter. In contrast, fraction D3 (density 1.5-2.0 g/cm^3) comprised predominantly irregularly-formed, vesicular particles with some well-formed individual spheres. The vesicular particles are principally partly combusted carbon. Among the glassy spherical particles some showed crystalline surface features, probably associated with mullite crystallization.

The majority of particles in fraction D5 (density 2.5-2.75 g/cm^3) were very irregular in form and appeared dense and partly fused. Many showed high Si concentrations by EDX analysis and are considered to be fire-polished detrital quartz.

The two high-density fractions, D7 and D8, contained many spheres with a wide range of particle sizes. Most of the spheres, however, had pronounced surfacial growths (Fig. 5). EDX analysis indicated a predominance of Fe associated with these particles and it is presumed that much of the observed surface structure resulted from the presence of crystals of exsolved iron-spinel and hematite.

Similar surface features were apparent in micrographs from the "magnetic" fraction (Fig. 6) where a variety of surface growth patterns are seen.

Chemical Speciation

Chemical analysis (Table I) showed little chemical speciation to accompany separation by particle size. The most notable chemical differences

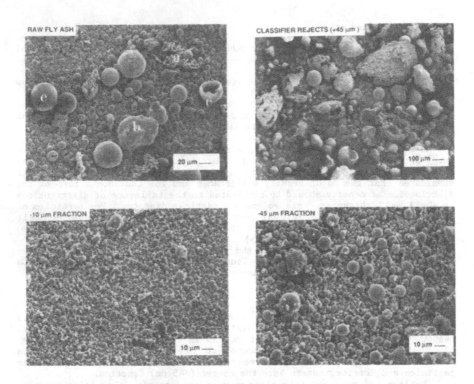

Figure 4. SEM photomicrographs of raw and size-fractionated fly ash.

between the raw ash and its size fractions is in the Fe content, which is enhanced in the +45 µm fraction (Al and Si being correspondingly lower). This is considered to be partly due to an artifact of the air separation process which was seen to cause some segregation of high-density (i.e. high-Fe), fine-sized (-10 µm) particles into the "reject" stream. This is also reflected in the marked bi-modal form of the particle size distribution (Fig. 2) of the +45 µm fraction.

More obvious chemical differences are apparent between the density fractions (Fig. 7), the most striking of which is the increase in Fe concentration with true particle density. Chemically, the density fractions can be divided into two distinctly different groups, fractions D1-D7 and D8. The lighter fractions are predominantly aluminosilicate compositions of decreasing Al and increasing Fe and Ca contents as density increased. These observations are similar to the findings reported for a high-Ca ash in which as "modifier" content (e.g. CaO) increased so did true particle density [14]. Potassium showed some speciation, being most concentrated (4.5% as K_2O) in the lightest fraction, while falling to 2.5% at higher densities. Fraction D8, although clearly containing some aluminosilicate constituents, is largely composed of iron oxides (70% as Fe_2O_3).

Mineralogical Speciation

XRD analysis of the raw ash (Fig. 8) showed quartz, mullite, magnetite (ferrite spinel) and hematite to be the principal crystalline phases, together

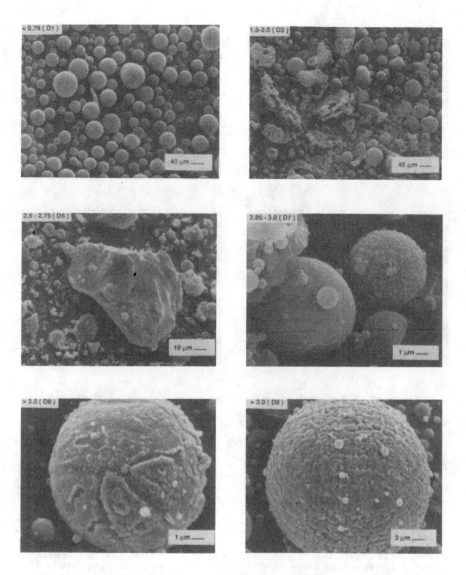

Figure 5. SEM photomicrographs of density fractions of fly ash.

with a glassy component as evidenced by the pronounced "halo" in the XRD pattern centred close to 25° deg.

Comparison of the XRD data of the raw ash and the three size fractions revealed relatively few differences in overall mineralogy (Fig. 8). Quantitative estimates of composition indicate an increase in glass as particle size decreased and a substantial concentration of quartz (19%) in the more coarse +45 µm fraction.

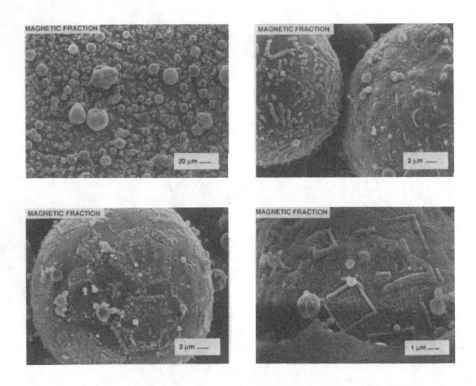

Figure 6. SEM photomicrographs of the magnetic fraction of fly ash.

The XRD data from the density fractions (Fig. 9) revealed extensive speciation of the mineral constituents. Estimated mineralogical compositions for selected density fractions are given in Table II. Again, as with the morphological and chemical differences, the data can be divided into two groups, D1–D7 and D8. Fractions D1 through D7 were found to be predominantly glassy with very little crystalline iron-containing phases, the crystalline phases in these fractions being principally quartz and mullite. The quartz is considered to be primarily present as free particles of detrital quartz (sand or silt size) from the coal, largely unaffected by combustion. The mullite, on the other hand, is most likely the result of crystallization from aluminosilicate melts. With regard to mullite formation, this high-Fe, low-Ca ash showed a marked difference when compared with the high-Ca ash studied previously [14]. For the latter ash, mullite was found only in the lower density fractions; as CaO content increased, Al_2O_3 appeared to show a greater tendency to remain soluble in the glass phase. For the present ash, mullite was crystallized from all of the aluminosilicate melts even when extensive Fe–oxide was present in the glassy phase. Thus, it would seem that iron acts as a less effective glass modifier than calcium in these ash-forming systems.

In this respect, it is of interest to note that the shape and position of the XRD "halo", attributed to glass, was not a function of particle density for the high-Fe ash. This is a marked distinction when compared with the properties of the high-Ca ash for which a considerable difference in the shape and position of the "halo" was found [14]. This observation is again consistent with lack of glass "modification" by Fe, and is particularly interesting

Table I

Chemical Analysis of Raw Fly Ash and Size-Classified Fractions
(% by mass)

Element Oxide	Raw Fly Ash	Rejects +45μm	Fine Prod. -45μm	Ultrafine Prod. -10μm
SiO_2	43.2	41.6	44.1	43.9
Al_2O_3	21.0	15.3	23.3	25.2
Fe_2O_3	24.2	35.1	21.7	18.1
CaO	1.6	1.3	1.5	1.5
MgO	1.0	0.7	1.0	1.2
Na_2O	0.5	0.3	0.6	0.7
K_2O	2.2	1.5	2.6	2.9
TiO_2	0.9	0.7	1.0	1.0
BaO	0.1	0.1	0.1	0.1
SO_3	1.3	0.6	1.6	2.2
C	1.9	1.9	1.5	1.3
LOI	3.3	1.3	3.2	3.7

in light of the large quantities of Fe apparently associated with non-crystalline phases in the high-Fe ash (see below).

By comparison, fraction D8 (density > 3.0 g/cm^3) was composed largely (64%) of crystalline iron-containing phases (spinel and hematite) with the remaining material being glassy.

These conclusions were supported in a general way by the vibrational spectra of the fractionated materials. The FTIR spectra (Fig. 10) of all fractions show similar principal features typical of aluminosilicates, in particular the band at 550 cm^{-1}, which is diagnostic and characteristic of an AlO$_4$ grouping. This band is relatively more intense in fraction D8 where the Al/Si is higher than the other fractions. The sharp band at about 800 cm^{-1} is diagnostic of quartz, which is seen to be most abundant in fraction D5, which is in keeping with the XRD data.

Considerable experimental difficulties were experienced in obtaining corresponding microfocus Raman spectral data. Only fractions D1 to D3 gave suitable spectra, and these were of low quality (Fig. 11). All other fractions apparently contained glasses that either totally absorbed the incident laser beam or resulted in a scattering level that was below the detection limits of the instrument. The Raman scattering from fractions D1 to D3 suggests that the glasses are nearly identical in composition being composed primarily of a silicate glass that is substituted by alumina. While the Raman data are generally consistent with fly ash glass spectra reported previously [17], clear spectral features associated with the function of Fe in the matrix have not been identified and are being investigated further.

Chemical Nature and Possible Origins of Glassy Components

The chemical analyses and XRD data from the density fractions were used to obtain calculated compositions of two types:

a. Glass compositions. Using the mineralogical compositions (Table II) for the density fractions and assigning exact stoichiometry to the crystalline phases (quartz, SiO$_2$; mullite, 3Al$_2$O$_3$·2SiO$_2$; magnetite spinel, Fe$_3$O$_4$; hematite, Fe$_2$O$_3$) and taking into account the presence of free

90

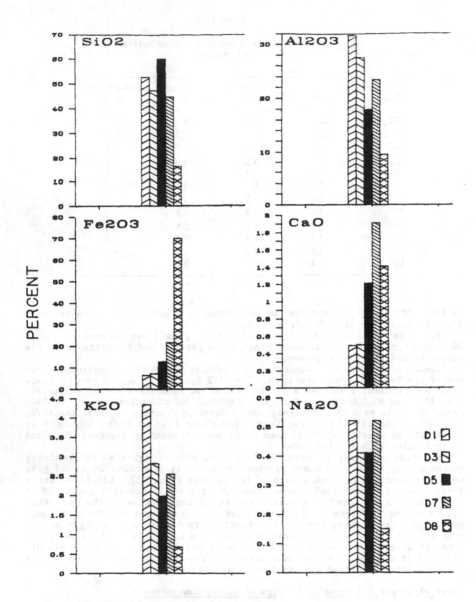

Figure 7. Distribution of major elements in density-fractionated raw
           ash. (See Fig. 1 for densities of D1–D8.)

carbon, calculations were made of the approximate compositions of the
non-crystalline constituents of each fraction. These are presented as
estimated glass compositions in Table II.

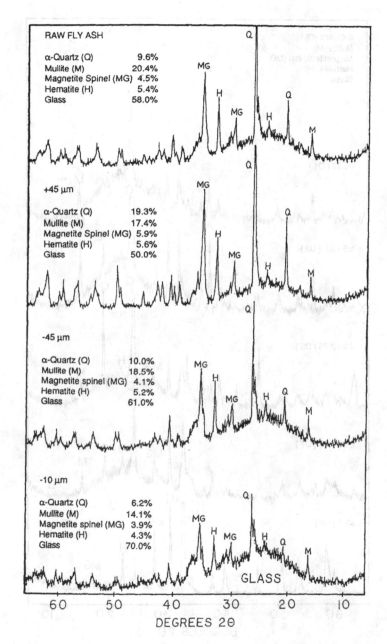

RAW FLY ASH

α-Quartz (Q)	9.6%
Mullite (M)	20.4%
Magnetite Spinel (MG)	4.5%
Hematite (H)	5.4%
Glass	58.0%

+45 μm

α-Quartz (Q)	19.3%
Mullite (M)	17.4%
Magnetite Spinel (MG)	5.9%
Hematite (H)	5.6%
Glass	50.0%

-45 μm

α-Quartz (Q)	10.0%
Mullite (M)	18.5%
Magnetite spinel (MG)	4.1%
Hematite (H)	5.2%
Glass	61.0%

-10 μm

α-Quartz (Q)	6.2%
Mullite (M)	14.1%
Magnetite spinel (MG)	3.9%
Hematite (H)	4.3%
Glass	70.0%

GLASS

DEGREES 2θ

Figure 8.   X-ray diffraction patterns and estimated mineralogical
            compositions of raw fly ash and size-fractionated ash.
            (See Fig. 1 for densities of D1-D8.)

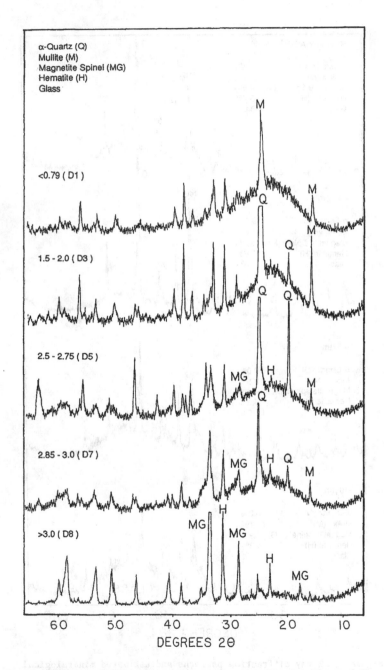

Figure 9. X-ray diffraction patterns from density-fractionated fly
ash. (See Fig. 1 for densities of D1-D8.)

Table II

Mineralogical Content of Density Fractions and Calculated Composition of Glass Phases

Mineralogical Composition (percent by mass)

Component	D1	D2	D3	D4	D5	D6	D7	D8
Glass	81	61	58	62	55	63	68	34
α-quartz	0	3	7	11	24	10	5	2
Mullite	19	36	35	27	18	24	22	0
Hematite	0	0	0	0	1	1	2	29
Spinel*	0	0	0	0	2	2	3	35

Glass Composition (Calculated percent by mass)

	D1	D2	D3	D4	D5	D6	D7	D8
SiO_2	59.4	66.7	67.0	66.9	60.8	54.7	51.1	44.0
Al_2O_3	22.7	9.8	4.5	8.4	9.8	12.1	11.4	29.9
Fe_2O_3	8.2	12.0	16.1	14.5	18.6	20.3	25.2	13.7
CaO	0.6	0.8	1.1	1.2	2.3	2.8	2.9	4.4
MgO	1.6	1.8	1.7	1.3	1.5	1.8	1.6	1.9
Na_2O	0.7	0.8	0.9	0.8	0.8	1.0	0.8	0.5
K_2O	5.5	6.2	6.3	4.8	3.9	4.7	3.9	2.1

*Calculated as Fe_3O_4

b. <u>Precursor melt compositions</u>. These were calculated by making two assumptions: first, for all fractions, all quartz was of detrital origin, and thus was not crystallized from a melt phase; and second, all other crystalline phases were the result of exsolution from melt compositions, with glass being the noncrystalline component.

From these assumptions it can be proposed that, for each glass/crystalline combination (i.e. each density fraction), the composition of the melt from which the glass and the crystals formed was the same as the bulk composition of the density fraction, less the quantity of SiO_2 associated with quartz.

To investigate the compositional relationships for the various density fractions it was decided to employ the system $SiO_2-Al_2O_3-FeO\cdot Fe_2O_3$ (Fig. 12) and, for simplicity, it was chosen to ignore the influence of CaO, MgO, K_2O and Na_2O on the system. Detailed descriptions of this system were presented by Muan [18] for various oxidation conditions expressed in terms of O_2 partial pressure.

The selection of an oxidizing condition (0.8 atm O_2) as representative of possible ash-formation conditions was to some extent guided by the work of Huffman et al. [9]. In their studies of the high-temperature behaviour of coal-ash, these workers reported that for high-Fe ashes under reducing conditions (60% CO/40% CO_2), melt components were identified as glass, wustite, a ferrite, fayalite, iron sulphide and hercynite. Most of these Fe-containing phases normally associated with the $FeO-SiO_2-Al_2O_3$ system were not observed in the XRD patterns of the ash from the present study. Rather, hematite and spinel phases were dominant; this was taken to indicate that ash formation had occurred largely under oxidizing (i.e. excess air) conditions. Again, this is consistent with the observations of Huffman et al. [9], who reported hematite and Ca-Mg-ferrite (i.e. spinel) as the phases resulting from ash formation in air.

When the compositional data for the glass components and those for the calculated precursors are plotted on the $SiO_2-Al_2O_3-FeO\cdot Fe_2O_3$ phase diagram,

94

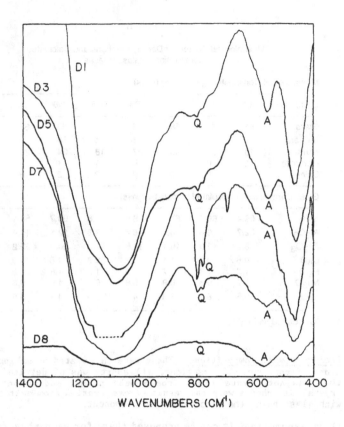

Figure 10. FTIR Spectra of density fractions. Band assignment:
A = AlO₄-bending; Q = quartz.

(Fig. 12) the following features are noted:

a. The glass compositions and the corresponding precursor compositions lie
   on two separate sets of tie lines. For fractions D1 through D7, tie
   lines connecting the precursor compositions to the mullite composition
   all extrapolate into the respective glass composition regions. This is
   consistent with crystal formation (i.e. mullite precipitation) from
   homogeneous melts. However, for fraction D8, in which no mullite was
   found, the corresponding tie line shows potential crystallization of
   hematite and spinel together with a glass of similar iron content to that
   of the other density fractions, but more concentrated in Al_2O_3. It would
   appear, therefore, that a substantial region of glass formation between
   SiO_2, Al_2O_3 and Fe_2O_3 is possible over a range of compositions, but that
   such glasses are not modified by Fe in the same manner as by Ca [14].

b. The precursor compositions are seen to lie along a single locus connecting $FeO \cdot Fe_2O_3$ with an aluminosilicate of approximately 65% SiO_2, 35%
   Al_2O_3, well in the region of composition typical of the common clay
   minerals.

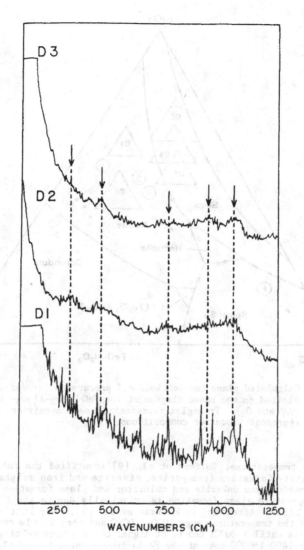

Figure 11. Microfocus Raman spectra of selected density fractions.

When ash is melted and quenched, the partitioning of Fe between glass and crystalline phases has been shown to be very dependent upon the formation (quenching) temperature [9]. At low temperature (below about 1000°C), about 20-25% of the iron is present in glass, the balance being in ferrite spinel (20%) and hematite (80%). It is presumed that at intermediate levels of oxidation an equilibrium between hematite and "magnetite" (spinel) will be established and a further Fe partitioning may be imposed upon that observed by Huffman et al. [9].

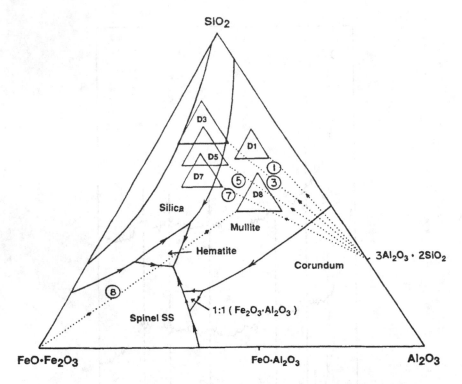

Figure 12. Calculated glass compositions and precursors for density fractions
plotted on the phase diagram of the FeO·Fe₂O₃-Al₂O₃-SiO₂ system
(0.8 atm O₂). Triangles represent glass compositions and circles
represent precursor compositions.

At low temperatures, Huffman et al. [9] identified the following reactions: hematite formation from pyrite, siderite and iron sulphates; Ca-Mg-ferrite formation from ankerite and chlorite; and glass formation from illite. The low-temperature glass formation is principally due to the fluxing action of K_2O in illite. According to Huffman et al. [9], very little melting is indicated in the temperature range 1000-1200°C and very little reaction of Fe-species occurs until 1300°C; hence, at higher temperatures melting is accelerated, and by 1400-1500°C most of the Fe is incorporated in the glass phase.

It is notable that for the fractions D1 through D7 in the present study, the calculated glass compositions indicate that as the Fe content in the glass increases, the K content decreases. This indicates perhaps the formation of different particle types from different precursors at different temperatures.

In addition to carbon and detrital quartz, the density fractions all contain particles that are to some degree glassy. These glass-containing particles can be distinguished on the basis of chemical and mineralogical composition into three types:

Type 1. Largely confined to density fraction D1, these are of relatively low Fe content (<10% as Fe_2O_3), with 80% glass and 19% mullite. The glasses associated with these particles are principally aluminosilicate in nature with %Al_2O_3 > %Fe_2O_3.

Type 2. Present in fractions D2 through D7, these contain substantial mullite (20-36%) and substantial Fe (Fe_2O_3 from 10 to 22%) but with only minor quantities of crystalline Fe-containing phases. The glasses associated with these particles are ferroaluminosilicate in nature, with $\%Fe_2O_3 > \%Al_2O_3$.

Type 3. Present in fractions of density >3.00 g/cm^3, they are composed of spinel/hematite crystals bonded or embedded in a glassy matrix (ca. 30% by mass).

In previous studies [14], it was proposed that two glass types are present in a high-Ca fly ash, originating from two different reactions:

Clay + minor fluxes and modifiers ---->   Glass I + mullite
Clay + CaO (or $CaCO_3$) and FeO ---->   Glass II

Some similarity can be seen for the high-Fe ashes with respect to the formation of Type 1 and Type 2 particles, namely:

Clay + minor fluxes and modifiers ---->   Glass I(f) + mullite
Fe-substituted clay ---->   Glass II(f) + mullite

Glass I (CaO system) and Glass I(f) are of similar composition and are both associated with low-density, hollow particles. Glass I(f) is distinguished from Glass II(f) in that in Glass I(f) $\%Al_2O_3 > \%Fe_2O_3$; whereas, a spectrum of Fe-substituted aluminosilicate compositions seems to be indicated for Glass II(f) (with an upper limit on Fe_2O_3 content in excess of 25%). The distinction between Glass II and Glass II(f) is the incorporation of significant levels of Ca and Fe, respectively, in otherwise aluminosilicate structures.

The Type 3 particles are considered to originate from oxidation reactions involving pyrite with minor associated clay or other aluminosilicates:

$$FeS_2 + clay \quad ----> \quad Fe_3O_4/Fe_2O_3 + Glass$$

It is noted that no mullite is found in the D8 fraction. This is consistent with a formation reaction from compositions high in Fe from which oxides of iron would precipitate. The glass associated with the Type 3 particles appears from the calculated composition and vibrational spectra to be of the Glass I(f) species and to be rich in Al_2O_3.

Physically, it may be considered that the glass material in the lower density particles provides a continuous phase in which are embedded mullite crystallites (as in a glass-ceramics). In the particles of highest density the glass is probably best considered as a continuous phase binding together crystals of spinel/hematite.

CONCLUSIONS

1. As was found in previous work on high-Ca ash, separation of the high-Fe ash into size fractions did not reveal major chemical or mineralogical speciation between particles.
2. Separation by density fractionation showed a continuous distribution of particles with most of the material being in a relatively narrow density range. A bimodal density distribution was found, indicating that a distinct group of particles of high-density was present in the ash.
3. Extensive chemical and mineralogical speciation was found between density fractions. In addition to quartz and carbon, three types of glass-containing particles were distinguished:
   Type 1 - low density aluminosilicate/mullite glass-ceramic particles;

Type 2 – intermediate density ferroaluminosilicate/mullite glass-
ceramic particles;
Type 3 – heterogeneous spherical particles composed of spinel/
hematite crystals and glass.

4. Two glass types were distinguished: Glass I(f) being aluminosilicate of
similar composition to the low-density glass found in other ashes; and
Glass II(f) being a ferroaluminosilicate.

5. The two glasses are considered to originate from pyrolysis of clays of
different iron/aluminum compositions.

6. The incorporation of Fe in Glass II(f) appears to result from substitution for Al in an aluminosilicate structure. No evidence of glass "modification" by Fe was seen from XRD or vibrational spectra.

7. It is proposed that the Type 3 particles originated from reactions between pyrite (framboids) and minor quantities of aluminosilicates such as
clays.

ACKNOWLEDGEMENTS

The authors wish to thank the Ontario Research Foundation for support of
this research and for permission to publish the findings of work in progress.

REFERENCES

1. P.J. Jackson, J. Appl. Chem. 7, 605 (1957).
2. H.S. Simons and J.W. Jeffery, J. Appl. Chem. 10, 328 (1960).
3. J.D. Watt and D.J. Thorne, J. Appl. Chem. 15, 585–594 (1965).
4. R.J. Lauf, Bull. Am. Ceram. Soc. 61 (4), 487–490 (1982).
5. L.D. Hulett and A.J. Weinberger, Env. Sci. Technol. 14, 965–970 (1980).
6. R.J. Lauf, L.A. Harris and S.S. Rawlson, Env. Sci. Technol., 16,
218–220 (1982).
7. C.C. Hinkley, G.V. Smith, H. Twardowska, M. Saporoschenko, R.H. Shirley
and R.A. Griffen, Fuel 59, 161 (1981).
8. F.E. Huggins, D.A. Kosmack and G.P. Huffman, Fuel 60, 577 (1981).
9. G.P. Huffman, F.E. Huggins and G.R. Dunmyre, Fuel 60, 585 (1981).
10. T.L.H. Aiken, J.D. Cashion and A.L. Oltrey, Fuel 63 1269–1275 (1984).
11. D.L. Biggs and J. Bruns, in Fly Ash and Coal Conversion By-Products:
Characterization, Utilization and Disposal I, edited by G.J. McCarthy
and R. J. Lauf, Mat. Res. Soc. Symp. Proc. Vol. 43 (Materials Research
Society, Pittsburgh, 1985), pp. 21–29.
12. S. Diamond, Cem. Concr. Res. 16, 569–579 (1986).
13. G.A. Norton, R. Markuszewski and H.R. Shanks, Env. Sci. Technol. 20, 409–
413 (1986).
14. R.T. Hemmings and E.E. Berry, in Fly Ash and Coal Conversion By-Products:
Characterization, Utilization and Disposal II, edited by G.J. McCarthy, F.
P. Glasser and D.M. Roy, Mat. Res. Soc. Symp. Proc. Vol. 65 (Materials
Research Society, Pittsburgh, 1986) pp. 91–104.
15. M. van Roode and R.T. Hemmings, CANMET Contract Report, No. ISQ83–00162,
July 1985.
16. B.E. Scheetz and W.B. White, in Fly Ash and Coal Conversion By-Products:
Characterization, Utilization and Disposal I, edited by G.J. McCarthy and
R.J. Lauf, Mat. Res. Soc. Symp. Proc. Vol. 43 (Materials Research Society,
Pittsburgh, 1985), pp. 53–60.
17. B.E. Scheetz, W.B. White and F. Adar, in Advances in Materials Character-
ization II, edited by R.L. Synder, R.A. Condrate and P.E. Johnson,
Materials Science Research Vol. 19 (Plenum Press, N.Y. 1985) pp. 145–154.
18. A. Muan, J. Am. Ceram. Soc. 40, 420–431 (1957).

SEM STUDY OF CHEMICAL VARIATIONS IN WESTERN U.S. FLY ASH

ROBERT J. STEVENSON and TIMOTHY P. HUBER
North Dakota Mining and Mineral Resources Research Institute, University of
North Dakota, Grand Forks, ND 58202.

Received 4 November, 1986; refereed

ABSTRACT

Scanning electron microscope/electron microprobe chemical analyses of individual grains of several western fly ashes have shown an inter-grain variation in composition for ashes derived from both lignite and subbituminous western coals. SRM 2689 (a Class F fly ash from a bituminous coal) and SRM 2691 (a Class C fly ash from a subbituminous coal) have been included in the study and also show inter-grain chemical variation. A classification scheme of fly ash grain compositions has been developed and has proven to be useful in illustrating the differences and similarities of the ashes. There is a positive correlation, for example, between grains rich in SiO_2 and Al_2O_3 and the Na_2O content of the grains in lignite and subbituminous coal derived fly ashes. There is also a positive correlation between grains rich in CaO and the MgO content of the grains in both classes of fly ashes.

INTRODUCTION

The first scanning electron microscope/electron microprobe (SEM/EMPA) grain analyses for this study were done in 1983 as part of research performed for the Gas Research Institute (GRI). That project included characterization of fixed-bed gasification ash as well as fly ash produced from the combustion of lignite. Results from the gasification ash study have been reported previously [1-3]. Chemical characterization of the grains of the three fly ashes sampled in that project (codes 83-319A, 83-342, and 83-523C) form the basis of the classification scheme presented and used in this work. The number of fly ash samples has been expanded in the subsequent years to include two of the three fly ash standard reference materials recently issued by the National Bureau of Standards (SRM 2689 and SRM 2691) [4]. Other sources of fly ash for this work include a follow-up project through GRI and the Western Fly Ash Research, Development and Data Center [5].

In order to evaluate the relatively large number of SEM chemical analyses, a scheme for the display and classification of the analyses was developed. This scheme is based on the major chemical components of ASTM C-618 Class C fly ash (CaO, SiO_2, Al_2O_3, and Fe-oxide) and can be illustrated using an oxide weight percent (wt.%) $CaO-SiO_2-Al_2O_3$ ternary diagram (Fig. 1). In these plots other oxides are ignored. The analyses represent the bulk composition of the exposed surface of each ash grain, including crystalline inclusions and the continuous glass phase.

EXPERIMENTAL

The ash samples were embedded in low viscosity epoxy in shallow molds. The resulting boats were then embedded in epoxy plugs at right angles to the bottom of the boats and subsequently cut and polished. This mounting method insured that the analyses of grains were unbiased as to density and size. Secondary electron image photographs of areas on the plugs at low (20x) and moderate (300x) magnification were taken and used for reference. An example of the resulting specimen is shown in Figure 2. Allocation into groups in the classification scheme used in this paper is also shown for some of the grains.

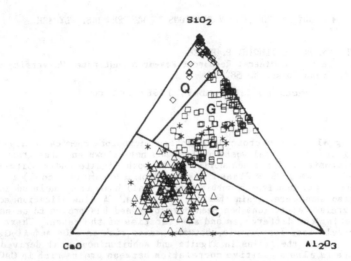

Figure 1. CaO–SiO₂–Al₂O₃ ternary wt.% oxide plot of Q-type (◇),
G-type (□), C-type (△), and F-type (✳) ash grain
compositions in ash 83-319A (L1). The divisions shown
are based on the classification scheme presented in
this paper.

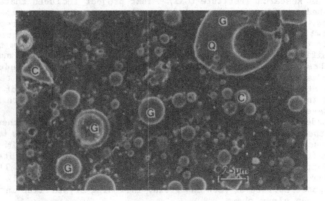

Figure 2. Secondary electron image of NBS SRM 2691 subbituminous
ash (S1). Scale bar is 25 micrometres long. The grain
types are represented by their respective letters.

Although only 175 to 400 grains were analyzed in each of the samples, the
authors believe that this is a large enough sample of grains to establish the
chemical trends observed. As a check on this assumption, the analyses from
sample 83-342 were plotted in groups of 100 analyses. The trends shown in the
groups were the same as those shown when all 431 analyses were plotted. The
analyzed grains were chosen from the reference photographs on the basis of
size. Grains ranging in size from 1 micrometre to over 200 micrometres were

selected. This way the complete size range of grains is represented on the plots.

The analyses and photographs were taken using a JEOL 35C scanning electron microscope/electron microprobe. The analysis operating conditions were changed over the course of the work. In the earlier portion of the work an accelerating voltage of 15 kV and a beam current of 1000 pA for a live time of 50 seconds was used. In the later portion of the work, the beam current was increased to 2000 pA and the live time was reduced to 10 seconds. The calculated minimum detectable limits at the later setting were all below 0.5 wt.% on an oxide basis for all oxides except for Na_2O which was 1.4 wt.%. This level of detection was considered acceptable for the planned use of the analyses. The grains were analyzed for Si, Al, Fe, Mg, Ca, Na, K, Ti, Mn, P, S, and Ba. The EDS spectra were reduced to oxide weight percents by using a Bence-Albee correction program [6].

There are two potential errors in the analysis of grains embedded in an epoxy matrix. One relates to the range in size of the grains. Analyses of the smaller sized grains, can lead to erroneous chemical analyses. The authors have assumed that for the scope of this study the error in analysis induced by the small volume grains can be ignored. Another possible source of error is the inability to determine the proximity of other grains below the polished surface of the sample. Figure 2 illustrates that the grains are sufficiently dispersed so that the potential problem of grain proximity does not exist.

RESULTS AND DISCUSSION

Fly Ash Classification Scheme

Five types of grains were used as categories in this classification scheme. They are FeO-rich (F), CaO-rich (C), glass-rich (G), SiO_2-rich (Q), and unclassified (U). These are defined as follows: F-type grains contain more than 50 wt.% FeO (on a normalized to 100% basis); C-type grains are those that have more CaO than SiO_2; G-type and Q-type grains both have more SiO_2 than CaO and are distinguished on the basis of the ratio $SiO_2/(Al_2O_3+SiO_2)$; with ratios >0.84 being assigned to the Q-type. The divisions used between C-type and Q-type and C-type and G-type are somewhat arbitrary but are based on observations made on a pilot-plant scale lignite fly ash (83-319A) (Fig. 1). All grains not assigned to these categories are designated U-type (for unclassified). The spheres that fall into this latter category either contain more than 15 wt.% of K_2O, MnO, P_2O_5, TiO_2, BaO, or SO_3, or contain between 15 and 50 wt.% FeO.

Chemical Trends of Grain Types

The bulk oxide chemistries of the ashes analyzed in this work are shown in Table I. Most analyses only total to about 95 wt.%. These totals are thought to be the result of analyses which do not include such oxides as TiO_2 and BaO as well as assumptions in the conversion of atomic absorption spectroscopy elemental analysis to oxide weight percent.

The bituminous (SRM 2689) and the bituminous-subbituminous mixed (86–008) coal derived ashes would fall into the Class F specification using the ASTM C-618 classification. The other ashes in Table I would fall into the Class C specification. For the purpose of this paper, the three lignite lime scrubber ashes (83-523C, 84-633, and 86-434) are considered to be Class C ashes that have been chemically altered by the addition of lime as a scrubber additive and sulfur from the scrubbing process.

Figure 1 shows the distribution of C-type, G-type, Q-type, and F-type grain chemistries for a lignite ash (83-319A). Note the general trend of the G-type and the C-type grains in which their chemistries form a band of compositions from near the CaO apex to a point on the SiO_2-Al_2O_3 join. This point on

Table I

Chemical Compositions (in wt.%) of Fly Ash Samples
(Analyses by Atomic Absorption Spectroscopy)

Sample Code[a]	SRM 2689 B	SRM 2691 S1	85-122 S2	85-600 S3	86-105 S4	83-319A L1	83-342 L2	83-523C L3	84-633 L4	86-434 L5	86-008 M
SiO_2	51.5	35.8	38.8	32.8	34.1	28.9	26.1	22.1	30.7	36.5	49.2
Al_2O_3	24.4	18.1	17.5	16.5	18.3	11.1	11.6	9.5	11.4	12.1	18.4
Fe_2O_3	13.3	6.4	5.5	5.8	5.9	11.1	6.8	4.7	5.1	5.0	8.9
CaO	3.0	24.7	22.4	28.9	26.0	18.9	23.0	28.0	21.1	19.8	15.4
MgO	1.0	5.3	5.6	5.6	5.0	6.0	7.3	6.0	5.6	5.7	3.5
Na_2O	0.3	1.4	1.8	1.7	2.7	8.0	8.8	7.6	5.3	5.4	1.0
K_2O	2.6	0.4	0.5	0.3	0.3	0.9	0.5	0.4	1.0	1.1	1.2
SO_3	0.70	2.10	2.13	2.98	3.12	6.83	9.10	14.62	12.70	10.20	2.32
LOI	1.80	0.20	0.17	0.26	0.41	0.42	0.25	2.91	0.70	0.90	1.48
Moisture			0.04	0.06	0.05	0.12	0.06	0.44	0.87	1.00	0.12
Total	98.6	94.4	94.4	94.9	95.9	92.3	93.5	96.3	94.5	97.7	101.5

NOTES:
a = Code used in figures.
SRM 2689 and SRM 2691 - NBS standard reference fly ashes. The source for SRM 2689 is a bituminous coal from
    Kentucky. The source for SRM 2691 is a subbituminous coal from Wyoming. Samples are the material before
    undergoing crushing and other standard reference material processing.
85-122, 85-600, and 86-105 - The sources for all three of these ashes are different subbituminous coals from
    Wyoming.
83-319A - The source for this ash is a lignite coal from North Dakota. The ash was produced at the University
    of North Dakota Energy Research Center in their particulate test combustor unit.
83-342, 83-523C, 84-633, and 86-434 - The sources for these ashes is a lignite coal from North Dakota from the
    same combustion plant. The ashes represent ash collected before a lime scrubber was on line (83-342),
    ash collected using a lime scrubber a few months after ash 83-342 was collected (83-523C), ash collected
    using a lime scrubber in 1984, one year after ash 83-523C was collected (84-633), and ash collected using
    a lime scrubber in 1986, three years after ash 83-523C was collected (86-434).
86-008 - The source for this ash is 80 wt.% subbituminous coal from Montana and 20 wt.% bituminous from
    Illinois.

the SiO_2-Al_2O_3 join is near the locations that compositions of kaolinite and
montmorillonite would plot [7]. Although most of the Q-type grains fall near
the SiO_2 apex, some of this grain type are close to the boundary of the G-type
area. Most of the Q-type grains contain over 90 wt.% SiO_2. The F-type grains
plot in the G-type and C-type regions of the diagram with most of these grain
compositions falling into the G-type classification.

    The grain compositions of a lignite lime scrubber ash (86-434) are plot-
ted in Figure 3. The general trend of the C-type, G-type, and Q-type grains
is the same as that of the lignite ash in Figure 1. That is, there is no discernible shift of individual grain compositions between the lignite ash (83-
319A) and the lime scrubber ash (86-434). This similarity is thought to be
due to the similarity of the coal used, as well as the presence of discrete
calcium sulfate grains. The calcium and sulfur are apparently not combined
with the fly ash grains. These discrete calcium sulfate grain compositions
(U-type category) plot very close to the CaO apex.

    Figure 4 is a CaO-SiO_2-Al_2O_3 ternary oxide plot of a subbituminous coal
derived ash (86-105). All subbituminous ashes listed in Table I were derived
from coals mined in Wyoming. The distribution of the subbituminous ash grain
compositions is different in detail from that of the lignite and lignite lime
scrubber ash trends. The differences observed in Figure 4 are the increased
number of Q-type grains close to the C-type/G-type boundary and the cluster of
G-type grains near the midpoint of the SiO_2-Al_2O_3 join. These two groups of
grain compositions give rise to a generally broader band of C-type and G-type
grain compositions than are present in the lignite and the lignite lime scrubber ash chemistries (Figs. 1 and 3). The overall trend, however, is the same.
The few F-type grains that are present tend to be on the CaO apex side of the
center of the ternary plot.

    The CaO-SiO_2-Al_2O_3 ternary oxide plot of the C-type, G-type, Q-type, and
F-type grain compositions for a bituminous ash (SRM 2689) is shown in Figure
5. Although this fly ash is designated as SRM-2689, it is the as-produced
material from the electrical generating plant that was used to make the NBS

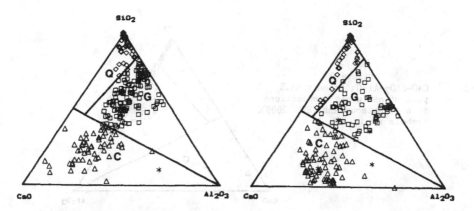

Figure 3
$CaO-SiO_2-Al_2O_3$ ternary wt.% plot
of ash grain compositions in
lignite scrubber ash (86-434).
Symbols as in Fig. 1.

Figure 4
$CaO-SiO_2-Al_2O_3$ ternary wt.% plot
of ash grain compositions in
subbituminous ash (86-105).
Symbols as in Fig. 1.

standard reference material and not SRM 2689 itself [4]. As can be seen in Figure 5, the plot of these grain compositions is significantly different than those shown for the other three ashes (lignite, lignite with lime scrubber, and sub-bituminous). In this ash the grain compositions for the G-type cluster near the $SiO_2-Al_2O_3$ edge of the ternary plot (near the composition of kaolinite and montmorillonite [8]) and there are very few C-type grains. The band of grain compositions seen for the lignite and subbituminous ashes is not present. The F-type grains all plot well within the G-type grain boundaries indicating that the non-FeO material in the F-type grains is similar to that found in the bulk of the ash grains. Most of the F-type grain compositions correlate with those of the high density fly ash grains studied by Hemmings and Berry [7].

Figures 6 and 7 are $CaO-Na_2O-MgO$ ternary plots of grain compositions of lignite ash (83-319A) and a lignite lime scrubber ash (86-434). The two plots are essentially identical. Both exhibit a narrow band of compositions of the Q-type and G-type grains from the Na_2O apex to the CaO-MgO tie line. This trend indicates that these grain types have a CaO/MgO ratio that is relatively constant. In contrast, the CaO/MgO ratio in the C-type ash grains in these plots varies over a large range. The relative amount of Na_2O present in the Q-type and G-type grains varies considerably; in contrast, there is almost no Na_2O in the C-type grains.

Figure 8 is a three-dimensional plot of the non-U-type grain compositions of a lignite ash (83-319A). The base of the plot is the $CaO-SiO_2-Al_2O_3$ ternary diagram used in Figure 1 and the vertical axis is the amount of Na_2O present (with total oxides normalized to 100%). Note that not only is there a proportionally large amount of Na_2O in the G-type and Q-type ash grains compared to CaO and MgO contents (Fig. 6), but there is also a large absolute amount in the G-type and Q-type ash grains. The pattern of compositions for the lignite scrubber ash (86-434) is the same as that for the lignite ash in Figure 8.

Figure 9 is a $CaO-Na_2O-MgO$ ternary oxide plot of a subbituminous derived ash (86-105). The trend of the Q-type and G-type grain compositions is very similar to that of the lignite and lignite scrubber ash grains (Figures 6 and 7). The C-type grain trend is somewhat different. The range of CaO/MgO ratios covers a much narrower range than that for the lignite fly ash and

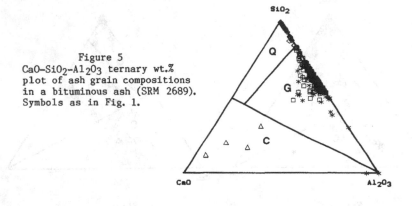

Figure 5
CaO–SiO$_2$–Al$_2$O$_3$ ternary wt.%
plot of ash grain compositions
in a bituminous ash (SRM 2689).
Symbols as in Fig. 1.

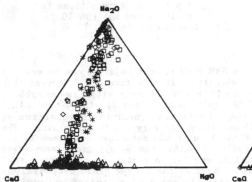

Figure 6
CaO–Na$_2$O–MgO ternary wt.%
plot of ash grain compositions
in a lignite ash (83–319A).
(83–319A). Symbols as in Fig. 1.

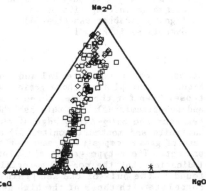

Figure 7
CaO–Na$_2$O–MgO ternary wt.%
plot of ash grain compositions in
a lignite scrubber ash (86–434).
Symbols as in Fig. 1.

lignite scrubber ash grain compositions. Figure 10 is a three-dimensional
plot of the non–U–type grain compositions for this subbituminous derived ash
(86–105) and has the same axes as Figure 8. As in Figure 8, the Na$_2$O content
of the Q–type and G–type grains is markedly higher than in the C–type. All of
the subbituminous ashes in this study showed the same trend of a higher Na$_2$O
content in the G–type and Q–type grains than in the C–type grains.

Figure 11 is a CaO–Na$_2$O–MgO ternary oxide plot of the bituminous derived
ash (SRM 2689). In this fly ash there is no CaO/MgO ratio trend of the Q–type
and G–type ash grains as seen for the other ashes. The spread of the CaO/MgO
ratio is as wide as that for the C–type grains in the lignite, lignite scrubber, and subbituminous derived ashes and the CaO/MgO ratio is generally lower
(a shift to the right on Fig. 11 as compared to Figs. 7 and 9) than that seen
for those ashes when Na$_2$O is present in the ash grains. In the ash grains
analyzed, the highest Na$_2$O content was below 2.0 wt.%, whereas in the other
ashes, the Na$_2$O content was as high as 16 wt.%. This low level of Na$_2$O

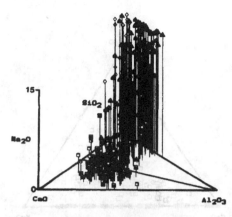

Figure 8
Three dimensional plot of Q-type
(diamond), G-type (pyramid), C-
type (cube), and F-type (cylinder) ash-grain compositions in
a lignite ash (83-319A). The
CaO-SiO$_2$-Al$_2$O$_3$ ternary wt.%
plot is the base and the normalized Na$_2$O content is the
vertical axis.

Figure 9
CaO-Na$_2$O-MgO ternary wt.% plot
of ash grain compositions in a
a subbituminous ash (86-105).
Symbols as in Fig. 1.

content in the bituminous derived ash is possibly the reason for the more
scattered appearance of the plotted grain compositions in Figure 11.

Distribution of Grain Types

Table II is a tabulation of the categories of the five grain composition
types used in the classification scheme presented herein. Listed in Table II
are the total number of grains analyzed, the number of U-type grains for each
sample, and the percentage of non-U-type grains (calculated for only the non-
U-type grains in the samples). These percentages of C-type, G-type, Q-type,
and F-type grains are plotted in Figure 12. Note that these are number based,
not weight or volume based, percentages.

The two lignite ashes (83-319A and 83-342) shown on Figure 12 (L1 and L2)
have similar proportions of non-U-type grains. The other three ashes from a
lignite source shown in Figure 12 are the lime scrubber ashes from the same
electrical generating facility as ash 83-342. The proportions of the grain
types shown in Figure 12 for these three scrubber ashes are consistent in
their order of abundances. The differences between the lignite ash after the
lime scrubber became operational (L3) and before operation began (L2) is an
increase in the proportion of Q-type and G-type grains and a decrease in the
proportion of C-type grains with the start-up of the scrubber. The decrease
with time (L3 is from 1983, L4 is from 1984, and L5 is from 1986) of the
relative amount of C-type grains and the increase with time of the G-type and
Q-type grains is at least partially caused by a change in the coal feed.
During this period of time the Great Plains Coal Gasification plant came on
line and the coal being sent to the companion electrical generating facility
was enriched in clay partings.

The grain-type proportions of the four subbituminous derived ashes shown
in Figure 12 are different from those of the lignite derived ashes. Relative
to the lignite proportions, the grain types of the subbituminous ash have a

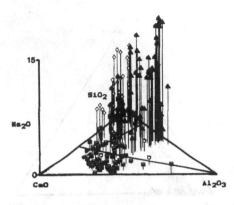

Figure 10
Three dimensional plot of ash grain compositions in a sub-bituminous ash (86-105). The $CaO-SiO_2-Al_2O_3$ ternary wt.% is the base and the normalized Na_2O content is the vertical axis. Symbols as in Figure 8.

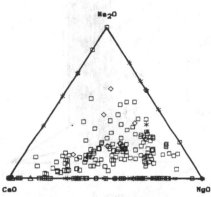

Figure 11
$CaO-Na_2O-MgO$ ternary wt.% oxide plot of ash-grain compositions in a bituminous ash (SRM 2689). Symbols as in Fig. 1.

Table II

Percentages of Grain Types

Grain type Sample	Total[a]	U[b]	F[c]	C[c]	G[c]	Q[c]
SRM 2689	249	13	16.1	1.7	69.1	13.1
SRM 2691	258	8	2.8	43.6	23.2	30.4
85-122	214	5	0.5	31.1	32.6	30.6
85-600	216	4	0.0	30.2	29.5	38.2
86-105	206	5	2.5	37.8	26.0	26.9
83-319A	300	7	8.5	42.0	37.0	9.9
83-342	431	17	5.1	41.6	39.2	11.1
83-523C	183	8	1.7	36.6	43.7	16.0
84-633	183	7	1.1	30.7	49.1	17.1
86-434	244	29	1.4	25.1	48.7	18.6
86-008	217	4	14.1	19.2	46.6	15.0

a. Total number of grains analyzed.
b. Number of grains in the U-type category.
c. Percent of the non-U-type grain total.

higher proportion (by a factor of at least 2) of Q-type grains and a slightly lower proportion of the C-type and G-type grains. The subbituminous ash grain proportions are not intermediate between the lignite and the bituminous derived ash proportions in this work, but have a signature of their own.

The grain-type proportions of the bituminous derived ash shown in Figure 12 are different from the proportions in both the lignite and subbituminous derived ashes. Relative to the lignite fly ash (83-319A, L1), the bituminous

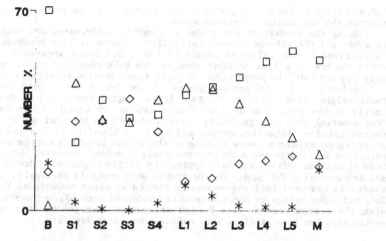

Figure 12. Percent of the non-U-type grain types in each ash sample.
Symbols as in Figure 1. The ashes are designated by the
codes listed in Table I.

fly ash has a much larger proportion of F-type and G-type grains (factor of
about 2 for each), a lower proportion of C-type grains (factor of one-twentieth), and about the same proportion of Q-type grains. Relative to the subbituminous derived ashes, the bituminous fly ash has a much larger proportion
of G-type and F-type grains (factors of 2 and 10, respectively), a lower proportion of Q-type (a factor of one-half), and a much lower proportion of C-
type grains (factor of one-twentieth).

Figure 12 also includes an ash (86-008, M) that has as its coal source,
20 percent by weight Illinois bituminous coal and 80% by weight Montana subbituminous coal. The resulting grain proportions are intermediate between the
relative amounts in the bituminous and subbituminous derived ashes of this
study.

The proportions of the grains are indicative of the coal source as well
as any processing by a collection system of the ash. The proportions are also
indicative of the ash if it comes from a mixture of coal types.

CONCLUSIONS

The composition of the grains of lignite and subbituminous coal derived
ashes in this work show generally similar trends, whereas the grains of the
bituminous derived ash show a distinctly different trend. There is a general
tendency for the C-type and G-type grains in the lignite and subbituminous
derived ashes to have a fixed SiO_2/Al_2O_3 ratio, which is illustrated on CaO-
SiO_2-Al_2O_3 ternary diagrams by a band of compositions from near the CaO apex
to an area on the SiO_2-Al_2O_3 join. This trend is probably from the clay
component (SiO_2 and Al_2O_3 rich) associated with Ca derived from the organic
material [8] in the coal. In lignite and subbituminous derived ashes, Na_2O is
generally associated with G-type and Q-type (Fig. 8 and 9) grains, and the
ratio of CaO/MgO is relatively constant in the lignite and the subbituminous
derived ashes. The C-type grains in the lignite and subbituminous derived
ashes have almost no Na_2O and a variable ratio of CaO/MgO. The range of

CaO/MgO ratio in the C-type grains of the lignite derived ashes is higher than that in the subbituminous derived ashes.

Using the classification scheme presented, differences and similarities in ashes and the effects of such collection devices as lime scrubbers can be categorized by the relative amount of the grain types present in the ash. While the use of a lime scrubber changes the bulk chemistry of the ash, notably SO_3 and CaO, the proportions of grain types remains relatively unchanged.

Relatively low CaO content grains of the G-type are thought to have as their origin, clay particles and to a lesser degree quartz and feldspar originally in the coal. The high CaO C-type grains are thought to be dominantly from material such as organically bound Ca and Mg in the coal and to a lesser degree calcite, dolomite, gypsum and clay. A blending of low and high CaO starting materials appears to make up the majority of the C-type and G-type particles in lignite and subbituminous derived ashes. However, G-type grains from bituminous derived ash have apparently little organically bound material and are generally CaO poor. Q-type grains were probably originally grains of quartz in the coal that subsequently picked up minor amounts of Na_2O, K_2O, CaO, and other oxides as these materials condensed onto [9] or came in contact with the quartz grains. The F-type grains are thought to originate from pyrite that has been oxidized into sulfur-poor, iron-rich grains.

ACKNOWLEDGMENTS

This research was partially funded by Gas Research Institute Contract No. 5086-253-1283 through the North Dakota Mining and Mineral Resources Research Institute. This research was also partially funded through the Western Fly Ash Research, Development and Data Center. The AAS analyses were performed by D. J. Hassett.

REFERENCES

1. R.J. Stevenson, Cem. Concr. Res. 14, 485-490 (1984).
2. R.J. Stevenson and R.A. Larson, in Fly Ash and Coal Conversion By-Products: Characterization, Utilization and Disposal I, edited by G.J. McCarthy and R.J. Lauf, Mat. Res. Soc. Symp. Proc. Vol. 43, (Materials Research Society, Pittsburgh, 1985) pp. 177-186.
3. R.J. Stevenson and G.J. McCarthy, in Fly Ash and Coal Conversion By-Products: Characterization, Utilization and Disposal II, edited by G.J. McCarthy, F.P. Glasser and D.M. Roy, Mat. Res. Soc. Symp. Proc. Vol. 65, (Materials Research Society, Pittsburgh, 1986) pp. 77-90.
4. H.M. Kanare, in Fly Ash and Coal Conversion By-Products: Characterization, Utilization and Disposal II, edited by G.J. McCarthy, F.P. Glasser and D.M. Roy, Mat. Res. Soc. Symp. Proc. Vol. 65 (Materials Research Society, Pittsburgh, 1986) pp. 159-160.
5. G.J. McCarthy, O.E. Manz, R.J. Stevenson, D.J. Hassett and G.H. Groenewold in Fly Ash and Coal Conversion By-Products: Characterization, Utilization and Disposal II, edited by G.J. McCarthy, F.P. Glasser and D.M. Roy, Mat. Res. Soc. Symp. Proc. 65, (Materials Research Society, Pittsburgh, 1986) pp. 165-166.
6. A.E. Bence and A. Albee, J. Geol. 76, 382-403 (1968).
7. R.T. Hemmings and E.E. Berry, in Fly Ash and Coal Conversion By-Products: Characterization, Utilization and Disposal II, edited by G.J. McCarthy, F.P. Glasser and D.M. Roy, Mat. Res. Soc. Symp. Proc. Vol. 65 (Materials Research Society, Pittsburgh, 1986) pp. 91-104.
8. F.R. Karner, S.A. Benson, H.H. Schobert and R.G. Roaldson, in The Chemistry of Low Rank Coals, edited by H.H. Schobert, ACS Symp. Ser. 264, (American Chemical Society, Washington, 1984), pp. 175-193.
9. R.J. Lauf, Am. Ceramic Soc. Bull. 61, 487-490 (1982).

CORRELATIONS OF CHEMISTRY AND MINERALOGY OF WESTERN U.S. FLY ASH

G.J. McCARTHY*, O.E. MANZ**, D.M. JOHANSEN*, S.J. STEINWAND* and R.J. STEVENSON**
*Department of Chemistry, North Dakota State University, Fargo, ND 58105
**Mining and Mineral Resources Research Institute, University of North Dakota, Grand Forks, ND 58202

Received 2 December, 1986; Communicated by D.M. Roy

SUMMARY

Fly ashes derived from low-rank coals mined principally in Montana, Wyoming and North Dakota are being studied by the Western Fly Ash Research, Development and Data Center [1]. Previous studies of the mineralogy of western U.S. fly ash by McCarthy et al. [1-3] using x-ray diffraction (XRD) form the framework of the present study. A database of chemical, mineralogical and physical properties, along with precursor coal characteristics, is being assembled. Based on studies to date of several hundred fly ash samples derived from lignite and subbituminous coals, as well as from several bituminous ashes, correlations of chemistry and mineralogy have been hypothesized and are being tested. These correlations are discussed below.

Analytical Alumina. Lignite ashes generally have Al_2O_3 below 16 wt.%; subbituminous ashes have between 18 and 24%. Bituminous coal fly ashes have Al_2O_3 contents greater than 23%. This trend is a result of the quantity and type of clay minerals, the principal source of Al_2O_3 in fly ash, in the original coal. The most important clay minerals in coal, their nominal chemical compositions (using Na-montmorillonite for a smectite) and the Al_2O_3/SiO_2 ratios for these compositions are:

Smectite	$Na_{0.33}(Al_{1.67}Mg_{0.33})Si_4O_{10}(OH)_2 \cdot nH_2O$	0.35
Illite	$K_{0.5}Al_2(Al_{0.5}Si_{3.5}O_{10})(OH)_2$	0.61
Kaolinite	$Al_4Si_4O_{10}(OH)_8$	0.85

Among the three major ranks of coal, lignite contains the greatest proportion of smectite, the clay mineral with the lowest ratio of Al_2O_3 to SiO_2, and bituminous coal with no smectite contains kaolinite and illite, minerals with higher ratios of Al_2O_3 to SiO_2. These ratios, combined with the dilution of both Al_2O_3 and SiO_2 by the CaO in high-calcium fly ashes, result in the observed lower Al_2O_3 contents of the low-rank coals.

Analytical CaO. In high-calcium ashes (those with CaO >17%), most subbituminous ashes have somewhat higher CaO content than lignite ashes. The sources of the Ca are Ca in the organic portion of the coal along with calcite and gypsum.

Analytical MgO. Lignite ashes have MgO greater than 4% while most subbituminous ashes have less than this level of MgO. The principal source of Mg is in the organic portion of the coal, along with montmorillonite and detrital ferromagnesian minerals and, rarely, dolomite.

Analytical alkali. $Na_2O > K_2O$ in both lignite and subbituminous ashes, in contrast to bituminous ashes where $K_2O > Na_2O$. In the low-rank coals, Na is present in smectites and on organic ion exchange sites. It combines with sulfate from gypsum and SO_x from oxidation of pyrite and organic sulfur.

Periclase. Crystalline MgO is more abundant in lignite ashes than subbituminous ashes.

C_3A and Merwinite. In x-ray diffractograms, the major characteristic peaks of C_3A and merwinite overlap [2,3]. Detailed studies of the diffractograms indicate that C_3A occurs in subbituminous ashes; the presence of merwinite has not been confirmed. In lignite ashes, with their higher MgO and lower Al_2O_3 contents, merwinite is dominant over C_3A; here, both phases appear to occur.

Alkali Sulfates. Phases such as thenardite and aphthitalite are common in the alkali-rich lignite ashes.

Table I

Partitioning of Chemical Components Among Phases* in High–Calcium Fly Ash

SiO_2:	GLASS > Quartz, Silicates (Mw,Ml,C_2S,Mu,So)
Al_2O_3:	GLASS > C_3A, Mu, Ml*, So, Sp
Fe_2O_3 (+FeO):	Sp > Glass, Hm, Ml*, Mw*
CaO:	GLASS > Lm, Ah, C_3A, Mw, Ml, C_2S
MgO:	Pc > GLASS, Mw, Ml*, Sp*
Na_2O:	GLASS, AS, So
SO_3:	Ah, AS, So, GLASS

Qz = Quartz, SiO_2
Mw = Merwinite, $Ca_3Mg(SiO_2)_2$
Ml = Melilite solid solution, $Ca_2(Mg,Al)(Al,Si)_2O_7$
C_2S = Dicalcium silicate-related, Ca_2SiO_4
Mu = Mullite, $Al_6Si_2O_{13}$
So = Sodalite-structure, $(Ca,Na)_{6-7}(Al,Si)_{12}O_{24}(SO_4)_{1-2}$
Sp = Ferrite Spinel, $(Mg,Fe)(Fe,Al)_2O_4$
Hm = Hematite, Fe_2O_3
C_3A = Tricalcium Aluminate, $Ca_3Al_2O_6$
Lm = Lime, CaO
Ah = Anhydrite, $CaSO_4$
Pc = Periclase, MgO
AS = Alkali Sulfates, Na_2SO_4 and $(Na,K)_2SO_4$

*Includes insights from SEM studies on low-rank coal fly ash
and lignite gasification ash by R.J. Stevenson [5-8]

Glass. The correlation of the XRD diffuse maximum with analytical CaO, as described by Diamond [4] is followed.

Anhydrite. Due to their higher CaO content, lignite and subbituminous fly ashes exhibit a SO_2 scrubbing activity during their formation and transport in the hotter parts of the furnace that results in anhydrite as a ubiquitous crystalline phase.

Qualitative estimates of the relative abundance of phases have been used to assemble Table I, which summarizes the current hypotheses for the partitioning of major oxide constituents of western U.S. fly ash among crystalline phases and glass. As this study progresses, efforts are underway to make the XRD analyses quantitative. When knowledge of the quantities of crystalline phases is available, additional insights on the chemical composition and structure of the glass phases can be derived. One example of the usefulness of such information would be in understanding sulfate resistance of these ashes. Dunstan [9] has proposed a factor (the "R-factor") for predicting resistance of concrete containing fly ash to attack by sulfate-containing solutions. This R-factor is based solely on the content of CaO, Al_2O_3 and Fe_2O_3. As Mehta [10] has pointed out, the relevant factors for predicting sulfate resistance of concrete are: (1) the mineralogy of alumina-containing crystalline and glass phases; (2) the hydration reactivity of these phases; (3) the availability of sufficient sulfate to combine with the aluminate phases to form ettringite and avoid monosulfoaluminate hydrate phases in freshly formed concrete. With the database being developed in this study, and the implementation of quantitative XRD, one will be able to study factors (1) and (3).

Lime and the Soundness of Concrete Containing Fly Ash

One crystalline phase occurring in western U.S. fly ash that is being monitored by quantitative XRD is lime (CaO). Schlorholtz and Demirel [11] have

Table II

Data From Tests of High-Calcium Lignite Fly Ash from Spain

BULK COMPOSITION (wt.%) AND ASTM C 618 RESULTS			
SiO_2 42.0	Al_2O_3 16.2	Fe_2O_3 8.7	CaO 24.9
MgO 1.0	Na_2O 0.3	K_2O 1.6	SO_3 4.6
LOI 0.17	Moisture 0.02	Fineness 26.2%	S.G. 2.60
Cement Pozzolanic Activity Index 99% Water Req. 94%			

ASTM SOUNDNESS TEST (AUTOCLAVE EXPANSION)

Bulk	Failed
+325 mesh	Failed
-325 mesh	0.3%

XRD LIME DETERMINATION (wt.%)

Bulk Ash	4.7 $\pm$0.7
-325 mesh	4.1 $\pm$0.6
+325 mesh	7.2 $\pm$1.0

indicated that lime is the expansive phase (forming portlandite, $Ca(OH)_2$, with a specific volume expansion of 96%) responsible for the occasional failure of a high-calcium fly ash in the ASTM soundness test. They suggested that when total crystalline lime content is greater than ~1 wt.% in a blended cement, soundness of concrete made from this material can be a problem [11]. A correlation of fly ashes failing the soundness test and high crystalline lime content has also been noted in our studies. We have shown that it is the lime in the +325 mesh (+45 μm) size fraction that is the problem. For a number of ashes that failed the test, separations with the 325 mesh sieve have been made; in each case the -325 mesh fraction passes the test. Consider the results given in Table II and Figure 1 for a lignite fly ash from Spain that has a high analytical CaO concentration and an exceptional crystalline CaO content (~7%). (Typical maximum lime contents of western U.S. ashes are ~3%.) XRD an-alysis of both fractions of the Spanish ash (Fig. 1) showed that the high lime content was concentrated in the coarse fraction. (It was also seen that anhydrite was concentrated in the fine fraction.) Sieving removed about 26% of the coarser ash particles, raising lime to 7.2% in the +325 mesh fraction and reducing lime in the -325 mesh fraction to 4.1%. Mass balance based on the fineness (Table II) indicates that the bulk ash should have 4.9% lime, in good agreement with the XRD value of 4.7%. Even though the -325 mesh fraction had almost as much lime as the bulk ash, it easily passed the soundness test. This indicates that it is the size of the lime particles, and not just the total amount of lime, that is responsible for the expansion. A third factor that should figure in delayed expansion of concrete is the reactivity of the lime. "Soft-burned" lime should hydrate and expand while the concrete is still plastic, whereas "hard-burned" lime would have hydration delayed until after the concrete has hardened. The lime in the present case was apparently hard-burned. Unfortunately, whether lime is hard- or soft-burned is not measurable with chemical analyses or XRD.

ACKNOWLEDGMENTS

The Western Fly Ash Research Development and Data Center is supported by Cooperative Power Association, Northern States Power, Otter Tail Power and Nebraska Ash Company. Chemical analyses were performed by D.J. Hassett.

112

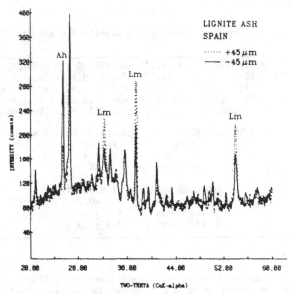

Figure 1. X-ray diffractograms of a lignite fly ash from Spain having an exceptionally high free lime content. [Lm = Lime (CaO); Ah = Anhydrite (CaSO$_4$)]

REFERENCES

1. G.J. McCarthy, O.E. Manz, R.J. Stevenson, D.J. Hassett, G.H. Groenewold, in Fly Ash and Coal Conversion By-Products: Characterization, Utilization and Disposal II, edited by G.J. McCarthy, F.P. Glasser and D.M. Roy, Mat. Res. Soc. Symp. Proc. Vol. 65, (Materials Research Society, Pittsburgh, 1986) pp. 161-162.
2. G.J. McCarthy, K.D. Swanson, L.P. Keller, W.C. Blatter, Cem. Concr. Res. 14, 471-478 (1984).
3. G.J. McCarthy and S.J. Steinwand, in Proc. 13th Biennial Lignite Symposium: Technology and Utilization of Low-Rank Coals, edited by M.L. Jones, DOE/METC-86/6036(Vol.2), (U.S. Department of Energy, Washington, 1985) pp. 600-608.
4. S. Diamond and J. Olek, this volume.
5. R.J. Stevenson, Cem. Concr. Res. 14, 485-490 (1984).
6. R.J. Stevenson and R.A. Larson, Fly Ash and Coal Conversion By-Products: Characterization, Utilization and Disposal I, edited by G.J. McCarthy and R.J. Lauf, Mat. Res. Soc. Symp. Proc. Vol. 43, (Materials Research Society, Pittsburgh, 1985) pp. 177-186.
7. R.J. Stevenson and G.J. McCarthy, in Fly Ash and Coal Conversion By-Products: Characterization, Utilization and Disposal II, edited by G.J. McCarthy, F.P. Glasser and D.M. Roy, Mat. Res. Soc. Symp. Proc. Vol. 65, (Materials Research Society, Pittsburgh, 1986) pp. 77-90.
8. R.J. Stevenson and T.P. Huber, this volume.
9. E.R. Dunstan, Jr., ASTM Cem. Concr. Aggreg. 2, 20-30 (1980).
10. P.K. Mehta, ACI Journal, November-December 1986, pp. 994-1000.
11. S. Schlorholtz and T. Demirel, in Fly Ash and Coal Conversion By-Products: Characterization, Utilization and Disposal I, edited by G.J. McCarthy and R.J. Lauf, Mat. Res. Soc. Symp. Proc. Vol. 43 (Materials Research Society, Pittsburgh, 1985) pp. 177-186.

SOME PHYSICAL, CHEMICAL AND MINERALOGICAL PROPERTIES OF
SOME CANADIAN FLY ASHES

R.C. JOSHI and B.K. MARSH
Department of Civil Engineering, University of Calgary, Calgary, Alberta,
Canada

Received 20 October, 1986; refereed

ABSTRACT

This paper gives physical and chemical properties of some Canadian fly ashes. Specific surface area, magnetic fraction, water soluble fraction and fraction finer than 45 µm were determined as part of the physical tests. Thermo-gravimetric analyses (TGA) in oxygen and nitrogen were conducted on raw ash samples. The change of pH with time in suspensions of the different ashes in water was also determined. Pozzolanic activity of the ashes with lime for all the ashes was evaluated to measure ash reactivity.

The ash activity seems to be related to fineness of the ash measured by the Blaine air permeability method, but not to the fineness measured by nitrogen sorption. Generally the greater the specific surface area, the higher the reactivity of the ash. The correlation was, however, not strong and no other physical or chemical parameter measured in this investigation seems to be related to pozzolanic activity.

The results of pH and TGA tests indicated that the ashes differ in many respects from each other. The TGA data suggest that loss-on-ignition in many of the ashes is not entirely due to the presence of unburned carbon. Specific surface area determined by various methods seems to provide different values. No characterization parameter was found that was uniquely related to coal type.

INTRODUCTION

Characteristics of any fly ash depend, amongst other factors, upon the coal type from which it is produced. The particular coal type used in power generation depends, in turn, to a large extent upon the geographical location of the plant. Across Canada there are many power plants using coal from many different sources. This paper compares physical, chemical and mineralogical properties of fourteen different ashes from Eastern, Central and Western Canada.

The purposes of the investigation were to determine:

1. the various physical and chemical properties of fly ashes produced in different parts of the country,

2. the amount of ash within fractions of low specific gravity, particularly the ash fraction lighter than water,

3. the magnetic fraction within the ashes,

4. the nature of the loss-on-ignition of the various fly ashes by thermogravimetric analyses, and

5. correlations, if any, between the physical and chemical properties and the pozzolanic activity of the ashes.

Table I. Fly Ash Types and Sources

Fly Ash No.	Coal Type	Sources
1	Sub-Bituminous	Alberta
2	Sub-Bituminous	Alberta
3	Sub-Bituminous	Alberta
4	Sub-Bituminous	Alberta
5	Lignite	Saskatchewan
6	Sub-Bituminous	Saskatchewan
7	Lignite	Saskatchewan
8	Lignite	Saskatchewan
9	Bituminous	Ontario
10	Bituminous	Ontario
11	Bituminous	New Brunswick
12	Bituminous	New Brunswick
13	Bituminous (?)	Nova Scotia
14	Sub-Bituminous	Nova Scotia

MATERIALS AND EXPERIMENTAL METHODS

Laboratory tests included chemical analysis, and tests to determine specific surface area, fraction finer than 45 μm, magnetic fraction, loss-on-ignition, fractionation of particles of different specific gravity, pH development of suspensions of ash in water, and pozzolanic activity with lime.

Fly Ash Source

Fourteen fly ashes from different power plants across Canada were studied (Table I). Fly ashes No. 1 through 4 from Alberta, fly ash No. 6 from Saskatchewan, and fly ash No. 14 from Nova Scotia, are produced from sub-bituminous coal while fly ashes No. 5, 7 and 8, from Saskatchewan, are produced from lignite coal. Fly ashes No. 9 and 10 from Ontario and fly ashes No. 11 and 12 from New Brunswick are produced from bituminous coal. Fly ash No. 13 may also be classified as marginally bituminous ash.

Available data indicate that a large proportion of the coal ash in Canada is produced in the Western Provinces, particularly Saskatchewan and Alberta. If the present trend of increased use of thermal coal for power generation continues, more and more fly ash is likely to be produced by the Western Provinces. This explains why 8 of the 14 samples selected for this study were from this area. The 14 ashes selected also reflect the three basic types of coal used in Canada, i.e. bituminous, sub-bituminous and lignite. Some of the coal sources, however, could be given dual classification.

The fly ashes were collected in electrostatic precipitators except ashes No. 6 and 7 which were collected by mechanical precipitators. The samples studied were obtained from the ash stockpiles of silos which are normally reserved for concrete work and/or structural fill in the respective provinces.

Test Procedures

Some of the chemical and physical properties of the 14 ashes are presented in Table II. Elemental analysis was done by atomic absorption spectrometry.

The water soluble fraction was determined by mixing 10 g samples of ash in 500 mL of distilled water. The ash suspension was stirred for at least one hour, allowed to settle for 24 hours, and restirred for one hour to ensure complete dissolution of the water soluble fraction.

Table II. Summary of Some Physical and Chemical Properties of Ashes

Fly † Ash No.	Weight Percentage										Magnetic** Fraction	Water Soluble Fraction
	Al_2O_3	Fe_2O_3	TiO_2	CaO	MgO	SO_3	Na_2O	K_2O	SiO_2	LOI*		
1	21.97	4.10	1.82	10.03	1.57	0.23	0.32	0.62	58.36	0.54	≤1	1.10
2	21.79	3.58	1.33	8.97	1.17	0.16	3.34	0.60	58.46	0.37	≤1	1.65
3	21.70	4.09	0.91	10.13	1.96	0.22	2.82	1.04	55.70	0.72	≤1	1.68
4	23.31	6.81	1.24	6.34	1.52	0.12	3.68	0.65	55.56	0.33	≤1	0.43
5	24.25	4.49	1.78	13.69	4.47	1.25	0.68	1.33	47.17	0.69	≤1	2.97
6	20.36	3.96	1.90	2.82	1.13	0.10	0.48	1.62	55.20	11.73	≤1	0.78
7	18.97	5.46	1.26	8.98	1.58	0.46	1.10	1.12	60.12	0.56	≤1	1.06
8	19.49	4.17	1.88	13.85	3.10	0.15	6.85	0.52	48.02	1.07	≤1	1.52
9	20.50	4.03	1.80	13.21	3.41	0.21	6.75	0.60	48.65	0.56	≤1	2.05
10	23.16	14.96	2.75	4.20	0.93	0.99	0.62	1.27	44.46	6.46	85	2.55
11	23.20	10.59	2.46	6.76	1.09	0.68	0.63	0.72	48.07	4.99	69	3.37
12	15.11	34.66	1.47	4.29	0.61	0.55	1.61	1.60	39.18	0.24	96	1.22
13	13.52	35.05	1.34	2.85	0.68	0.23	1.53	1.50	40.94	2.07	87	1.04
14	21.48	19.22	1.95	1.98	1.25	0.72	0.59	1.84	50.09	0.67	94	2.08

†   See Table I for coal type and ash source
*   Loss-on-ignition
**   Percentage of sample attracted to a magnet. It does not necessarily indicate the amount of magnetite.

The magnetic fraction was determined by suspending a known mass of ash in water in a test tube and placing the tube at the end of a strong horse-shoe magnet. Particles that were not held in place by attraction to the magnet were washed away and the procedure was repeated until no more non-magnetic particles could be removed.

Specific surface area of the ashes was found by the Blaine Air Permeability method, ASTM C204, and by nitrogen sorption using a single point B.E.T. method. At least three tests were performed on each sample. To examine the influence of carbon, or other combustible matter, on the specific surface, all the ashes were heated in air to 550°C and the specific surface was again determined by nitrogen sorption.

The fractions of each ash with different specific gravities were determined by measuring the amount floating on liquids of various densities (water, chloroform and 2-propanol of specific gravity 1.0, 1.47 and 0.78 respectively). A 10 g sample of ash was mixed with each liquid, and the floating particles were separated by decantation and filtration. Sediments from lighter fluids were tested for floating particles in the heavier liquids to ensure separation of particles of different specific gravity from one sample of ash.

Thermogravimetric analyses were performed at a heating rate of 20°C/min, from ambient to 1000°C. Samples were tested in an atmosphere of flowing oxygen; further tests were made in an atmosphere of flowing nitrogen. The sample size was approximately 100 mg.

Development of pH in suspensions of ash in water was monitored using a universal glass pH probe and pH meter. Fly ash samples weighing 2 g were suspended in 100 mL of water by continuous stirring. The indicated pH of the solution immediately on addition of the ash and every 0.25 seconds for a period of 250 seconds, was read and stored by computer.

Tests to determine pozzolanic activity of the ashes with lime were conducted in accordance with ASTM C 311.

Table III. Specific Surface Area of Fly Ashes

Fly Ash No.	By Air Permeability Method	Specific Surface Area (m²/g) By B.E.T. Method Ambient	Specific Surface Area (m²/g) By B.E.T. Method Heated to 550°C	Difference or Change	% Retained on 45 μm Sieve
1	0.42	1.52	0.99	-0.53	32.0
2	0.46	1.61	0.64	-0.97	26.0
3	0.43	1.61	0.65	-0.96	22.0
4	0.59	1.64	0.78	-0.86	9.8
5	0.50	1.14	1.23	+0.09	2.8
6	0.22	6.70	1.32	-5.38	20.4
7	0.17	1.01	0.60	-0.41	44.8
8	0.22	3.47	0.36	-3.11	26.6
9	0.28	0.55	0.28	-0.27	24.0
10	0.25	3.28	1.69	-1.59	27.0
11	0.31	2.68	1.37	-1.31	21.4
12	0.18	0.43	0.53	+0.10	26.4
13	0.36	0.85	0.52	-0.33	28.2
14	0.38	0.67	0.77	+0.10	34.2

average difference   1.10 m²/g
standard deviation of mean of 14 readings   1.44 m²/g

RESULTS AND DISCUSSION

Fineness Variation

An examination of the data in Table III indicates that the specific surface area, determined by the air permeability method, of the ashes collected by electrostatic precipitators is, for the most part, higher than the specific surface area of the ashes collected in mechanical precipitators (ashes No. 6 and 7). Nevertheless, even some of the ashes collected in electrostatic precipitators had very low specific surface area.

Table III also includes data on the weight fraction of each ash coarser than 45 μm. Comparison of these results with those in Table I shows that there does not appear to be a relationship between the percentage retained on the 45 μm sieve and the coal type or ash source. There is also no clear difference between those ashes collected mechanically (ashes No. 6 and 7) and those collected electrostatically. It is, however, interesting to note the large variation between the ashes - from the very small amount of coarse particles found in ash No. 5, to the ash No. 7 where coarse particles comprise almost half the weight of the ash.

Specific surface area was measured both by the Blaine air permeability method and the single-point B.E.T. nitrogen sorption method. Comparison of the results, in Table III, from air permeability with those from the B.E.T. method on unheated ash samples show considerable differences. The indicated specific surface area by B.E.T. is in all cases larger than that by air permeability. The amount of this difference varies from a factor of 1.8, for ash No. 14, to 30.5 for ash No. 6.

It was noted that some of the ashes showing very large differences between the two estimates of specific surface area also showed high values for loss-on-ignition. In particular, this was true of ashes No. 6, 10, and 11, although ash No. 8 proved to be an exception. The possibility of this effect being caused by porous carbon particles was investigated by repeating the nitrogen sorption tests on the ashes after they had been heated in air

Table IV. Data on Particles of Various Specific Gravities

| Fly Ash No | Overall Apparent Specific Gravity | % Floaters on Liquids of Density | | |
		1.47	1.00	0.78
1	2.19	1.7-2.7	0.5-0.65	0.25-0.35
2	1.92	2.1-3.2	0.45-0.55	0.25-0.30
3	1.91	0.7-0.8	0.1-0.11	0.07-0.09
4	2.03	0.35-0.55	0.2-0.22	0.08-0.12
5	2.54	0.3-0.5	0.2-0.23	0.13-0.17
6	2.15	1.3-1.6	0.3-0.6	0.3-0.5
7	2.37	2.5-2.8	0.5-0.75	0.4-0.6
8	2.39	1.1-1.4	0.6-0.8	0.2-0.4
9	2.46	0.8-1.0	0.3-0.5	0.1-0.3
10	2.31	1.2-1.5	0.25-0.4	0.2-0.3
11	2.94	0.7-0.9	0.2-0.5	0.08-0.10
12	2.87	0.8-1.2	0.6-0.8	0.3-0.5
13	2.53	1.8-2.4	1.1-1.4	0.7-0.1
14	2.44	0.05-0.07	0.01-0.04	0.003-0.01

to 550°C. This heating would result in the removal of any such carbon particles.

The specific surface areas of the ashes after heating to 550°C are given in Table III. Also given are the differences between these readings and those obtained from the unheated ash. Most of the ashes showed a significant reduction in specific surface area due to the heating. The average reduction for all the ashes was 1.10 m^2/g which, statistically, is significantly different from zero at the 1% level. Moreover, the greatest reduction in specific surface area was experienced by ashes No. 6, 8, 10 and 11. It would thus appear that the specific surface area measured by nitrogen sorption is greatly affected by the presence of high surface area particles of unburnt carbon. Such high surface area particles do not, however, influence the air permeability results to such an extent as ashes No. 6, 10 and 11 do not show high specific surface by this method.

Even after heating the ash to 550°C, however, there is still poor correlation between the specific surface area of ash measured by nitrogen sorption and the specific surface area of unheated ash by air permeability.

The high surface area carbon particles would not affect the reactivity of an ash, other than to reduce the amount of reactive material, but may affect workability and air-entrainability of any concrete in which the ash is used.

Ash Fractions of Different Specific Gravities

The data in Table IV includes values for the apparent specific gravity of the overall ash sample. The values range from 1.91, for ash No. 3, up to 2.94 for ash No. 11. Nevertheless, all the particles within a particular ash sample are not necessarily of the same density; often an ash will contain a significant fraction of very light particles. The proportion of particles of low specific gravity was determined from the amount of ash found to float on liquids of different densities. The results are given in Table IV.

All the particles of low specific gravity must have trapped gas bubbles, since all the glassy and crystalline matter in ash has a true specific gravity higher than 2.65. The majority of the light particles were coarse and could be easily discerned by the naked eye.

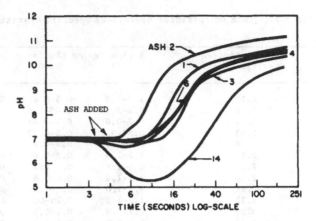

Figure 1. pH development for sub-bituminous ashes

The lightweight particles, sometimes known as cenospheres or floaters, can be used as lightweight fillers in a variety of applications. Ash No. 13 could be used as an economic source of floaters that could be separated using water.

Magnetic Fraction

The magnetic fraction, as reported in Table II, represents the fraction or percentage by weight of the ash attracted to a magnetic field. Many fly ash particles are very small and some may become trapped between the magnetic particles during separation; as a result, the reported data for the magnetic fraction may include some non-magnetic particles. Nevertheless, the data still clearly indicates large amounts of magnetic particles in ashes No. 9, 10, 11, 12 and 13. Apart from ash No. 14, for which no data is available, all the other ashes contain very few magnetic particles.

pH Development

The data on development of pH with time of solutions of ash in water are presented in figures 1 to 3. Each figure shows the results for a particular ash type. The pH development does not appear to be characteristic of the ash type. There are, however, different types of behaviour between different ashes.

Some ashes cause the solution to first become acidic before finally becoming basic. In the cases of ashes No. 12 and 14, the pH of the solution drops to below 5.5. In the cases of ashes No. 4, 6, 8 and 9, the initial drop in pH is evident but not large. Ashes No. 1, 2, 3 and 11 show only a rise in pH.

The rate of increase of basicity varies considerably between ashes. Ash No. 11 causes the pH to rise from pH 7 to approximately 10.5 in less than 6 seconds, whereas ash No. 12 causes the pH to fall to a minimum of approximately 5.5 after about 16 seconds and by 4 minutes the pH has risen only to 7.

When used as a partial replacement for cement, the effect of the fly ash on the pH of the mixing water will be largely masked by the rapid rise in pH

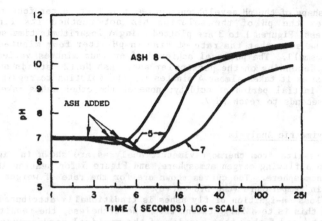

Figure 2. pH development for lignite ashes.

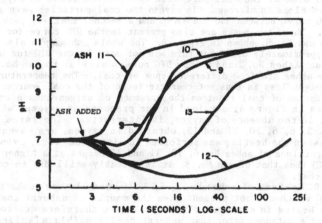

Figure 3. pH development for bituminous ashes.

resulting from the reaction of the cement grains. Nevertheless, the study of
the pH development of ash in solution is interesting from the point of view of
characterization of the ash itself. The rapid change in pH as measured in
this investigation is probably indicative of the nature of the soluble components existing on or near the surface of the ash.

The initial acidity experienced by some of the ash solutions was likely
caused, at least in part, by dissolution of sulphates that had been deposited
on the surface of the ash particles during formation. There is, however, no
direct relationship between the degree of acidity initially achieved and the
sulphate content of the ash, as given in Table II. Most of the ashes with low
sulphate contents show little or no acidic behaviour, whereas most ashes with
higher sulphate contents (approximately 0.5% and above) show some acidic
behaviour but ashes No. 11 and 13 are exceptions to this rule.

The shape of the pH development curves show that, after four minutes, for most ashes the pH of the solution has not reached its final value. Nevertheless, Figures 1 to 3 are plotted using a logarithmic time scale and it can thus be seen that the rate of rise in pH after four minutes, for most ashes, is small. The pH level achieved after four minutes varied between 7 and 11.5, depending on the particular ash. Ash No.12 stands out from the other ashes as it takes almost 4 minutes for the solution to regain neutrality after the initial period of acidity; none of the other ashes takes more than about 30 seconds to reach pH 7.

Thermogravimetric Analysis

The results from thermogravimetric analyses are shown in Figure 4, for heating in a flowing oxygen atmosphere, and Figure 5 for heating in a flowing nitrogen atmosphere. The curves shown are for the rate of weight change with increase in temperature (the DTG curve).

The loss-on-ignition of fly ashes is traditionally attributed to unburnt coal, and this is to a large extent true. Nevertheless, the results presented in Figures 4 and 5 show that burning of the residual coal occurs at temperatures that may differ between ashes. The results also show that loss-on-ignition is not always entirely due to unburnt coal.

Coarse particles of visibly unburnt coal were removed from a sample of ash No. 8; some of the coal particles were heated alone in oxygen and some were heated alone in nitrogen. In oxygen the coal particles gave a sharp peak on the DTG curve between 350 and 450°C and a second, smaller, peak between 450 and 500°C. These two peaks are also present in the DTG curve for a sample of the whole ash No. 8, shown in Figure 4. The whole ash sample also exhibited a peak at approximately 650°C but this was not seen in the DTG for unburnt coal particles. Ashes No. 2 and 9 show DTG curves similar to ash No. 8 although all these ashes are from different types of coal. The temperature distribution of weight loss is thus not characteristic of the coal source.

Combustion of coal requires the presence of oxygen thus a comparison of the curves in Figure 4, obtained in the presence of oxygen, with those in Figure 5, in the absence of oxygen, is interesting. The large DTG peak of ashes No. 3, 6, 8, 10, 11 and 13, obtained in oxygen, are absent in curves obtained when the heating was performed in nitrogen. Thus, even though the major weight loss of ashes No. 6, 10, 11 and 13 occurs at a higher temperature (600-650°C) than that of ash No. 8, it is probably still due to combustion of residual coal.

The DTG curves of ashes No. 2, 8, 9, 10, and 11 show peaks in the range 400-450°C which are not present when the sample of unburnt coal, from ash No. 8, is heated in nitrogen. These peaks are therefore due to the decomposition of substances other than unburnt coal possibly alkalines or other condensates deposited on ash particles during quenching on their exit from the flame in the furnace. At this stage, however, it is not possible to identify these other substances; further work, possibly using evolved gas analysis, is necessary.

The Relationship Between Pozzolanic Activity and Various Other Parameters

Generally loss-on-ignition, specific surface area, chemical composition and fineness are regarded as the parameters which have some effect on the pozzolanic activity of an ash. Joshi and Rosauer [1] and Joshi and Lam [2] have suggested that the type and amount of glass (or amorphous matter), fineness and chemical composition of the glass, in that order, control the pozzolanic behaviour of an ash. No test has yet been developed which allows accurate determination of glass or amorphous matter in fly ash. As a result of this, the relationships between pozzolanic activity of an ash, with hydrated lime, and combinations of parameters were evaluated.

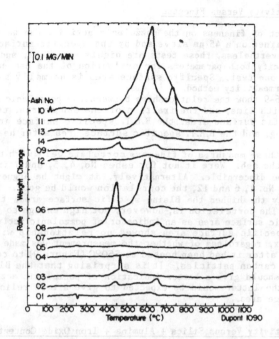

Figure 4. Derivative thermogravimetric curves for ashes heated in oxygen

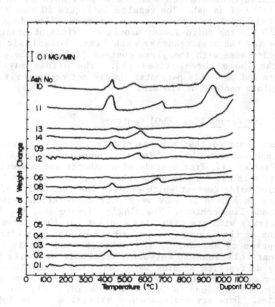

Figure 5. Derivative thermogravimetric curves for ashes heated in nitrogen

Pozzolanic Activity Versus Fineness

The effect of fineness on the pozzolanic activity can be related to the fraction retained on a 45 μm sieve and by the specific surface area of fly ash [3,4]. Nevertheless, these tests are highly empirical, and many agencies prefer to specify both parameters for evaluation of the ash' activity rather than rely on one test. Specific surface area is normally assessed by the Blaine air permeability method.

Figures 6-9 show the relationships between the pozzolanic activity index of the ash, with lime, and the fraction of the ash coarser than 45 μm, the Blaine specific surface area. the B.E.T. specific surface area of the ash before heating, and the B.E.T. specific surface area after heating to 550°C respectively.

None of these measures of fineness correlate well with the pozzolanic activity of the ash. Were it not for ashes No. 4, 5, and 7 in Figure 6, no trend would be discernible. Alternatively, it might be argued that were it not for ashes No. 1, 6 and 12, the correlation would be good. Figure 7 shows that generally the higher the Blaine specific surface area, the greater the reactivity. The correlation is, however, not high enough to justify use of Blaine specific surface area as an indicator of pozzolanic activity.

B.E.T. specific surface area showed no relationship with pozzolanic activity index, regardless of whether the measurement was made on the raw ash (Figure 8) or after it had been heated to 550°C (Figure 9) to remove the high surface area carbon particles. It is surprising that the Blaine specific surface area should give a better indication of pozzolanic activity than the B.E.T. area; the latter would be expected to give a more reliable estimation of true surface area.

Pozzolanic Activity Versus Silica + Alumina + Iron Oxide Content

Oxides of silicon, iron and aluminum comprise the bulk of fly ash. The sum of these oxide contents has sometimes been used as an indication of pozzolanic activity of an ash. The results in Figure 10 show that, for the 14 ashes within this test programme, such an assumption is quite invalid. Indeed, ash No. 5 has the third lowest amount of oxides of silica, alumina and iron, but is by far the most reactive with lime. Pozzolanic activity of an ash is said to increase with the glass content [1,2] and the glass is mainly composed of the above three oxides [5]. The oxides may also exist in crystalline form and thus this parameter should not necessarily be taken as an indication of glass content of the ash.

Pozzolanic Activity Versus Lime (CaO) Content

It has been reported that ash reactivity increases as the calcium oxide content of the ash increases [6]. This view, however, is not universally held and Joshi and Rosauer [1], from a study of synthetic and commercial fly ashes, concluded that it is the form in which the calcium compound is present, not the total calcium oxide content which affects the reactivity of the ash.

The results in Figure 11 show very poor correlation between pozzolanic activity index and lime content. The single lignite ash, No. 5, shows a high degree of reactivity with its high lime content but this ash also has a very small retention on the 45 μm sieve and a high Blaine specific surface area. With the exception of ash No. 5, the ashes tested within this programme exhibit no clear relationship between pozzolanic activity index and lime content. X-ray diffraction studies were conducted to evaluate the types of minerals or compounds present in these ashes. However, except for the lignite ashes, the calcium compounds in most of the Canadian fly ashes could not be easily identified. Data are therefore not available on the form in which the lime existed within the ashes so it is not possible to speculate on the

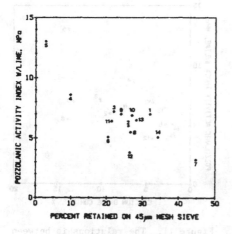

Figure 6. The relationship between
pozzolanic activity index
and proportion greater
than 45 μm

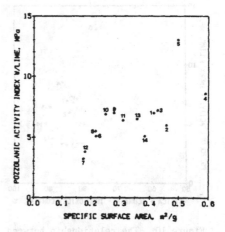

Figure 7. The relationship between
pozzolanic activity index
and Blaine specific
surface

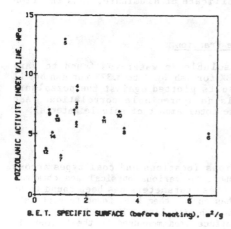

Figure 8. The relationship between
pozzolanic activity index
and B.E.T. specific
surface before heating
the ash

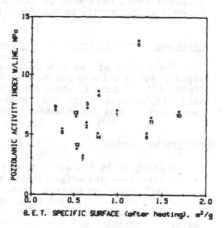

Figure 9. The relationship between
pozzolanic activity index
and B.E.T. specific
surface after heating
the ash

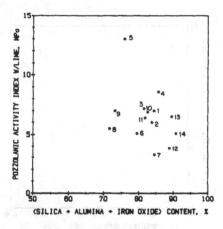

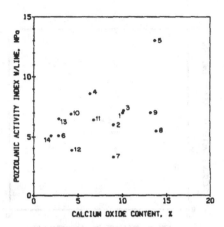

Figure 10. The relationship between pozzolanic activity index and the major oxide content

Figure 11. The relationship between pozzolanic activity index and the lime content

proposal that, only when present as silicate or aluminate, does the lime impart high reactivity to the ash [5].

Pozzolanic Activity Versus Water Soluble Fraction

The amount of the ash that was soluble in water was found to vary significantly between ashes, from 0.43% for ash No. 4 to 3.37% for ash No. 11 (Table II). Figure 12 shows these results plotted against the pozzolanic activity index of each ash; there is no appreciable correlation. Ash reactivity is thus not dependent upon the total amount of soluble material.

CONCLUDING REMARKS

A group of 14 fly ashes from various locations and coal types across Canada have been investigated with respect to various physical and chemical characterization parameters. None of these parameters has been found to be uniquely related to the coal type; nor has any of them been found to correlate well with the pozzolanic activity of the ash with lime. The best indicator of pozzolanic activity was the specific surface area measured by the Blaine air permeability method; even so, the correlation was insufficient to allow prediction of pozzolanic activity from knowledge of this parameter.

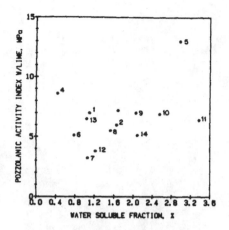

Figure 12. The relationship between pozzolanic activity
index and water soluble fraction

ACKNOWLEDGEMENTS

The work reported here was performed within the Calgary Fly Ash Research
Group supported by a Strategic Research Grant funded by the Natural Sciences
and Engineering Research Council of Canada.

REFERENCES

1. R.C. Joshi and E.A. Rosauer, Am. Ceram. Soc. Bull. 52, 459–465 (1973).
2. R.C. Joshi and D. Lam, Sources of Self-Cementitious Properties in Subbituminous Fly Ashes, University of Calgary Research Report No. CE81-6,
   1981.
3. P.T. Sherwood and M.P. Ryley, The Use of Stabilized Pulverized Fuel Ash in
   Road Construction, Road Research Laboratory Report No. 49, 1966.
4. G.G. Fowler and R.W. Styron, in Proceedings of the Fifth International Ash
   Utilization Symposium, edited by J.D. Spencer and C.E. Whieldon, Jr. (U.S.
   Dept. of Energy, Morgantown, WV, 1979) pp. 587–610.
5. J.P. Capp and J.D. Spencer, Fly Ash Utilization – A summary of Applications
   to Technology, U.S. Bureau of Mines Information Circular No. 8485, 1970.
6. Pacowski, in The Application of Brown Coal Fly Ash to Road Base Courses,
   FHWA-RD 79-101 (Federal Highway Administration, Washington, DC, 1979)
   pp. 79–101.

THERMAL CHARACTERIZATION OF COAL ASH POWDERS: HEAT CAPACITIES AND MINIMUM SINTERING TEMPERATURES

R. LEDESMA, P. COMPO and L. L. ISAACS
Department of Chemical Engineering, The City College of CUNY, New York, New York 10031

Received 15 October, 1986; refereed

ABSTRACT

Ashed mineral matter from ground coals were prepared by combustion at 973 K. Five coal samples ranging in rank from lignitic to high-volatile bituminous coals were used as starting materials. The heat capacities of the coal ash materials were measured between 140 K and 900 K by differential scanning calorimetry. The minimum sintering temperatures of each of the ash powders were determined using a high temperature dilatometer.

INTRODUCTION

The use of coal either as a fuel for power generation or as a raw material in chemical processes is usually accompanied by generation of ash. The ash comes from the mineral content of the coal and may amount to as much as 15% of the total weight for soft coals. The ash contribution to the heat load must be accounted for in an energy balance.

Most of the processes proposed for the conversion of coal to synthetic fuels (and feedstocks) require the use of fluidized bed type reactors. Since the agglomeration of ash particles in such reactors is an operating problem, the value of the minimum temperature for ash sintering becomes an important operating parameter.

Because of the wide variation in the composition of coal mineral matter, it is necessary to ascertain how ash properties vary with the composition and the type of the coal. In this communication, the results of thermal characterization experiments on ashes derived from low temperature (973 K) combustion of five different coals, specifically the results of heat capacity and dilatometry measurements, are reported.

EXPERIMENTAL

Ash Preparation

A suite of five coals was obtained from the Pennsylvania State University Coal Bank. Their rank and other pertinent properties are listed in Table I. In Table II, the major mineral constituents of some similar type coals [1] are listed. The coals were ground with a ball mill and sized by sieving. The 250 mesh sieve fractions were used for the preparation of the ash powders. These were prepared by placing five gram portions of the coals into muffle furnaces which were preheated to 825 K, and then raising the temperature slowly to 973 K. The ashes were kept at 973 K for two to three hours to ensure total combustion. No indication of particle size coarsening was observed. The chemical compositions of the prepared ashes were not determined directly. Spectroscopic analysis of ashes obtained from coals according to ASTM procedures, reported as oxides, for the coals whose mineral constitution breakdown is given in Table II, indicate that they contain: 33-55 wt% SiO_2, 10-25 wt% Al_2O_3, 5-15 wt% Fe_2O_3 and 2-20 wt% CaO as their major constituents [1].

Table I. Properties of Ashes and Coals.

Sample Identification	PSU Number	Total Mineral Matter Content as % (dry)	Pyritic Sulfur as %(dry)	FeS_2 as %(dry)	C (KJ/Kg-$^{\circ}$K)			T_s (K)
					300°K	600°K	900°K	
No. Dakota Lignite	PSOC-246	10.99	0.03	0.06	0.63	0.54	0.46	(853) 1035
Wyoming Subbit. B	PSOC-241	8.13	0.46	0.86	0.74	0.60	0.82	1036
Illinois #6	PSOC-022	11.88	0.89	1.67	0.63	0.72	0.93	1084
Illinois #6 HVB	PSOC-282	11.78	0.64	1.20	0.60	0.72	0.66	1066
Virginia HVA	PSOC-265	17.98	0.02	0.04	0.68	0.67	0.35	(927) 1101

Table II. Mineralogical Analysis of the Low Temperature Ash of Three Coals (Taken from Conn and Austin [1].)

Mineral (wt.%)	Keystone	Illinois #6 HVC	North Dakota Lignite
Quartz	25	20–25	25
Pyrite	10	10–15	10
Calcite	–	1–5	2–5
Gypsum	5	1–5	1–5
Kaolinite	30	20–25	5
Illite	20–40	10–20	–
Feldspar	–	1–5	–
Coquimbite	–	5–10	–

Heat Capacity Measurements

Ash heat capacities were determined by differential scanning calorimetry (DSC) [2,3]. A DuPont 1090 thermal analyzer system was used for these measurements. The temperature range involved was between 140 K and 900 K. However, some of the measurements were done only between 300 K and 900 K. Sample sizes of the order of six to ten milligrams were used. Heat capacities were calculated by comparing the heat flow curves for the samples with heat flow curves for a sapphire standard. Each sample run was accompanied by a standard run. The data were reproducible (multiple runs) and the estimated precision was 2%.

Dilatometry Measurements

One of the methods for finding the minimum sintering temperature, T_S, of a granular material is to use elongation-contraction versus temperature data obtained by the use of push-rod dilatometry [4]. In these experiments the powder to be tested is placed into a quartz tube and piston assembly, and the sample is compressed by an adjustable load. The experimental assembly is then heated at some preselected rate and the expansion or contraction of the sample is detected by a linear variable differential transducer and recorded as a function of temperature. In general, the powder dilates on heating due to thermal expansion. Eventually, a temperature is reached where the surface of the particles begin to deform due to viscous flattening and/or sintering at the intergranular contact points resulting in contraction of the sample. The temperature at which this phenomenon occurs is the minimum sintering temperature, T_s, which is characteristic for each powder and is the temperature at which particle agglomeration will first occur.

In this work, the minimum sintering temperatures of the five coal ash powders listed in Table I were determined by dilatometry. The samples were tested in air without sieving. They were loaded into a six millimeter diameter quartz tube and compressed by a 23 gram load. The assembly was heated at a 12 K/min. rate from ambient temperature to approximately 1250 K. The heating rate was comparable to the heating rate (10 K/min.) used for the DSC measurements. The change in length of the sample is reported as ΔL/Lo where Lo is the initial length of the sample.

It should be noted that the value of the piston weight used to compress the sample was chosen so that the intergranular force was kept small, thereby simulating a zero compression sintering process. Some minimum load is necessary; if too small a load is used, frictional forces become important and the results become irreproducible. Intergranular forces resulting from sample compression loads on the order of 20-30 grams have been shown to lower the minimum sintering temperature of various materials by 2-4% [5].

RESULTS AND DISCUSSION

Heat Capacity of Ashes

Figures 1-5 show the heat capacity of ashes prepared at 973 K, T_{Ash}, as a function of temperature. Measured heat capacities at 300 K, 600 K and 900 K are tabulated in table I. The data are rather similar to each other irrespective of the source of the ash. The magnitudes of the observed heat capacities are in good agreement with literature data [6] on ashes obtained from the combustion of tar sands and other ash like materials.

The non-monotonic behavior of the heat capacity as a function of temperature is real and reproducible. The conversion of coal minerals to ash involves several processes; dehydration, thermal decomposition, dehydroxylation of clays, oxidation, and phase transformations. Certain features of the obtained heat capacity curve are explainable in terms of these processes.

In the ashing process the coal minerals were dehydrated and formed a material that was porous in character and rather hydroscopic. Water from the air was adsorbed on the ashes. It has been observed [3] that in high porosity materials the adsorbed water behaves as two distinct entities, nonfreezable (molecular) water, adsorbed in the pores, and freezable (bulk) water adsorbed on the surface. The onset of molecular motion for the pore adsorbed water molecules is endothermic and an anomalous increase in the matrix heat capacity, between 150 K and 250 K is observed. The melting and vaporization phase transformations of the adsorbed bulk water, instead of occurring over the usual narrow temperature ranges, now spread out over a wide temperature range. The broad maxima in the heat capacities of the ashes between 250 K and 450 K are associated with the desorption of water.

Iron pyrite (FeS_2) is one of the minerals associated with the coals used in this study. The heating of iron pyrite in air leads to the formation of the oxide, Fe_3O_4, and of the nitride, Fe_4N, and may leave a range of compounds with formulae Fe_yS_x. A number of the compounds are magnetic. The Curie temperatures of the above mentioned compounds are respectively 858 K, 763 K and 573 K. When the ash was heated a series of magnetic to non-magnetic phase transitions took place. Each of these transitions was observed to be a lambda type discontinuity in the heat capacity.

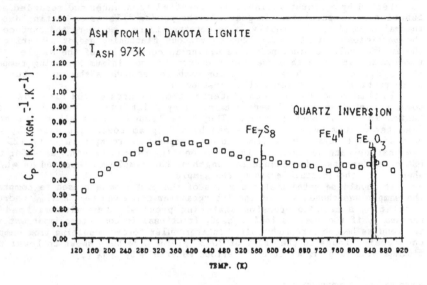

Figure 1. Ash Heat Capacity vs. Temperature

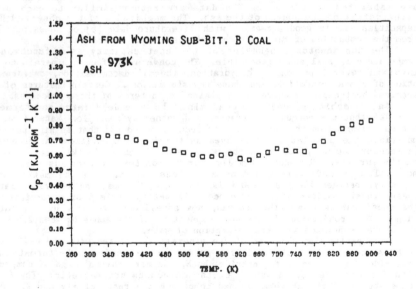

Figure 2. Ash Heat Capacity vs. Temperature

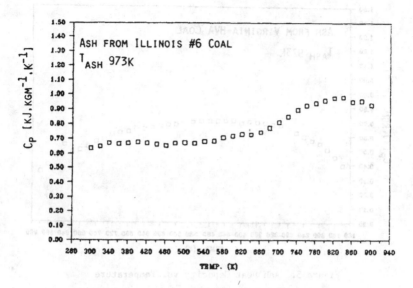

Figure 3.   Ash Heat Capacity vs. Temperature

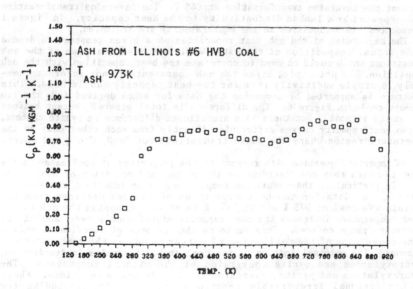

Figure 4.   Ash Heat Capacity vs. Temperature

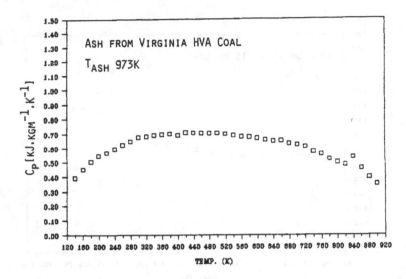

Figure 5. Ash Heat Capacity vs. Temperature

Quartz is a major mineral constituent of the coals used in the study. While the quartz remained unaffected by the low temperature ashing process, it underwent the inversion transformation at 845 K. The inversion transformation is accompanied by a lambda discontinuity in the heat capacity. In Figure 1 the expected discontinuities in the heat capacity are indicated.

The magnitudes of the ash heat capacities at a given temperature depend on the actual composition of the ash. We do not have analyses of the ash compositions which could be used to correlate the heat capacities with the ash composition. In principle, since the ash components do not interact chemically, a simple additivity rule for the heat capacity should hold. This conjecture is supported by comparing the data for ashes derived from the two Illinois coals in Figure 6. The difference in total mineral content between these coals is small, but there is a significant difference in pyrite content. The two heat capacity curves differ significantly from each other only in the temperature region where the phase transitions for Fe_4N, Fe_3O_4 and quartz occur.

An important question with respect to the properties of coal ashes is the extent to which they are dependent on the method and condition of ash formation. In particular, the combustion temperature is an important variable. In Figure 7, the data for the heat capacities of the ash obtained from the Virginia HVA coal at 973 K and at 1373 K is shown. Exposing the ash to the higher temperature increases its heat capacity significantly over most of the temperature range covered. This reflects the sequence of reactions involved in the conversion of the coal's mineral matter to ash as the temperature is increased. The original heating of the coal to 973 K results in dehydration, dehydroxylation and pyrite conversion of the mineral components. The dehydroxylation and pyrite conversion are irreversible reactions. Above 1073 K additional irreversible reactions are possible, including the decarbonation of calcite, dolomite and siderite and the high temperature phase transformations of clays. Therefore it is not surprising that ashes heated to the higher temperatures show changed heat capacities.

133

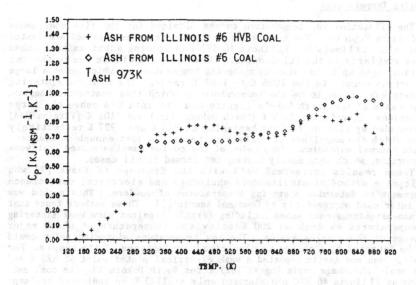

Figure 6. Ash Heat Capacity Dependence
on Composition

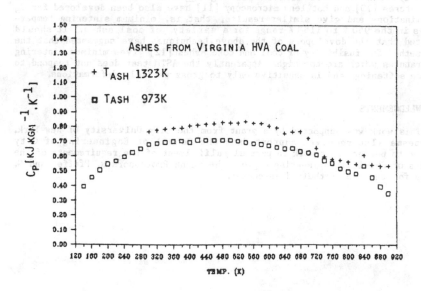

Figure 7. Dependence of Ash Heat Capacity
on Ashing Temperature

Sintering Temperatures

The dilatation vs. temperature curves obtained for the five coal ashes are shown in Figure 8. The minimum sintering temperatures obtained are listed in Table I. Illinois #6, Illinois #6 HVB and Wyoming subbituminous ashes behave similarly in the dilatometer. These samples dilate due to thermal expansion right up to the minimum sintering temperature at which point a large contraction occurs in the 1036 K to 1084 K range. The minimum sintering temperature is taken to be the temperature at which this contraction begins. In the cases of the North Dakota lignite and Virginia HVA ashes the large contractions occurring at 1035 K (North Dakota lig.) and 1101 K (Virgina HVA) are preceded by plateau regions beginning at 853 K and 927 K respectively during which the magnitude of surface softening is just enough to counterbalance thermal expansion. Inspection of the cooled samples showed that weak agglomerates, which were easily broken, had formed in all cases.

These results correspond well with the findings of Raask [7] who developed a method of simultaneous shrinkage and electrical conductance measurement to determine sintering temperatures of powders. This method was also later used successfully by Conn and Austin [1]. These authors found that some non-bituminous coal ashes including certain lignites show weak sintering at temperatures as much as 200 K below the temperature at which major shrinkage of the samples begins. The low temperature sintering point could only be detected by a change in electrical conductance of the sample. For example, Conn and Austin reported a weak electrical sinter point of 873 K and large scale shrinkage beginning at 1023 K for North Dakota lignite coal ash, whereas an Illinois #6 HVC ash sintered only at 1123 K as indicated by large changes in both conductance and the length of the sample. Dilatometry has also been used by others to measure coal ash sintering temperatures [8,9]. Siegell [8] measured values for T_s ranging from 975 K to 1148 K for various ashes which were obtained from actual boilers.

Additional methods such as sieving agglomerates formed at various temperatures [10] and hot lens microscopy [11] have also been developed for T_s determinations and give similar results, that is, minimum sintering temperatures in the 950 K to 1150 K range for a variety of coal ashes. It should be noted that the developers of the above techniques have suggested that the ASTM method for fusibility of coal and coke ashes [12] gives minimum sintering temperatures which are too high. Apparently the ASTM test does not respond to surface softening and is sensitive only to gross particle deformations.

ACKNOWLEDGEMENTS

This work was supported by a grant from the City University Of New York. R. Ledesma also received support from the School of Engineering of City College to perform this work in partial fulfillment of the requirements of the Masters of Chemical Engineering degree. We thank Professors R. Pfeffer and G. Tardos for the use of their dilatometer.

135

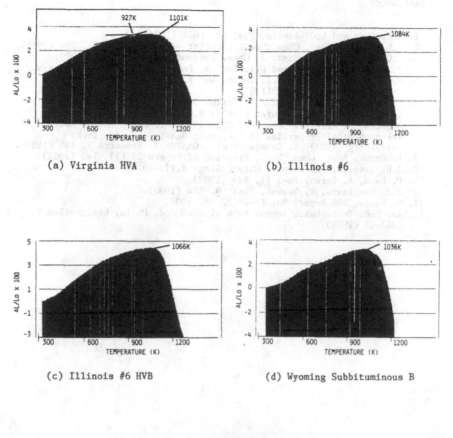

(a) Virginia HVA

(b) Illinois #6

(c) Illinois #6 HVB

(d) Wyoming Subbituminous B

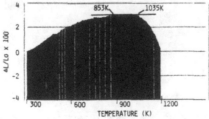

(e) North Dakota Lignite

Figure 8. Elongation – Contraction of Coal Ashes.

REFERENCES

1. R.E. Conn and L.G. Austin, Fuel 63, 1664 (1984).
2. L.L. Isaacs, Amer. Chem. Soc., Preprint of Papers, 29 (6), 234 (1984).
3. S.C. Mraw and D.F. Nass, J. Chem. Thermodynamics 11, 567 (1979).
4. P. Compo, R. Pfeffer and G.I. Tardos, Powder Tech., 1987, In Press.
5. P. Compo, D. Mazzone, G.I. Tardos and R. Pfeffer, Particle Characterization 1, 171 (1984).
6. M. Gomez, J.B. Gayle and A.R. Taylor Jr., U.S. Bur. Mines Res. Invest. No 6607 Pittsburgh (1965); N.C. Majumdar and T.J. Stevens, J. Austral. Cer. Soc. 28, (1966); I. Barin and O. Knake, Thermochemical Properties Of Inorganic Substances, New York: Heidelberg (1973); R. Cassis et al., OASTRA J. Research 1, 163 (1985).
7. E. Raask, Amer. Chem. Soc., Preprint of Papers 27 (1), 145 (1982).
8. J.H. Siegell, PhD thesis, City College N.Y., 1976.
9. P. Basu, A. Sarka, Fuel 61, 924 (1983).
10. J.J. Stallmann, R. Neavel, Fuel 59, 584 (1980).
11. S. Katta, DOE Report No. FE-19122-38, 1984.
12. Am. Soc. Test. Mat., Annual Book of ASTM Std. Phila, Designation D, 1857-68 (1978).

PART III

Reactions, Microstructure
and Modeling

HYDRATION REACTIONS IN CEMENT PASTES INCORPORATING FLY ASH
AND OTHER POZZOLANIC MATERIALS

F.P. GLASSER[a], S. DIAMOND[b] and D.M. ROY[c]
[a]Department of Chemistry, University of Aberdeen, Old Aberdeen AB9 2UE, U.K.
[b]School of Civil Engineering, Purdue University, Lafayette, IN 47907
[c]Materials Research Laboratory and Department of Materials Science and
Engineering, The Pennsylvania State University, University Park, PA 16802

Received 16 March, 1987; Communicated by G.J. McCarthy

ABSTRACT

A model for reactions that occur in hydrating portland cement is now
generally well developed. Incorporation of various by-products to form blended cements modifies both the hydration reactions and the physical properties
of the resulting pastes. A review of recent progress in understanding the
effects of blending agents on these reactions is presented. The blending
agents considered are low-calcium (Class F) fly ash, high calcium (Class C)
fly ash, blast furnace slag, silica fume, biosilica and natural pozzolans.
Effects of the blending agents on physical properties such as rheology are
also considered. Particular attention is given to the essential role of
alkalies in pore solutions and the beneficial reactions that occur with high
silica content blending agents.

INTRODUCTION

The use of coal combustion by-products presents economic opportunities,
but also a number of technical challenges. A principal use of fly ash (FA) is
in the formulation of blended cements, either "formal" blended cements produced and marketed by a cement manufacturer, or more commonly, "informal"
blended cements batched as part of the concrete making operation by concrete
producers. The technical challenges involved in optimizing such use may be
considerable.

The behavior of cements blended with fly ash is in many respects similar
to the behavior of cements blended with other blending components, such as
ground blast furnace slag (BFS), natural or artificial pozzolans, and silica
fume (SF). In this paper we have attempted to consider relevant aspects of
cement hydration reactions as influenced by some of these alternative components, notably BFS and SF, as well as by fly ash.

These blending components are ordinarily available as a consequence of
some other industrial process and their characteristics and compositions tend
to be governed by constraints imposed by that process, rather than by potential use in concrete. For example, metallurgical requirements control the
bulk composition of BFS, and power generating requirements control many of the
characteristics of fly ash. When selecting materials for use as blending
components in concrete, it is thus important to have adequate procedures in
place to ensure their suitability.

Previous symposia [1,2] have presented significant advances in the characterization of fly ashes. These studies, as well as those reported in this
volume, disclose the chemical and mineralogical nature of FA to be highly
variable. The variation occurs on different scales: combustion conditions,
ash collection procedures and coal source affect ash constitution. However,
even within one plant, much interparticle variability occurs. This is not
necessarily a disadvantage; portland cement (PC) is itself mineralogically
inhomogeneous.

Slags, on the other hand, are more homogeneous. They are obtained as
fully-molten material which, with rapid cooling, retains much of the former

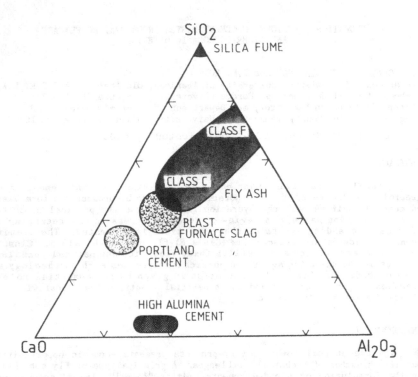

Figure 1. Ternary plot, showing compositions of cements and blending agents.

melt as glass. Some crystallization may occur during cooling, but the glassy matrix normally comprises 50 to more than 90% of the whole. The particle shapes are more uniform than FA; granulation and grinding produces shards with conchoidal fracture surfaces, typical of broken glass. The metallurgical requirements such as desulfurization potential ensure that the bulk chemistry of the slag is closely controlled, although the target compositions also reflect the ore source, nature of the fluxes available and the composition and content of the non-carbon constituents of the coal.

Figure 1 shows some characteristic composition ranges, projected on a $CaO-Al_2O_3-SiO_2$ ternary grid. The compositions are averages, and the method of projection ignores the presence of MgO, often at the 8-12% level in BFS, and iron oxide, often at similar and sometimes even higher levels in low calcium FA. Nevertheless, the figure is useful (1) in depicting the wide range of variation in FA compositions, and (2) in visualizing the potential changes in system chemistry which arise as a consequence of adding blending agents. The incorporation of blending agents is sometimes regarded as a cheap way of extending cement. However, economic and performance considerations go together, because blending agents can also markedly improve the performance of cement-based systems. This potential is not, however, automatically obtained but arises from careful formulation and optimization of mix properties, adequate curing and good design and workmanship. In any comparisons made in this paper, we assume good practice.

Table I

Some Properties of Cementitious Systems Affected by the Presence of
Blending Agents

Approximate Time Scale of Interest	Characteristic
0–1 h	Rheology
0–10 h	Initial set time
0–10 d	Heat evolution, initial strength gain
0–180 d	Strength gain, shrinkage, evolution of microstructure and relationship with porosity and permeability
>1 yr.	As above, plus enhancement of durability; resistance to adverse environmental factors (carbonation, sulfate and chloride attack, freeze-thaw, etc.)

Table I summarizes some of the system properties which may be affected by
the presence of blending agents. We classify these according to time scale.
This is of particular importance in the short term, up to roughly 1–3 months,
because blended cement systems often mature at different rates than PC systems. This time factor is especially important in comparing system properties. For some engineering purposes, e.g. strength gain, it may be essential
to make comparisons at fixed times. But for other types of assessment, e.g.
of potential durability, it is perhaps more important to conduct tests only
after steady-state conditions are established within the cemented system.
Since maturation may occur more slowly in some blended systems relative to PC,
it is important that this be allowed for in designing comparative tests.

Both FA and BFS are pozzolanic in the sense that when suitably activated,
they react with water to give products which are cementitious. However, the
choice of blending agent, the potential benefits to be obtained and the quan-
tification of the performance and durability of blended cement systems are
areas in which engineers need advice and guidance. This is particularly true
of those factors that relate to chemistry and microstructure of the blended
material and the implications for assessing performance. One of the purposes
of this paper is to describe some of these features and relate them to performance. Major emphasis is placed on FA, BFS and SF, but brief mention is
made of natural pozzolans and biosilicas (e.g. rice-husk ash).

COMPONENTS AND CERTAIN PROPERTIES OF BLENDED CEMENTS

Cement, Aqueous Phases and Cement Hydration

Cement is a manufactured product obtained by reacting naturally-occurring
raw materials, which must have closely-defined chemical compositions, at high
temperatures. Composition and temperature are chosen so that after loss of
volatile material (organic matter, CO_2 from carbonate, water from clays, etc.)
an indurated polymineralic product termed "clinker" is obtained. Table II
shows idealized compositions of the phases obtained, and their reactivity with
water. This reactivity is achieved by grinding the clinker to a relative high
specific surface area, typically to Blaine fineness values of about 4×10^3
$cm^2 g^{-1}$. Gypsum is universally added to the clinker at the grinding stage to
regulate the setting behavior. In modern cements set occurs typically in 4–6
hours (h), and significant strength gain is accomplished within the first day.

Table II

Constitution of Portland Cement and Hydration Characteristics
of its Constituents

Nature of Phase	Reactivity Towards Water
Calcium silicate, Ca_3SiO_5, "C_3S"	Relatively rapid
Calcium silicate, Ca_2SiO_4, "C_2S"	The β-polymorph is less reactive than C_3S
Calcium aluminate, $Ca_3Al_2O_6$, "C_3A"	Reacts rapidly with water; reaction retarded by soluble SO_4^{2-}
Calcium aluminoferrite, $Ca_2(Fe,Al)_2O_5$, "C_4AF" ($Fe:Al = 1:1$)	Reacts rapidly with water; reaction retarded by soluble SO_4^{2-}
Sulphates: $(K,Na)_2SO_4$, $CaSO_4 \cdot 2H_2O$ (gypsum), $CaSO_4 \cdot 0.5H_2O$ (hemihydrate), $CaSO_4$ (anhydrite)	Present on surface of clinker; calcium sulfates usually added to control set time

Provided that water continues to be available, strength gain continues indefinitely, although at a decreasing rate. Many concretes will have reached almost 3/4 of their ultimate compressive strength potential in 28 days (d), and the strength at this age is usually taken as a bench mark for the performance of the particular material.

Differences in the rate of strength gain for different cements at early ages reflect differences in specific surface area and in mineralogical composition; finer ground cements and cements richer in tricalcium aluminate (C_3A) and tricalcium silicate (C_3S) tend to gain strength more rapidly.

The cement hydration reactions are exothermic; unfortunately, the heat released can prove troublesome in mass concrete structures, where large temperature differences can develop in different parts of the structure. The high and rapid heat evolution during the early hydration of cements rich in C_3A, and to some extent C_3S, is particularly troublesome, and for mass concrete structures cements lower in these components may be specified. Historically, much of the early use of blending components such as fly ash has been in mass concrete structures, the replacement of some of the cement with the blending component serving to reduce the rate of early heat evolution.

Chemical studies of the aqueous phase in hydrating cement systems disclose that the alkali sulfates present as minor components in cement undergo rapid and complete dissolution. The cement phases dissolve incongruently, liberating calcium but retaining most of their anionic (silicate and aluminate) components. The solution quickly becomes supersaturated with respect to calcium hydroxide, and precipitation of this phase begins almost immediately. Apparently amorphous deposits, often of compositions resembling ettringite (AFt), appear on the surfaces of many cement grains, and may partly protect them against further hydration, leading to a so-called dormant stage during which hydration occurs slowly. However, after a few hours, it is considered that the protective film becomes unstable and ruptures, leading to a renewed burst of hydration.

The relationship among the OH^- ion concentration (or pH), the calcium ion concentration, and the various reaction processes undergone by the different cement components have been difficult to establish. It is observed that (1) pure calcium hydroxide dissolves congruently and generates a saturated solution that is approximately 0.04 N and has a pH of ~12.5, but that (2) the aqueous phase in cement systems has a significantly higher pH than this from the earlier stages, and after some days of hydration the pH may reach

13.5 or greater, i.e. a 10-fold higher concentration of OH^-. It appears that the earliest step in the cement hydration reactions involves attack on the silicate structure of the calcium silicates by protonation, the protons being derived from water molecules. Each water molecule providing a proton (for all practical purposes, a H^+ ion) leaves over an OH^- ion as a product of the dissociation of water; this provides the primary source of OH^- ions. As mentioned previously, the attacked silicates release Ca^{2+} ions but retain a residue enriched in silicates. Similar responses may be generated in the calcium aluminates. Alkali- and alkali calcium sulfates present as minor components dissolve quickly and contribute relatively high concentrations of potassium, sodium and sulfate ions; additional alkalies present in solid solution with certain of the cement phases (primarily C_3A and C_2S) are liberated more slowly as these phases hydrate. As a result of these processes, typical solution concentrations after a few hours are approximately 0.35 M in alkalies (mostly as K^+) and about 0.022 M in Ca^{2+}; the balancing anions are of the order of perhaps 0.2 M in SO_4^{2-} and perhaps 0.17 M in OH^-, for ionic strengths on the order of 0.5 M [3]. The silica content is in the ppm range, and both alumina and iron contents are lower than this. The pH level is of the order of 13.

Under these conditions, the solution is clearly supersaturated with respect to calcium hydroxide, which precipitates continuously, and also with respect to gypsum. The solid gypsum (and hemihydrate which may have been produced from it on clinker grinding) dissolves relatively slowly, and gypsum does not ordinarily re-precipitate. Rather, despite the paucity of aluminate species in the bulk solution, ettringite precipitates at or near the cement grain surfaces, practically from the earliest stages of hydration.

It appears that sulfate deposited into solution from progressive gypsum dissolution approximately keeps pace with sulfate depleted by ettringite precipitation, until the solid gypsum reservoir is fully depleted, usually by about 5 or 6 h after mixing. At that point the dissolved sulfate concentration starts to diminish rapidly. The OH^- ion concentration increases in step with the removal of sulfate, maintaining the total anion concentration approximately constant, but raising the pH significantly. Sulfate concentration drops to levels of the order of 10% or less of their original values, and remain low. Calcium ion concentrations diminish to even lower levels (on the order of 0.001 M), partly in response to the increase in OH^- concentration.

Subsequent to these changes over the first day or two, the composition of the aqueous phase remains more or less stable, subject to slight fluctuations in alkali and sulfate contents. Table III provides typical analyses for pore solutions of pastes made from a moderately high alkali cement, after periods of hydration ranging from 7 to 84 d. Similar data ranging up to almost two years have been secured at Purdue University. The steady-state aqueous phase can be described as essentially a concentrated solution of potassium and sodium hydroxide. It may contain, in addition, small contents of sulfate, still smaller contents of calcium, silica concentrations of only a few ppm, and almost negligible contents of aluminum and iron.

Replacement of some of the cement by blending components can lead to some change in aqueous phase chemistry, but changes are often surprisingly small. Vapor-deposited alkali sulfate coatings on some fly ashes dissolve quickly and augment the early concentrations of these components. Some high calcium fly ashes may release calcium, sulfate and aluminate ions on dissolution of anhydrite, C_3A and other soluble crystalline components. However, alkalies in the glass of most low calcium fly ashes and in slags are released only very slowly, and do not appear to influence aqueous phase chemistry to any significant degree.

The changes in solid phases occurring as cement hydrates are even more complex and difficult to summarize adequately, but an attempt will be made here to sketch the main features as presently understood. As mentioned above, early formation of amorphous-appearing deposits around cement grains have been reported by various authors. Some of these deposits have compositions resembling ettringite. Gelatinous appearing material, and the presence of hollow

Table III

Pore Fluid Analyses of Cement Paste

Cure time at 22°C (d)	Ion concentration, mmole/L				
	Na	K	Ca	OH⁻	SO_4^{2-}
7	263	612	1	788	23
28	271	629	1	839	31
56	332	695	3	839	44
84	323	639	2	743	27

(Cement alkali content: 0.41 Na_2O, 1.18 K_2O; w/c = 0.50)
Source: C.L. Page and Ø. Vennesland, Materiaux et
Constructions No. 91, 19-24 (1983).

sacs have been observed by Barnes et al. [4], and by Pratt and Jennings [5]. Deposits of very small crystals of calcium hydroxide, ettringite, and fibrous C-S-H on cement grain surfaces are reported to arise from the initial filmy deposit by Dalgleish et al. [6].

Subsequently, each cement grain appears to form a complete or nearly complete shell of hydration product around it [7]. These shells are limited in thickness to about 2 µm or less. Later hydration may deposit hydration product between the shell and the residual cement grain in an "inner deposition mode", or outside of the shell entirely, leading to the formation of a hollow hydration shell or "Hadley grain".

In either event, more or less elongated particles of Type I C-S-H gel and of longer ettringite needles deposit radially outward from the outer surfaces of the shells. These eventually intersect and combine with similar depositions from adjacent shells, and with calcium hydroxide precipitated into the free space between grains to develop a skeletal structure. The rheology of the paste reflects the extent of progress toward this structure, with increases taking place in both yield stress and plastic viscosity. When a complete network of the required rigidity is attained, the paste sets.

Subsequently calcium hydroxide continues to deposit, as does C-S-H gel, the latter primarily of the Type III variety. The earlier skeletal deposits and the shells tend to be covered up, and depending on the original water to cement (w/c) ratio, the original space between grains may be entirely filled. Mature fracture surfaces of hardened pastes tend to show primarily these latter stage deposits, although in local areas of higher original water content, the early morphology can still be recognized.

The deposition of calcium hydroxide may not be quite proportional with the extent of hydration, as would be expected from stoichiometric equations. Late deposition and perhaps a process of redissolution and precipitation in different areas may occur. There is a tendency to develop massive calcium hydroxide deposits in regions rich in formerly empty space, such as capillary pores, zones influenced by proximity to aggregate surfaces and even air voids. Ettringite also tends to deposit as a secondary process in such spaces.

Once initial set is achieved cement hydration continues over a period of months or even years. Thus, after about 28 d, much of the ferrite and aluminate phases have hydrated but some unhydrated silicate, especially β-C_2S, still persists. As a rule of thumb, about 2/3 of the total anhydrous phases will have been consumed at about 28 d. The continuation of hydration reactions over long periods of time helps explain the importance of good curing, defined as prevention of loss of unbound water, the continued presence of which is essential to the completion, or near completion of hydration. Figure 2 shows diagrammatically how the balance between unhydrated clinker, hydration

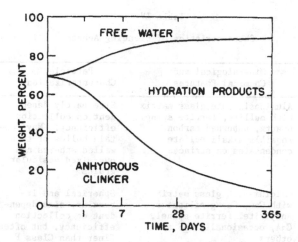

Figure 2. Schematic, showing the distribution of phases in a hydrating
cement paste, initially to a w/c ratio -0.45.

products, and free water (assuming no loss by evaporation) change during
hydration. Of course, the amount of free water is rather sensitive to the w/c
ratio used in mixing. The critical w/c ratio required for complete hydration
is -0.24; in practice, w/c ratios substantially exceed this, so pores may
remain largely filled with aqueous pore solution.

Blending agents are thus faced with a local environment which is characterized by containing a number of reactive phases. The most important is the
aqueous phase which contains much Na and K,, essentially as hydroxides.
Although Ca concentration in solution is low, Ca remains sufficiently soluble
so that the aqueous phase apparently can act as a bridge, transporting Ca from
the high specific surface area, Ca-rich hydration products of cement to the
pozzolanic constituents of blended cements. Silica fume and Class F fly ash
are notable consumers of Ca; Class C fly ash and BFS consume rather less Ca.
The composition of the aqueous phase is subject to fairly large changes during
the first day, but thereafter settles down to relative constancy [3]. Of
course, as water is withdrawn to satisfy the hydration demands of blending
agents and any remaining unhydrated cement, the more soluble constituents,
especially Na and K, tend gradually to increase in concentration.

Characterization of Reactive Blending Materials

The influences of the various classes of blending components on the processes of cement hydration and structure formation depend to some degree on
their individual particle characteristics. Here we include size distribution,
shape and chemical nature of the exposed surface.

The size distributions of portland cements depend somewhat on the clinker
grinding process employed, but typically have mean diameters of the order of
15 μm, with comparatively few very large grains, i.e. those over 75 μm (No.
200 sieve), being present. Only a small weight percentage is smaller than 2
μm, although here the number of particles may be considerable.

Table IV

Characteristics of Blending Agents

Designation	Mineralogical and Chemical Features	Particle Characterization	Activator
Fly Ash (Class F)	Aluminosilicate glass matrix with mullite, ferrite spinel, quartz, unburned carbon. Possible alkali sulfate condensates on surfaces	Size partly dependent on collection efficiency. May contain hollow or thin-walled spheres and oversized vesicular grains	(1,2)
Fly Ash (Class C)	Calcium-rich glass matrix with CaO, Ca-Mg silicates, anhydrite, ferrite spinel, C_3A, occasionally C_2S and others	Spherical and ir-regular, size dependent on collection efficiency, but often finer than Class F	(1)
Slag	Mainly Ca-Mg-Al-Si-rich glass. Contains reduced S	Glassy shards with conchoidal fracture	(1,2,3,4)
Silica Fume	Mainly SiO_2 glass, may have traces of silicon carbide and some unburned carbon	Mainly extremely fine high surface spheres	
Rice Husk Ash	Mainly SiO_2 glass, some unburned carbon	High surface, complex shapes reflecting biological origins	(1,2)
Natural Pozzolan	Very variable; active constituents may be silicate glass, zeolites, clay or mixtures thereof	Polymineralic nature yields ground particles having sizes and shapes similar to cement	(1,2)

Key to activators: (1) = PC; (2) = $Ca(OH)_2$; (3) = alkali hydroxides, carbonates; (4) = gypsum.

Some of the general chemical features of important blending agents were presented in Figure 1. However, some additional chemical features, and particlarly their physical form, require comments. Table IV summarizes these features. Fly ash may also contain substantial amounts of unburned carbon in addition to the components shown in the table.

Silica Fume, Biosilicates. Chemically, these are relatively pure SiO_2, at least ideally. SiO_2 fume used with cement is apt to contain <5% total non-SiO_2 material, although fumes of lower Si content are also available from the production of ferrosilicon and other products. Silica fume tends to be almost entirely composed of extremely fine spheres, with a mean size typically of the order of 0.1 to 0.2 μm with BET surface area of about 20 m^2/g. Occasional oversize spheres may be found, and fragments of wood or coke may occasionally be encountered, depending on the degree to which the fume may have been bene-ficiated. Some silica fume particles occur in clusters that may be difficult or impossible to disperse chemically, the latter implying that a certain degree of fusion may have taken place at points of contact.

Rice husk ash is an extremely high surface area material with a characteristic dendritic particle structure. This complex "shape" is difficult to characterize, but clearly increases the water demand in a suspension containing a significant proportion of such grains.

Fly Ash. The chemical compositions of this blending agent reflect the geological setting of the coal: some ashes are aluminum-rich or aluminum- and iron-rich silicates (Class F) while others may contain much Ca (Class C). The Class F fly ashes are characterized by relatively high glass contents. It is this glass which contributes to the pozzolanic reaction. The glass also surrounds crystalline minerals, some of which are residual unreacted mineral impurities in the coal, e.g. quartz. Other crystals develop from high-temperature reactions occurring in the combustion process (e.g. ferrite spinel and mullite), and are derived from pyrite oxidation, thermal dissociation of clays, etc. These mineral phases, as well as unburned carbon, are unreactive. The more Ca-rich Class C ashes often contain substantially less glass and a different suite of minerals which are developed during combustion. The minerals reflect the high Ca content of the ash and may themselves be reactive, e.g. CaO, C_3A, anhydrite, and C_3S and C_2S.

Fly ashes tend to have roughly similar size distributions, although they are influenced considerably by the collection process used. In a suite of 14 different fly ashes from Indiana, mean particle sizes were found to range from extremes of 3 μm to 78 μm [8]. However, both these end-member values are highly atypical; 11 of the remaining 12 fly ashes showed mean sizes between 10 μm and 32 μm.

The similarity of size distributions does not, however, tell the whole story. Commercial fly ash is the finer residue of a collection process in a coal-burning plant that typically captures large particles ("bottom ash"), and intermediate size particles ("economizer ash") upstream of the electrostatic precipitators that capture the fly ash, thus limiting the upper size. The upper size limit of any given fly ash assemblage captured in a given installation depends somewhat on the efficiency of operation of these preliminary separation steps. It may range up to several hundred μm in poorly functioning installations.

At the other end of the scale, the content of finest sizes (~1 to 2 μm) also tends to be limited. Since there is no grinding step involved in fly ash production, there is no opportunity to split off small chips of this size range as there is in clinker grinding. The finest sizes of fly ash tend to be solid spheres.

Fly ash particles come in a variety of shapes and surface characteristics. Most are solid spheres, but fly ashes also contain residual coal fragments, imperfectly rounded ellipsoids, vesicular solids of irregular shape, and various hollow or partly hollow grains. The content of "cenospheres" (thin walled hollow particles without smaller interior particles) tends to be quite small in most fly ashes, but there may be a more appreciable content of "plerospheres" (hollow spheres with smaller spheres enclosed), and even occasional two-level plerospheres (spheres within smaller hollow spheres within primary hollow spheres).

Slags. Slags occupy a rather narrower range of compositions than FA. Moreover, they have a different shape; granulation of slag produces shards bounded by conchoidal fractures of the type which, on a larger scale, are seen on fragments of broken glass bottles. Since they are deliberately ground to meet a standard for surface area (or "fineness"), the size distribution is more or less controlled, and tends to be similar to that of portland cement (or finer). The individual glassy particles show the expected conchoidal fracture and tend to be quite jagged. Fine "chips" are not uncommon. Two additional chemical features of slag deserve comment, their content of MgO,

Table V

Short Term Influence of Blending Agents on the Properties of Cement Systems

Property	Comment
Contribute to aqueous phase composition	Alkali may be present in readily soluble form, as on surfaces of FA grains
Removal of ions from solution by sorption, etc.	Silica fume and to a lesser extent FA uptake of alkali from aqueous solution
Flocculation	Particle size, shape, zeta potential, etc., all influence the flocculation behavior of mixed particle systems
Nucleation	Particles of blending agents may act as sites for preferential nucleation of hydration products

which distinguishes them from cement and fly ash, and their content of species reflecting the strongly-reducing regime in blast furnaces. Their content of reduced sulfur, as S^{2-}, and of colloidal iron are examples of reducing species. The crystalline components of slag may be strongly reactive with cement, and there is some evidence that partial crystallization may enhance somewhat the strength gain of slag-cement blends.

Natural pozzolans generally show some similarity to slags, in that they tend to be at least partly vitreous and are ground to a specified fineness. However, they also typically contain crystalline components, which may affect particle shape to some degree.

One important point should be made in connection with the use of these blending agents. Silica fume and biosilicas have a high potential water demand, and are usually blended at up to a 10-15% level. Fly ash and slag, on the other hand, are broadly neutral in their water demand, or may reduce it, and fly ash is often incorporated at levels of 10-30%. Blast furnace slags are frequently used in even greater amounts; blends containing up to 70% slag are employed in Europe.

Physical Interactions Between Cement and Blending Agents

The initial reactions between cement and blending agents may be summarized with the aid of Table V. Any readily-soluble components of the blending agents contribute to the short-term behavior of the system. These arise mainly from the presence of vapor phase condensates on the surfaces of grains, as occurs especially in some FA. However, during the first few minutes or hours of hydration, extensive hydroxylation of the particle surfaces of blending agents occur. As a consequence, the sorptive characteristics of the surface of the blending agents may be changed. In general, Class F FA and silica fume develop negative surface charges, and can sorb cations, while Class C FA tends to develop positive surface charges, and may well be able to sorb anions, although this has not been investigated to our knowledge. The presence of fine particulate matter, other than the cement itself, is also liable to influence the flocculation behavior of the system. This, in turn, influences the packing properties of the partially-hydrated system and may well have considerable influence on the pore size distribution and pore interconnectivity, even at longer ages such that hydration approaches completion. Finally, the blending agent itself may act as preferential sites for the nucleation of cement hydration products, especially in the period immediately following the initial burst of supersaturation. Many of these features are not readily investigated by conventional techniques, so most studies are confined to indirect measurements of the effect of blending agents on system properties, e.g. rheological changes. It is nevertheless desirable to extend studies in this area because the blending agents do not simply behave as inert materials during the initial stages of cement hydration.

REACTIONS IN BLENDED CEMENT SYSTEMS

Details of reactions that occur when the various blending agents are included in cement systems vary considerably, and have only recently come under extensive investigation. It appears that the more or less casual generalization that "pozzolans react with calcium hydroxide produced during cement hydration to produce additional or secondary C-S-H gel" is a grossly inadequate statement of what actually takes place.

The primary reaction step very likely is formation of a shell of C-S-H (derived from the hydrating portland cement) around the mineral admixture grain; dissolution of glass at the surface of the grain within the shell may then take place, the dissolution being effected by the alkali hydroxide-rich pore solution usually present in cement-based systems. Combination with dissolved calcium and precipitation of a reaction product may take place elsewhere within the system, although the details are still obscure.

In general, the "average" C-S-H gel produced in such systems shows a lower Ca:Si ratio than is found in a plain portland cement system, and the skeletal structure of linked silica tetrahedra within the C-S-H gel may be different in some details.

The content of calcium hydroxide, as determined for example by TGA, is less than would be found in a corresponding plain portland cement system, and an absolute reduction in the calcium hydroxide content with time in the later stages of hydration is usually demonstrated; this is of course due to the the the classic pozzolanic reaction.

The details of the processes naturally vary with the nature of the mineral admixture considered and, of course, with the degree of substitution for cement and the nature of the cement itself.

Fly Ash Reactions

The results of incorporating fly ash into cement pastes and concretes are still being explored, and no fully definitive summary can yet be made. It is far more difficult to generalize about reactions in fly ash bearing systems than in those with other components, because of the variations between different fly ashes and between different particles within a single fly ash. Fly ashes are assemblages of particles of individually varied compositions and crystalline contents, some of which are reactive.

One major influence of fly ash incorporation does not necessarily stem from chemical reaction at all, but rather derives from physicochemical effects. Fly ash inclusion may have a significant effect on the size distribution and structure of the individual flocs comprising fresh cement paste. Observed decreases in water demand most likely stem from such floc structure modifications, rather than from supposed "ball-bearing effect" of the spherical particle shape.

The changed floc structure is ordinarily carried over into changes in primary structure of the hardened paste, including especially its pore size distribution and permeability. However, evaluation of such changes by the usual mercury porosimetry techniques is still in a state of some controversy.

Chemical reactions in fly ash cement systems do clearly occur, and these reactions strongly influence the final properties of the resulting hardened cement paste or concrete. However, because of the wide range in chemistry among fly ashes, some distinctions among them need to be made when discussing chemical reactions.

In low calcium fly ashes, the reactive component is entirely the glass. The crystalline phases included within fly ash spheres, quartz, and hematite and magnetite-like crystalline compounds, which may make up most of the bulk of a separate class of iron-rich spheres, are not reactive in concrete. The glass is relatively slowly reactive at "ordinary" temperatures, but reaction is much faster and more extensive at higher temperatures developed briefly within thick concrete sections, or in steam-cured or autoclaved concretes.

In high calcium fly ashes, both glass and certain crystalline components may be reactive. The glass appears to be of at least two types, one showing an x-ray diffuse scattering maximum of symmetrical character near 27° 2θ (Cu radiation), the other an extremely asymmetrical band peaking near 32° 2θ. The latter is thought to be richer in calcium and aluminum, and to be more reactive. Reactive crystalline components, some of which may be readily available to mix water, can include free CaO, anhydrite, C_3A, and others. Certain fly ashes containing large proportions of such components may auto-hydrate when added to water, producing ettringite, monosulfate, and eventually C-S-H gel even in the absence of cement.

Occasionally fly ashes may influence the course of early hydration by depositing significant contents of alkali sulfates into the mix water; the long-term alkali hydroxide concentrations of pore solutions of cement systems containing such fly ashes may be substantially raised. Despite significant alkali contents in the glass of FA, most fly ashes seem to do little more than dilute the cement in terms of pore solution effects, reducing the concentrations of dissolved components more or less proportionately to the amount of cement replaced. Apparently the alkalies released by slow long term dissolution of the glass are effectively absorbed into reaction products.

The processes of reaction and deposition of the resulting reaction products are not at all well understood. According to Diamond et al. [9] a "duplex film" type skin may be deposited around fly ash grains; next, a dense shell similar to that formed around hydrating clinker grains is formed; subsequent reaction involves etching and dissolution from within this shell.

Micrographs showing the "skeletons" of unreactive crystalline inclusions within the original glass spheres left undisturbed where the glass has been entirely removed by dissolution, have been published by various authors, and the generality of this phenomenon seems established. However, what happens to the dissolved glass is not at all clear. Presumably it is precipitated elsewhere as "hydration" product, but the details of the precipitation process, and exactly where such products precipitate are yet unclarified.

As mentioned earlier, the overall $Ca(OH)_2$ content decreases with time more or less concomitantly with the dissolution of the glass, especially with low calcium fly ashes. At least one worker [10] has suggested that this process may leave empty spaces in the structure of the hardened product, especially in the vicinity of sand grains or coarse aggregate particles, but this has yet to be independently confirmed.

Reactions in BFS-Cement Blends

Nature of Slag. The chemical composition of blast furnace slags (BFS) was depicted in Figure 1 in terms of three components—CaO, Al_2O_3, and SiO_2; their compositions are controlled largely by metallurgical requirements. In addition to its CaO, Al_2O_3, and SiO_2 components, BFS contains substantial MgO and sulfur. Various proportioning formulae exist, but in general the sum of the "acidic" oxides (SiO_2, Al_2O_3, TiO_2) divided by the sum of the "basic" oxides (CaO, MgO, MnO, FeO) gives a ratio close to unity. Alkalies are damaging to the furnace refractories, so they are usually not allowed to exceed 1-2% Na_2O equivalent. The resulting slag is tapped molten, or very nearly so, from the blast furnace; if quenched or quickly chilled it yields a largely glassy product. BFS for use as an admixture is granulated and ground. However, not all slags are equally suitable for cement blending. Reactivity is generally assessed by determining the compressive strength gain of slag-cement blends; in general, high-calcium slags give the best reactivity. The MgO content is normally present mainly in the glass phase although some may be combined in crystalline components, e.g. melilite or merwinite, but not normally as free MgO (periclase). Thus, the type of problems associated with high MgO clinkers do not occur. The Al_2O_3 content of slags is, however, quite variable; while many slags contain ~10–15% Al_2O_3, it is not uncommon, depending on ore source, to find Al_2O_3 contents ranging up to 24%. Vitreous high-alumina slags give satisfactory strength gain, but it is likely that the resistance of cement blends to sulfate attack is lowered by very high Al_2O_3 contents.

The reactive nature of slag in slag-cement arises primarily from the glass content, and only to a small extent to certain of the crystalline phases present; β–Ca_2SiO_4 is an obvious example of a desirable crystalline phase. The glass content is usually determined by petrographic analysis using a polarizing microscope, or by x-ray diffraction. Using XRD the glass content can be determined semi-quantitatively from the integrated area under the broad amorphous glass scatter maximum centered at ~30° 2θ (Cu radiation). The crystalline phases are best characterized by XRD, although this is not normally done for routine quality control.

Portland cement is not the only activator which can be used for slag cements; aqueous solutions of alkali hydroxides, soluble sulfates, e.g. gypsum, soluble anhydrite or alkali sulfates, also act as activators. It is, however, customary to use some (20–30% minimum) portland cement to achieve adequate early strengths.

Hydration. No comprehensive picture of the hydration kinetics of cement-slag blends is as yet available. In the short term, the kinetics can be inferred from calorimetric data on the rate of heat release, assuming that the rate of heat release per unit mass is constant. Wu et al. [11] have recently compared the heat evolution of cement-BFS blends at two temperatures: their data, reproduced in Figure 3, show the very marked dependence of slag hydration rates on temperature. The marked temperature dependence of slag hydration is in contrast to portland cement, the hydration rate of which is less sensitive to temperature changes in this range. The actual heat release under adiabatic conditions (the maximum possible), however, is often reduced by granulated slag substitution for cement, as seen in the mortars of Figure 4. However, self heating, as occurs in most real constructions, may create substantial differences between the apparent reactivity of slag unless allowance is made for its thermal activation.

In the longer term, slag cement blends, particularly those containing large amounts of slag ($\geq$40%) appear to show an initial burst of hydration followed by a dormant period, during which little slag hydration occurs. The amount of slag which has reacted may be determined by selective dissolution methods, so some uncertainty is attached to the absolute values; however, Luke and Glasser [12] made a critical analysis of existing methods following which they analyzed a suite of aged samples, with the results shown in Figure 5. After an initial burst of hydration, during which ~40% of the slag was consumed in the first 28 d, little further reaction occurred until nearly one year, when hydration activity appeared to resume. This reconciles, at least qualitatively, two apparently contradictory features: rapid initial reactivity, suggesting that slag is very reactive, and experience of old slag cement blends which, although admittedly having an uncertain moisture regime, contain much unhydrated slag.

The products obtained from slag cement hydration are, of course, complicated by wide variations in the amount of slag used (5 to 70% or more), as well as by variations in slag composition and the extent to which hydration reactions have reached completion. It is possible however, to provide some guidelines concerning the nature of the solid phases obtained. Most slags

152

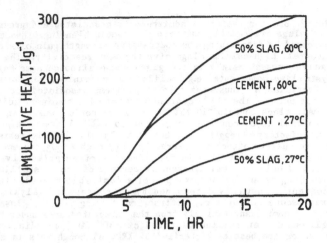

Figure 3. Isothermal cumulative heat evolution of cement
(ASTM Type II) pastes with and without slag
additions at 27°C and 60°C.

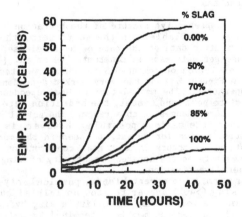

Figure 4. Adiabatic temperature rise of ASTM Type I cement
mortars with different proportions of slag;
adiabatic curing, 1:2.75 mortar, base temperature 25°C.

consume $Ca(OH)_2$ during hydration and, in mature cement-activated blends, the
amount of $Ca(OH)_2$ present decreases faster than would be expected upon simple
dilution of the cement by an inert material. Correspondingly, hydration of
slag gives rise to a C-S-H phase having characteristically lower chemical
CaO/SiO_2 ratios than would be obtained from portland cement hydration, with
ratios from 1.2 to 1.5 being common. Indeed, immature blends often yield
C-S-H which exhibits large local variations in CaO/SiO_2 ratio, with high ratio

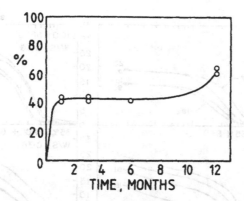

Figure 5. Experimentally determined kinetics of slag hydration. Blend
composition: 30% ground, granulated BFS, 70% OPC;
w/c = 0.6; cured at 100% relative humidity and 18°C.

C-S-H forming from cement and low ratio material from slag. The balance
between AFt and AFm phases during the early stages of hydration appears to be
controlled by the more rapid hydration of the cement component and the availability of sulfate; early-formed AFt tends to be replaced by AFm as hydration
progresses. None of the Ca-containing hydrate phases are capable of taking up
appreciable magnesium, and as Mg becomes increasingly available from hydration
of the slag component a hydrotalcite-like phase, a magnesium equivalent of
AFm, develops. It is not known whether the latter makes any contribution to
the strength of the blend.
     The internal environment generated by slags has also been analyzed in
terms of pore fluid chemistry [13,14]. The most striking feature is the
change from a relatively oxidizing environment, as obtained in portland cements or blends with low BFS contents, to a reducing environment [13]. The
redox (shorthand for reducing or oxidizing) conditions can be measured quantitatively on a scale, termed the $\bar{E}_h$ scale. The reference point of this scale,
assigned a value of 0.00 volts, is taken as that potential necessary to discharge H_2 from water at unit activity and 25°C. Relative to this, most portland cements or blends containing fly ash have E_h values in the range +100 to
+200 mV, whereas slags may lower the E_h to -200 to -400 mV and, moreover,
appear to buffer the E_h within this reducing range. The buffering couples
appear to involve the sulfur content of slag. Unlike cement, in which S is
present in oxidized from as SO_4^{2-}, slags contain S in its most reduced form,
as S^{2-}. During hydration, S^{2-} is liberated and is partially soluble in the
pore fluid. Some of the S^{2-} is oxidized to thiosulfate, $S_2O_3^{2-}$, and the
generally slightly lower pH relative to PC, as well as the presence of these
two reduced species, enhances the solubility of sulfate. Pore fluids in slag
cement are typically less alkaline than those of portland cements, with
typical pH's in the range 12-12.2 for slag-rich formulations compared with
12.5-13.2 (portland). It is of interest to note that the shift towards lower
pH's with increased slag contents appears to be gradual, whereas the transition to lower E_h's appears to be abrupt, requiring a certain minimum slag
content. At 28 d, this minimum is high, corresponding to ~70% replacement
but it probably shifts towards lower slag contents with longer cure times.

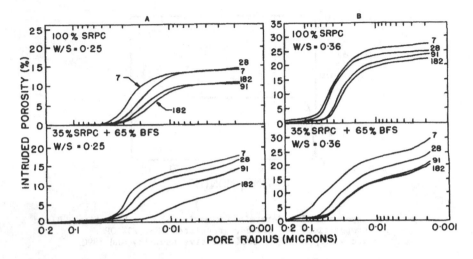

Figure 6. Cumulative pore size distribution curves for sulfate resistant
portland cement (SRPC) and BFS pastes; (A) w/s = 0.25,
(B) w/s = 0.36.

Pore Structure and Permeability

Hooton [15] has determined cumulative pore size distributions in hydration products of blends of cement made with increasing content of BFS and FA.
The MIP pore size distribution was determined after 7, 28, 91, and 182 d of
curing at ambient temperatures and two w/s ratios: 0.25, and 0.36. The conclusions (Fig. 6) are broadly representative of results obtained in other
studies; at w/s = 0.25, the cement matrix has lower intruded porosity up to 91
d, although the mean pore size was finer in the BFS blend, especially in the
<0.025 μm radius range. However, with increasing cure duration, by 182 d, the
BFS blend had markedly intruded lower porosity. Similar results were obtained
at w/s = 0.36, although the total pore volume intruded at comparable ages was
greater than at the lower ratio. Permeabilities were also measured; at w/s
0.25 and 182 d, the 65% BFS blend had a coefficient of permeability of
<0.1x10^{-13} m/s; while the cement-only control gave 0.3x10^{-13} m/s. Comparable
values at w/s ratio 0.36 were 0.4x10^{-13} (BFS) and 1.0x10^{-13} m/s (cement).
Correlations between intruded porosity and permeability were found to present
difficulties, and the nature of such correlations is still an area of uncertainty.

In other studies, it has been suggested that a critical pore radius
exists, taken as the radius corresponding to the point of maximum slope in the
differential pore size distribution curve, dV/dP. The value of this actual
radius has been related to permeability [16]. Similar concepts have been
advanced in other studies [17,18]. Although some correlation undoubtedly
exists between porosity and permeability, the data exhibit much scatter, partly due to drying effects during specimen preparation [19,20].

In general, although it has not yet been possible to define a quantitative relationship between porosity and permeability, high-slag blends (and to
differing extents silica fume and fly ash blends) do produce pastes with lower
porosities and permeabilities than do PC matrices, and hence greater durability may characterize such materials [21]. Slag blends and to a greater extent
Class F fly ash blends, mature more slowly, and moist curing for extended

periods is required to optimize the performance of blends; inadequate curing may result in disappointing performance.

Silica Fume Reactions

Ultra-fine particles of silica fume, when properly dispersed, tend to fill the interstices of the fresh cement paste structure, where they are available to react with the alkalies and $Ca(OH)_2$ liberated by the hydrating portland cement, to form cementitious compounds. When properly dispersed (usually with superplasticizer) silica fume used in proportions up to 10% replacement of cement also significantly reduces bleeding and segregation of the mixtures.

Aqueous slurries of SF already incorporating superplasticizer or other dispersing agents are commercially available and provide convenience in batching. Aqueous slurries of SF can be prepared without chemical admixture, but such slurries are unstable as well as difficult to produce, and usually require higher total water:cement ratios when used in concrete.

Silica fume, because of its high surface area, reacts much more quickly than other blending components. The overall heat of hydration of SF blends is frequently equivalent to that of high early strength cement, although the rate at which heat is liberated is a function of proportions and w/c ratio. Significant early reaction occurs, although the reactions continue for some weeks; "index" strengths at 28 d are usually much higher than those of conventional cement pastes.

Details of the microstructural effects produced by silica fume incorporation are still under investigation, but the overall effect is a much denser paste matrix with few large voids, and much better bonding to aggregates or other reinforcing materials. The interfacial zone near such reinforcements is much denser, and calcium hydroxide deposition in such areas is much reduced. SF incorporated into the narrow spaces between glass fiber filaments also prevents deposition of calcium hydroxide in these special zones, thus preserving the needed ductility in glass fiber reinforced cement products subject to wet exposure [22]. The finer microtexture characteristic of SF-bearing pastes, either with portland cement alone, or in conjunction with FA or BFS blends, can result in dramatic reductions in capillary porosity and in permeability, often with favorable consequences for expected durability of the products.

Alkali-Aggregate Reactions

It is generally accepted that cement blending agents including FA, BFS, and SF can, if properly used, mitigate the harmful effects of alkali-aggregate reaction (AAR). The basic causes and mechanisms of AAR are well documented [23-24]. Hydroxyl ions (OH^-) and their accompanying alkali counter-ions diffuse into certain reactive siliceous components of susceptible aggregates, depolymerize the silica structures, and generate an alkali silica (or more commonly, an alkali-calcium silica) gel of low density. This gel in turn absorbs additional fluid, swells, and creates expansive stresses sufficient to crack affected concrete. Progressive cracking, accompanied by secondary effects (carbonation, freezing damage, etc.) contributes to deterioration of the concrete. Exudation of diluted reaction gel at the surface through the cracks is a frequent occurrence, and the resulting unsightly appearance often makes the damage look worse than it is.

The onset of alkali aggregate reaction is a function of the concentration of alkali hydroxide developed in the aqueous phase, which in turn is a function of the alkali content of the cement used. However, alkalies are to some extent removed from solution by absorption into C-S-H hydration products, so that the long-term steady state concentration reflects this factor as well as the potential alkali supply contributed by the cement. It appears that Na^+ may be absorbed by C-S-H gel somewhat preferentially as compared to K^+, and that both may be taken up more easily by gels of lower Ca:Si ratio, i.e. those produced by blended cements.

In a sense, the major effect of blending components is to provide a sink for the removal of alkali hydroxide in pore solution competitive with that provided by the formation of alkali silica or calcium alkali silica gel in the process of alkali attack. To the extent that alkalies are removed from solution prior to alkali aggregate attack, the concentration of alkali hydroxide is reduced, and the attack mitigated. Except for alkali attack on opal or certain cherts, which have a highly porous structure and relatively labile silica, the rates of alkali attack on aggregates tend to be slow, and the process of reaction prolonged over many years, providing considerable time for the alternate sink for alkali hydroxide developed by blending component reaction product to operate.

Different blending components react with cement at very different rates; however, many fly ashes are only slowly reactive. On the other hand, silica fume, because of its fine particle size and high surface area, reacts relatively quickly, and reduces the alkali hydroxide concentrations to relatively low levels over a period of weeks or a few months [25,26].

BFS and high calcium FA both tend to have somewhat lower silica contents than the other components. This may be instrumental in reducing their effectiveness as sinks for alkali hydroxide in preventing alkali aggregate attack, or at least in requiring larger proportions of the blending component for a given degree of effectiveness. However, other mechanisms may be operating, especially with BFS. The greatly lowered permeability of BFS-cement pastes tends to reduce the alkali ion mobility and limit the alkali hydroxide availability at particular reaction sites. Local depletion of alkali in such systems may effectively halt alkali-aggregate reaction before extensive damage can occur.

Both types of FA, BFS, and SF all contain alkali in non-negligible levels, and it seems paradoxical that they serve as net sinks for alkali hydroxide in solution. The alkali content in silica fume is usually the smallest of any of the blending components, and it ordinarily is in the form of Na rather than K. At least one study [25] has shown a short-term augmentation of the alkali hydroxide content of the aqueous phase when silica fume was used, but this effect was quickly reversed, and alkali hydroxide depletion proceeded thereafter. BFS sometimes contains significant alkali contents, and most FA materials contain 2 or 3% combined alkali oxides, K predominating with low calcium FA and Na with high calcium FA. It appears that the rates of release of these alkalies from the blending components are sufficiently slow that they do generally serve as sinks for alkali hydroxide released from cements, rather than as augmenters of the alkali content of the pore solutions.

SUMMARY AND CONCLUSIONS

A large number of chemical and physical factors involved in the hydration reactions in cement pastes incorporating fly ash, slag, silica fume, and other pozzolanic materials are responsible for the course of reactions and microstructural development. The floc structure of fresh cement paste affects its workability, and the incorporation of blending materials such as fly or blast furnace slag affects the floc structure, generally interrupting the structure, imparting greater fluidity. Silica fume if adequately dispersed, can provide a similar effect and also, due to its ultra-fine particle size, fill the interstices and produce a denser paste structure.

Each of the materials affects the cement hydration kinetics because of a combination of physical and chemical characteristics. Silica fume accelerates the early hydration reaction, but much fume remains unreacted until a later stage. The overall extent of reaction after a few days tends to be greater with slag and Class C fly ash than with Class F fly ash. Pastes made from blends with Class F ashes, and to a lesser extent, Class C ashes or slag, will contain much residual unhydrated material even at later ages. However, pastes from blended cements will in general have finer pore structures than comparable plain pastes, the finest being from silica fume and slag blends. In such pastes, water permeabilities and molecular or ionic diffusivities are much reduced. In concrete, such pastes tend to give rise to greater durability on exposure to forces of degradation. This greater durability, of course, depends on adequate curing having been provided.

Additional chemical factors such, as the higher silica content of blended cement pastes, when combined with the development of a physically finer pore structure, enhance resistance of the pastes to alkali-aggregate (silica) reaction.

In summary, the microstructural development of these blended cement pastes is a complex process, different for each of the materials, and for compositional varieties within each. The difficulties of working with such supplementary cementitious materials arise from their complexity, the necessity for adequate characterization and of following guidelines for the proper selection and use of each material. When utilized within proper guidelines, such as control of early rheological properties, use in proper proportions, and assuring adequate curing, the resulting cement pastes will frequently impart superior properties to the matrices of the concretes which they form.

REFERENCES

1. Fly Ash and Coal Conversion By-Products: Characterization, Utilization and Disposal I, edited by G.J. McCarthy and R.J. Lauf, Mat. Res. Soc. Symp. Proc. Vol. 43 (Materials Research Society, Pittsburgh, 1985).
2. Fly Ash and Coal Conversion By-Products: Characterization, Utilization and Disposal II, edited by G.J. McCarthy, F.P. Glasser and D.M. Roy, Mat. Res. Soc. Symp. Proc. Vol. 65, (Materials Research Society, Pittsburgh, 1986).
3. S. Diamond, in Technology of Concrete When Pozzolans, Slags, and Chemical Admixtures are Used, (ACI - RILEM Symposium, Monterrey, Mexico, 1985) pp. 1-18.
4. P. Barnes, A. Ghosh and A.L. Mackay, Cem. Concr. Res. 10, 639-646 (1980).
5. P.L. Pratt and H.M. Jennings, Ann. Rev. Mater. Sci. 11, 123-149 (1985).
6. B.J. Dalgleish, A. Ghose, H.M. Jennings, and P.L. Pratt, Proc. RILEM Conf. Concr. at Early Ages, Paris, Vol. 1, pp. 137-143 (1982).
7. D.W. Hadley, Thesis, School of Civil Engineering, Purdue University (1972).
8. S. Diamond, Selection and Use of Fly Ash for Highway Concrete, Final Report, FHWA/IN/JHRP-85/8, Purdue University (1985).
9. S. Diamond, D. Ravina and J. Lovell, Cem. Concr. Res. 10, 297-300 (1980).
10. R.F. Feldman, Cem. Concr. Res. 16, 31-39 (1986).
11. X. Wu, D.M. Roy and C.A. Langton, Cem. Concr. Res. 13, 277-286 (1983).
12. K. Luke and F.P. Glasser, Cem. Concr. Res. 17, 273-282 (1987).
13. M.J. Angus and F.P. Glasser, in Scientific Basis for Nuclear Waste Management IX, edited by Lars and Werme, MRS Proc. Vol. 50 (1986) pp. 547-555.
14. M. Silsbee, R.I.A. Malek and D.M. Roy, in Proc. 8th Intl. Congr. Chem. Cement, Brazil, Vol. IV, pp. 263-269 (1986).
15. R.D. Hooton, in Blended Cements, ASTM STP 897, edited by G. Frohnsdorf, (ASTM, Philadelphia, PA, 1986) pp. 128-143.
16. D.M. Roy and K.M. Parker, in American Concrete Institute SP 79, Vol. 1, (1983) pp. 397-414.
17. B. Nyame and J.M. Illston, in Proc. 7th Intl. Conf. on the Chemistry of Cements Vol. 3, (Paris, 1980) pp. VI 181-185.
18. D.N. Winslow and S. Diamond, J. Materials 5, 564-585 (1970).
19. R.F. Feldman, Cement Technology 3 (1), 5-14 (1972).
20. R.F. Feldman, in Proc. 8th Intl. Conf. on the Chemistry of Cements Vol. 1, (Rio de Janeiro, 1986) pp. 336-356.

21. D.M. Roy, in Proc. 8th Intl. Congr. Chem. Cement, Brazil, V. I, pp. 362–380 (1986).
22. A. Bentur and S. Diamond, in Proc. PCI Intl. Symp. on Durability of Glass Fiber Reinforced Concrete, (Chicago, 1985) pp. 337–351.
23. L.S. Dent Glasser and N. Kataoka, Cem. Concr. Res. 11, 1–9 (1981).
24. S. Diamond, R.S. Barneyback Jr. and L.J. Struble, in Proc. 5th Intl. Conf. Alkali–Aggregate Reactions in Concrete, (Capetown, 1981).
25. S. Diamond, J. Am. Ceram. Soc. 66, C82–C84 (1983).
26. C.L. Page and O. Vennesland, in Materials and Structures (RILEM) 16, No. 91, 19–25 (1983).

LABORATORY MODELING AND XRD CHARACTERIZATION OF THE HYDRATION REACTIONS
OF LIGNITE GASIFICATION AND COMBUSTION ASH CODISPOSAL WASTE FORMS

P. KUMARATHASAN and GREGORY J. McCARTHY
Department of Chemistry, North Dakota State University, Fargo, ND 58105

Received 31 October, 1986; refereed

ABSTRACT

Cementitious reactions in gasification and combustion ash derived from
North Dakota lignite permit the fabrication of monolithic wastes forms from
nonhazardous ash by-products and hazardous liquid wastes from a coal gasification plant. To better understand such cementitious reactions, x-ray
diffraction has been used to characterize the hydration reaction products of
crystalline phases in the gasification and combustion ashes from the Beulah,
ND, complex and the crystalline reaction products formed. A cementitious
lignite fly ash was also studied. pH of solutions at a liquid to solid ratio
of 2.4 and times up to 14 days was measured, and compared to pH calculated
from leachate chemical analyses. Reactions were monitored for up to one year.
Among the principal crystalline phases in gasification ash, carnegieite
($Na_{1.5}Al_{1.5}Si_{0.5}O_4$) was the most reactive. Solution pH's for two samples of
gasification ash were in excess of 13; carnegieite reaction may be responsible
in part for these very high pH's. Hydration products included gaylussite,
($Na_2Ca(CO_3)_2 \cdot 5H_2O$), a carbonate-sulfate ettringite structure phase, calcite
and three or more zeolites (NaA, laumontite, faujasite and/or NaX).
Ettringite and calcite were the principal hydration reaction products of
scrubber ash, fly ash and the composite codisposal waste form. The formation
of ettringite may be one of the principal reactions responsible for
consolidation of the waste forms (along with noncrystalline calcium silicate
hydrate formation which could not be observed by XRD). Ettringite was
observed to decrease in abundance at long reaction times. Calcium aluminate
monosulfate hydrate formed in the later stages of the fly ash reaction.

INTRODUCTION

The Great Plains coal gasification plant at Beulah, North Dakota, presently owned by the US Department of Energy and operated by ANG Coal Gasification Company, has a number of solid and liquid waste streams. Gasification
ash is by far the most abundant of the solid wastes. After extensive evaluation, including tests specified in the Resource Conservation and Recovery Act
(RCRA), it has been determined that gasification ash is not a hazardous waste
[1-3]. Two liquid waste streams from this plant, liquid waste incinerator
blow-down (LWIBD) and multi-effect evaporator concentrate (MEEC), are hazardous wastes because of their content of several toxic trace elements [3]. The
feasibility of codisposal of these toxic wastes combined with gasification ash
and two solid wastes from the adjacent Antelope Valley electrical generating
station (AVS), owned by the Basin Electric Power Cooperative, is being explored in a program supported by the Gas Research Institute (GRI). Both plants
currently dispose of ash by burial in separate, licensed, clay-lined cribs at
the adjacent strip mine.

In Volume II of this series it was reported [4] that, because of cementitious reactions, mixtures of gasification ash, scrubber ash and bottom ash
could be fabricated, using procedures specified in ASTM C 593, into monoliths
having compressive strengths in excess of 1000 psi. For most of the toxic elements, this processing resulted in reductions in leachability compared to
that measured in individual ash components [3,4].

Solid-liquid reactions account for the consolidation of these ash materials and liquid wastes into monoliths, and one of the objectives of the GRI

Table I

Chemical Composition (wt.%) of Ash Samples

	GASIFICATION ASH GPGA-3 GPGA-4w		SCRUBBER ASH	BOTTOM ASH	CD MIX	FLY ASH[a]
SiO_2	25.2	21.9	30.7	36.8	29.2	45.6
Al_2O_3	12.9	11.7	11.4	13.7	11.8	15.5
Fe_2O_3	10.2	7.8	5.1	9.2	6.3	7.3
CaO	23.3	17.9	21.1	20.9	20.2	20.3
MgO	8.8	6.6	5.6	7.3	6.1	5.5
Na_2O	8.8	9.9	5.3	7.4	6.8	1.0
K_2O	0.6	0.6	1.0	0.5	0.8	0.7
SO_3	2.1	2.2	12.7	1.6	8.5	1.9
LOI	4.23	14.57	0.70	0.21	4.24	0.14
Moisture	0.07	4.19	0.87	1.52	1.82	0.04

a. ASTM C-618 physical properties: Fineness, 16.1% +325 mesh; Pozzolanic Activity Indices: 106% of control with Portland cement (28 d) and 1672 psi with lime (7 d); Autoclave Expansion, 0.09%; Water Requirement, 92% of control; S.G., 2.50.

study is to understand these reactions so that future mix ratios and processing conditions can be designed to produce maximum consolidation and fixation of toxic trace elements. This paper describes a laboratory simulation and x-ray diffraction (XRD) characterization of hydration reactions of individual ash components and of one reference mixture of three types of ash. For comparison to another cementitious material widely used as a mineral admixture in concrete, fly ash from a nearby power plant was also included. The objective of the study described here was to elucidate the crystalline phases in ash that react with pore liquids during a time frame of months as well as the crystal-line phases that form from these reactions. Once the reactive phases are known, experiments can be designed with synthetic versions of these phases to provide mechanistic and kinetic insights into the hydration reactions.

EXPERIMENTAL PROCEDURES

Ash Materials

Six types of ash, all derived from the same coal (Beulah-Zap North Dakota Lignite), were included in the study. Although codisposal waste forms are fabricated from as-received ash [3], for this study each ash was ground to -325 mesh (<45 μm) in order to make more surface area available for reactions. Chemical analysis and XRD characterization of crystalline phases in these ashes are summarized in Tables I and II. Ash composition and mineralogy have been discussed previously by Stevenson and McCarthy [5] and McCarthy et al. [4].

Gasification Ash. The fixed-bed Lurgi gasifiers produce a dominantly crystalline clinker. After being discharged from the gasifiers, the ash is sluiced, dewatered and hauled off for disposal. Because of the ash's hydraulic behavior (self-hardening in the sluicing system), NaCl is added to the sluice water to retard cementitious reactions. Two samples were provided by ANG Coal Gasification Company. GPGA-3 was collected dry below one of the gasifiers so that the reactions of ash that had not contacted aqueous solutions could be studied. The second gasification ash was GPGA-4w. It had also

Table II

Principal Crystalline Phases in Ash Samples

CODE	NAME	NOMINAL FORMULA	CHARACTERISTIC PEAK[a]
		GASIFICATION ASH	
Mw	Merwinite	$Ca_3Mg(SiO_4)_2$	33.3
Cg	Carnegieite	$Na_{1.5}Al_{1.5}Si_{0.5}O_4$	21.0
Ml	Melilite	$Ca_2(Mg,Al)(Al,Si)O_7$	31.1
Sp	Ferrite Spinel	$(Mg,Fe)(Fe,Al)_2O_4$	35.5
C_2S	Dialcium silicate	Ca_2SiO_4[b]	32.7
Pc	Periclase	MgO	42.8
Qz	Quartz	SiO_2	26.7
Cc	Calcite	$CaCO_3$	29.4
Gy	Gaylussite	$Na_2Ca(CO_3)_2 \cdot 5H_2O$[c]	13.8
		SCRUBBER ASH	
CSf	Calcium Sulfite Hemi-hydrate	$CaSO_3 \cdot 1/2H_2O$	28.2
Bs	Bassanite	$CaSO_4 \cdot 1/2H_2O$	14.2
Ah	Anhydrite	$CaSO_4$	25.5
Gp	Gypsum	$CaSO_4 \cdot 2H_2O$	11.6
Qz	Quartz	SiO_2	26.6
Pc	Periclase	MgO	42.8
Sp	Ferrite Spinel	$(Mg,Fe)(Fe,Al)_2O_4$	35.5
C_3A	Tricalcium aluminate	$Ca_3Al_2O_6$	33.1
Mw	Merwinite	$Ca_3Mg(SiO_4)_2$	33.3
		BOTTOM ASH	
Px	Pyroxene	$(Ca,Na)(Mg,Fe)(Si,Al)_2O_6$	
Ml	Melilite	$Ca_2(Mg,Al)(Al,Si)O_7$	
Pc	Periclase	MgO	
Sp	Ferrite Spinel	$(Mg,Fe)(Fe,Al)_2O_4$	
Ne	Nepheline	$NaAlSiO_4$	
Cg	Carnegieite	$Na_{1.5}Al_{1.5}Si_{0.5}O_4$	
Qz	Quartz	SiO_2	
		FLY ASH	
Qz	Quartz	SiO_2	
C_3A	Tricalcium aluminate	$Ca_3Al_2O_6$	
Pc	Periclase	MgO	
Sp	Ferrite Spinel	$(Mg,Fe)(Fe,Al)_2O_4$	
Lm	Lime	CaO	
Ah	Anhydrite	$CaSO_4$	
Mw	Merwinite	$Ca_3Mg(SiO_4)_2$	
Ml	Melilite	$Ca_2(Mg,Al)(Al,Si)O_7$	

a. Two-theta angle for CuK-alpha diffractograms.
b. Three compositions of C_2S have been observed [5]; the principal phase is similar to bredigite, $Ca_{14}Mg_2Si_8O_{32}$.
c. Gaylussite was observed only in GPGA-4w.

been collected dry but was treated with simulated sluice water (without the NaCl) for six hours in order to see if hydration reactions would be diminished by prior aqueous solution treatment. The NaCl addition was omitted because this treatment was considered to be only a temporary local measure at the time the study was planned.

Scrubber Ash is the by-product of the Antelope Valley Station flue gas desulfurization scrubbers. The AVS uses its high-calcium lignitic fly ash, with some supplemental hydrated lime, as the active scrubber material. This process gives a fine ash with an elevated sulfur content (Table I), present as calcium sulfite hemihydrate and $CaSO_4$ as anhydrite, bassanite and minor gypsum

(Table II). In spite of the fact that the original fly ash is slurried prior to atomization in the dry scrubber chamber, there was unreacted C_3A in the scrubber ash.

Bottom Ash is a partially melted clinker from the pulverized coal fired boilers at the AVS. It has higher silica, alumina and iron oxide contents than the other ashes derived from the same coal (Table I). This results in a crystalline phase assemblage containing the metasilicate augite-like pyroxene and lacking the orthosilicates, merwinite and C_2S (Table II). This ash also contains some admixed "economizer ash" that may contribute to hydration reactions. Nevertheless, it was assumed that this ash would contribute little to the overall consolidation of codisposal waste forms. Consequently, it was not studied separately for its hydration properties, and was used only in the codisposal reference mix.

Codisposal Mix (CD Mix) consists of a mixture of 26% GPGA-4w gasification ash, 61% scrubber ash and 13% bottom ash. These weight percentages represent the relative quantities of the three by-products produced at the adjacent gasification and power plants at Beulah, ND.

Fly Ash was obtained from the Coal Creek Station at Underwood, ND. It is used widely as a mineral admixture in concrete throughout the upper midwestern US. Its chemical analysis and ASTM C-618 test results are given in Table I. The mineralogy of this material (Table II) is typical for a lignite fly ash [6,7].

Hydration Experiments

The liquid to solid (L/S) ratio used in the initial codisposal waste forms (the optimum moisture content of ASTM C 596) varied from about 0.16 to 0.20 [4]. It was decided to use larger L/S ratios because of the higher surface areas and reactivities of the ground gasification and scrubber ashes. A value of 0.4 was selected as the reference L/S ratio. Experiments were also performed with the principal ash component of codisposal waste forms, scrubber ash, and with a gasification ash, to see whether L/S ratios twice and three times this value would have marked effects on the hydration reactions and consolidation.

The ground ash materials were mixed with distilled water and held in capped polypropylene vials. Groups of vials were placed in larger sealed containers and held at room temperature. For a minimum of 25 intervals over the one-year duration of the experiments, the reactions were monitored by opening the vials, checking for relative consolidation by noting resistance to scratching with a stainless steel spatula, and removing several milligrams of solid for XRD analysis. This was done under ambient conditions, so some loss of moisture and carbonation was possible at each monitoring. The monitoring sample was prepared for XRD analysis by grinding with a large excess of ethanol (so as to quench the hydration reactions).

Characterization of Reaction Products

Solution pH Measurement. To aid in interpretation of formation and reactions of the crystalline phases, pH established in an aqueous solution by each solid was measured. It was necessary to use a L/S ratio of 2.4, differing by factors of 6, 3 and 2 from the L/S ratios used in the hydration experiments, in order to provide sufficient solution for measurement. After trials with various meters and electrodes, a Corning pH meter, equipped with a Ag/AgCl combination electrode, that exhibited the least drift during measurement was chosen. Five grams of each of the −45 µm powders were mixed with 12 mL of distilled deionized water and held at room temperature (22–25°C) in capped polypropylene bottles. The meter was calibrated with pH 7 and 10 standards before each measurement. The pH of the solution over the solid was measured

Table III

Measured and Corrected pH in a 2.4 to 1 Liquid to Solid Ratio Mixture,
Solution Concentrations, and pH Calculated from Solution Concentrations

	SCRUBBER ASH	GASIFICATION ASH GPGA-3	GPGA-4w	BOTTOM ASH	CD MIX	LIGNITE FLY ASH
Measured pH at day:						
1	12.3	12.9	12.9	10.6	12.1	12.8
2	12.4	13.3	13.4	11.1	12.2	13.0
3	12.4	13.4	13.6	11.3	12.2	13.0
4	12.4	13.5	13.7	11.5	12.3	13.0
5	12.4	13.5	13.7	11.5	12.3	13.0
6	12.2	13.5	13.6	11.6	12.2	13.0
7	12.1	13.5	13.6	11.6	12.2	13.0
14	12.2	13.5	13.6	11.6	12.2	13.0
14 d Corrected pH[a]	11.8	13.2	13.4	11.2	11.8	12.5
14 d Solution Composition[b]						
Na	1318	7224	7191	1037	3579	38
K	129	207	187	37	211	32
Ca	277	<2	<2	27	21	285
Mg	<1	<1	<1	<1	<1	<1
SO4	2071	4669	3589	1905	7306	198
pH Calculated from 14 d Solution Composition[c]	12.37	13.21	13.25	11.79	11.87	12.00

a. from calibration procedure using NaOH (see text)
b. analyses by D.J. Hassett, Fuels Analysis Laboratory, University of North Dakota
c. NDEGM program [9]; pCO_2 = E-12.5; 25°C.

daily for one week and at the end of two weeks. The bottle was magnetically
stirred for approximately 5 min and allowed to settle for an additional 5 min
before each reading. Additional experiments (discussed below) were aimed at
complex equilibrium computer modeling of pH's and correction of measured pH's.
These were instituted near the end of the study when it appeared that the pH
readings were somewhat high.

XRD. Philips manual and automated powder diffractometers using CuK-alpha
radiation were used. All specimens were well-ground powders prepared as
slurries with ethanol on glass slides. In order to be able to estimate the
progress of reactions involving crystalline phases by monitoring peak intensities, every effort was made to maintain consistency in specimen preparation
and instrument operation. The intensity of the x-ray source at the chosen
power setting was monitored periodically, and corrections were made to compensate for a small loss in x-ray intensity (due to aging of the tube) over the
year-long course of the experiments.

RESULTS AND DISCUSSION

pH Behavior

Results from the L/S = 2.4 measurements are given in Table III. Readings
did not vary much after the second day. The highest pH readings were observed
for the gasification ashes, and the lowest for the bottom ash. The higher pH
values suggest high dissolved Na^+ concentrations without sufficient quanti-
tites of acid counterions (such as sulfate). Hassett et al. [1] found that
after as little as one day of contact of an ash similar to GPGA-3 with water

at L/S = 2.0, the Na$^+$ concentration of the solution was 0.36 M. The somewhat higher pH's produced by the GPGA-4w, compared to GPGA-3, ash may be due to increased leachable Na in the sluiced ash. On a LOI- and moisture-free basis, the Na$_2$O content of GPGA-4w is 12.6% compared to 9.6% for GPGA-3 (Table I). The high pH's measured for gasification ash solutions are similar to the pH's found in pore solutions squeezed from Portland cement pastes (see for example, Luke and Glasser [8]). To have such high pH values, the solutions must be rich in NaOH.

The scrubber ash, with its lower Na concentration and higher concentration of the acidic sulfate anion, established pH values about 1.0 unit lower than that of gasification ash. Bottom ash gave pH's about 1.0 lower than scrubber ash. It is likely that the fine-grained economizer ash mixed with the bottom ash controls much of this pH behavior. The pH of the codisposal mix solution appeared to be controlled by the scrubber ash, its principal component at 61%. This result was surprising because a weighted average of the pH's measured for its components is in excess of 13.

The pH's established by the lignite fly ash were somewhat higher than expected. Previous experience with pH's of lignite fly ashes suggested that the reading should be closer to 12.5. This results suggested that further work on pH was necessary.

At this point, the pH meter was checked with a saturated Ca(OH)$_2$ solution, measured in N$_2$, and a reading of 13.15 was obtained. Calculations with a complex equilibrium (geochemical) computer model [9] indicated that the correct value is 12.42 (25°C, pCO$_2$ = E-12.5, K$_{sp}$ of Ca(OH)$_2$ = 4.6 E-6). Rather than proceeding to correct all readings with this single calibration point, it was decided to prepare a calibration curve based on NaOH solutions, direct titration of (OH), and calculation of pH from the resulting NaOH concentration using the computer model. The resulting calibration called for subtraction of about 0.5 from each measurement. After applying the correction, the pH of the saturated Ca(OH)$_2$ solution became 12.6, in reasonable agreement with the calculated value of 12.4. Table III also gives the corrected 14 d pH's obtained using this calibration procedure. Finally, as a third indication of pH, the compositions of the major ions in the 14 d supernatant solution were determined and the computer model was employed to calculate pH. These results are also given in Table III.

The agreement between the corrected and calculated pH was good in the case of the three high-Na solutions, namely those from the two gasification ashes and the codisposal mix. The corrected and measured pH values differed by about 0.5 for the three low-Na solutions (scrubber ash, bottom ash and fly ash). Beyond this observation, no explanation for the difference is apparent.

Hydration Experiments

Gasification Ash. Heights of characteristic peaks of the reactants and products in the x-ray diffractograms were measured to give an approximation of the rates of hydration of reactants and formation of products. Data for GPGA-3 treated at three L/S ratios and for GPGA-4w are given in Table IV. It should be cautioned that these peak heights are derived from weak peaks of minor phases on a single diffractogram made from an aliquot of only a few milligrams, and thus are subject to considerable uncertainty. Also, because of the large number of phases, few of the peaks monitored are unique to one phase and in several cases, hydration product peaks grew at virtually the same angles that were being monitored for reaction of ash phases. This was a problem with both Cg and C$_2$S. Nevertheless, general peak height trends were useful to indicate the reactivity of the original crystalline phases in aqueous solutions and to identify the formation of crystalline reaction products.

The peak height results indicated that among the principal crystalline phases in gasification ash, carnegieite (Cg), dicalcium silicate (bredigite) (C$_2$S), merwinite (Mw), melilite (Ml), periclase (Pc) and calcite (Cc) were affected the most by hydration reactions. Quartz and ferrite spinel showed little or no reaction and are not included in Table IV. The crystalline

Table IV

Heights of Characteristic Peaks[a] of Crystalline Reactants and Products in the Gasification Ash Hydration Experiments[b]

TIME	GPGA-3 L/S = 0.4							GPGA-3 L/S = 0.8							GPGA-3 L/S = 1.2							GPGA-4w L/S = 0.4						
	Cg	C2S	Mw	Ml	Pc	Cc	Gl	Cg	C2S	Mw	Ml	Pc	Cc	Gl	Cg	C2S	Mw	Ml	Pc	Cc	Gl	Cg	C2S	Mw	Ml	Pc	Cc	Gl
0.00	24	41	72	37	47	13	0	24	41	72	37	47	13	0	24	41	72	37	47	13	0	24	41	52	40	43	24	10
0.04	18	35	63	37	47	19	0	19	38	68	30	47	29	0	19	37	64	30	47	31	0	20	43	54	44	44	29	10
0.17	18	39	60	37	47	31	5	20	41	62	33	49	40	8	18	37	64	29	53	48	3	19	40	53	41	48	33	12
0.38	18	39	59	37	47	27	7	16	41	62	30	49	30	6	15	39	63	29	50	37	2	21	43	53	41	45	36	12
0.67	17	39	59	37	43	26	7	16	39	62	30	55	34	5								20	46	54	39	45	37	10
2	13	37	54	31	43	35	7	13	39	62	30	53	33	4								18	46	54	39	45	37	11
3	12	35	50	31	43	40	7	13	39	52	26	43	30	6								17	44	54	38	48	37	10
5	10	35	47	31	43	61	7	11	33	52	30	40	63	6	14	37	58	29	48	69	6	17	42	52	39	44	41	12
7	10	35	47	30	43	65	8	11	39	52	30	43	40	7	13	38	57	29	49	48	7	15	44	49	38	46	41	12
21	6	35	37	30	39	58	9	9	34	40	30	41	72	6	11	31	43	28	37	52	4	10	39	36	36	43	70	12
27	6	33	37	30	39	59	6	7	34	39	30	41	73	5	8	29	39	28	37	63	3	8	31	30	36	39	70	11
38	8	29	32	30	36	81	4	5	23	33	39	38	94	3	4	26	35	34	46	92	2	7	28	30	37	40	72	8
48	9	27	33	30	40	74	4	3	23	31	39	35	83	4	4	24	30	35	37	85	5	5	28	31	36	38	73	7
55	9	27	31	30	33	89	5	2	23	31	39	39	100	4	4	24	30	36	39	111	6	3	29	31	36	38	60	8
83	8	26	25	35	38	90	9	4	19	26	35	33	106	3	4	22	26	31	31	105	5	8	31	31	36	37	59	8
131	4	28	38	36	43	59	5	5	19	26	32	26	75	11	4	22	23	30	24	73	7	8	33	31	31	38	40	7
155	4	28	37	37	38	37	4	6	17	26	32	23	47	7	5	20	22	28	26	70		7	31	31	31	39	45	7
202	6	28	31	33	31	36	4	8	17	24	31	26	66	3	8	19	22	28	24	57	7	6	26	25	30	22	48	6
244	5	21	27	31	24	32	4	9	17	24	26	26	63	3	8	14	22	28	25	64	3	6	22	22	31	28	47	5
315	5	17	24	26	22	33	4	11	15	20	24	16	45	4	8	14	22	27	22	65	1	5	15	16	28	16	80	4
363	5	15	22	26	28	79	3	4	14	20	24	26	61	3	8	13	22	27	24	89	1							

a. Heights in 0.1 inch on a strip chart for diffracometer conditions: 40 kV, 20 mA, 250 cts/sec full scale, TC = 2 sec.
b. See Table II for listing of crystalline phase codes.

reaction products in GPGA-3 were gaylussite ($Na_2Ca(CO_3)_2 \cdot 5H_2O$), additional calcite and an ettringite-like phase that had a cell parameter between sulfate ($Ca_6Al_2(SO_4)_3 \cdot 32H_2O$) and carbonate ($Ca_6Al_2(CO_3)_3 \cdot 26H_2O$) forms. The presence of the Na-Ca carbonate is consistent with a mechanism where high-pH pore solutions react with CO_2 (and available calcite) to form the carbonate. The sources of the CO_2 are presumed to be air and oxidation of residual organic material in the gasification ash. Gaylussite is already present in GPGA-4w from the reaction of dry ash with the simulated sluice water.

In GPGA-4w, additional calcite formed along with small amounts of perhaps four zeolite structure phases:

NaA	$Na_{96}Al_{96}Si_{96}O_{384}(H_2O)_{216}$
Laumonite	$Ca_4Al_8Si_{16}O_{48}(H_2O)_8$
NaX	$Na_{88}Al_{88}Si_{104}O_{384}(H_2O)_{220}$
	and/or
Faujasite	$Na_{9.6}Ca_{4.8}Al_{57.6}Si_{134.4}O_{384}(H_2O)_{220}$

The zeolite identifications were assisted by comparison to diffractograms given in von Ballmoos [10], and the formulae used are the unit cell contents of each zeolite as given in this reference. The NaX and faujasite are related structurally and have very similar x-ray patterns, which makes it difficult to distinguish between them. In these four zeolites, the high proportions of Na, Ca and Al compared to silica are consistent with the observed high reactivity of the silica undersaturated phases (phases that would react with silica to form more silica-rich phases) carnegieite and merwinite (cf. mineral formulae in Table II). The zeolites may have been observed only in GPGA-4w because of the added Na resulting from the sluice water treatment and the somewhat higher pH during reaction (Table III).

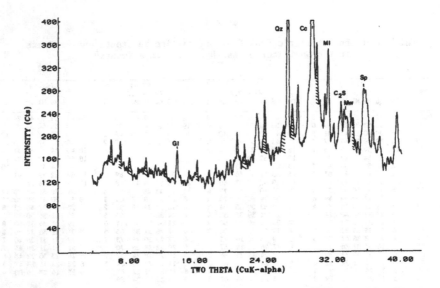

Figure 1. X-ray diffractogram of GPGA-4w after 173 days. Peaks of the various zeolite peaks are shaded.

Figure 1 is an x-ray diffractogram of GPGA-4w sampled at 173 days. It shows the high background and diffuse maximum characteristic of noncrystalline phases. The strongest or characteristic peak of each crystalline phases is labelled. The zeolite and gaylussite peaks are generally weak, but are un-ambiguous enough to permit identification of phases of the NaA, laumonite and faujasite/NaX structure types. There is also a broad peak centered at approximately 6.5 deg (13.6 A). This could be due to very poorly crystallized zeolites or a tobemorite-14A-like phase.

The data in Table IV also indicate that similar reactions occurred with each of the three L/S ratios. The most significant differences were that the rate of carnegieite reaction is decreased somewhat and that the quantity of calcite formed is increased, both with increasing L/S ratio. The calcite formed in much greater abundance in the L/S = 0.8 and 1.2 experiments. Peri-clase reaction did not begin until at least 50 days had elapsed. The role of magnesium in the hydration products is unknown, as no brucite $(Mg(OH)_2)$ or other crystalline phase containing major Mg was observed. Residual C_2S in the fly ash component of scrubber ash reacted steadily with time, but quantifying this trend was not possible because of the growth of reaction product peaks at the angle of the peak being monitored in the diffractograms. Among the phases included in Table IV, melilite reacts at the slowest rate.

The gasification ash-water mixtures showed the greatest ability to self-harden, as qualitatively assessed by scratching with a steel spatula. With GPGA-3, the solids had hardened by 16 h in the L/S = 0.4 mixture, at 13 d for the 0.8 and at 25 d for the 1.2. These results were discussed in the previous paper by McCarthy et al. [4].

Table V

Heights of Characteristic Peaks[a] of Crystalline Reactants and Products in the Scrubber Ash Hydration Experiments[b]

TIME (days)	L/S = 0.4						L/S = 0.8						L/S = 1.2					
	C3A/Mw	Gp	CSf	Pc	Et	Cc	C3A/Mw	Gp	CSf	Pc	Et	Cc	C3A/Mw	Gp	CSf	Pc	Et	Cc
0	16	7	18	25	0	0	16	7	18	25	0	0	16	7	18	25	0	0
0.01	16	14	19	25	0	0	16	11	18	30	0	0	16	9	18	29	0	6
0.04	15	13	19	25	0	0	16	11	18	32	0	47	16	9	18	29	0	16
0.17	15	13	19	25	0	9	14	10	18	28	0	48	15	10	19	28	0	50
0.67	13	12	19	24	4	10	14	8	18	29	3	28	15	9	19	28	3	45
2	13	11	20	26	4	9	14	8	19	29	8	26	15	9	19	29	6	25
3	13	10	20	26	8	9	14	7	19	29	8	24	15	9	19	28	6	22
4	12	8	20	24	8	7	13	7	19	27	8	17	15	6	19	28	9	20
8	11	7	19	26	8	9	12	6	19	28	10	9	6	7	18	26	9	25
10	20	4	19	24	9	10												
14	9	4	19	27	11	8												
18	9	0	19	25	13	5	13	4	19	28	17	7	6	0	20	23	13	18
21	9	0	19	25	17	6	14	2	19	29	19	7						
28	9	0	19	23	17	5	12	1	19	26	22	6	6	0	22	23	14	19
35	9	0	19	22	16	5	9	0	19	25	21	8						
39	9	0	20	25	16	5							9	0	23	25	19	20
49	7	0	21	26	12	18	6	0	20	22	20	10	9	0	23	26	20	18
56	6	0	21	27	12	14	6	0	17	21	18	10	8	0	23	23	23	20
63	6	0	21	27	12	16												
70	6	0	21	27	12	24												
84	7	5	21	25	10	40	6	0	14	19	15	11	7	0	16	19	19	18
105	7	4	21	26	14	22												
126	6	3	17	25	16	36												
142	4	2	14	19	14	40	6	0	14	18	22	8	7	0	16	20	21	18
172	4	2	14	18	13	47	5	0	14	16	23	8	7	0	13	19	23	15
203	4	2	12	16	12	48	5	0	9	15	24	20	7	4	13	19	19	22
245	6	2	10	14	11	25	5	0	9	14	20	21	7	7	6	18	9	90
275	7	4	7	13	10	55												
316	8	9	8	12	10	36	5	5	9	14	20	64	7	5	6	19	11	40
363	8	5	8	11	10	12	6	4	9	13	21	54	8	0	4	19	15	70

a. Heights in 0.1 inch on a strip chart for diffracometer conditions: 40 kV, 20 mA, 250 cts/sec full scale, TC = 2 sec.
b. See Table II for listing of crystalline phase codes.

Scrubber Ash. XRD results for scrubber ash hydration at three L/S ratios are given in Table V. The first reaction to begin was the hydration of anhydrite and bassanite to gypsum. Ettringite (Et) and calcite (Cc) were the only other crystalline reaction products identified by XRD. The decrease in the C_3A/Mw composite peak coincided with a decrease in the gypsum peak and the formation of Et early in the experiments. This suggests that the familiar (in Portland cement hydration) C_3A plus Gp reaction was at least one of the Et forming reactions.

Et was first observed in the 16 h diffractogram, and its characteristic peak increased in intensity until about day 40. In the L/S = 0.4 mixture, the Et intensity decreased to about two-thirds of its maximum intensity at day 84, increased again for another 40 days and then decreased back to its day 84 level by the end of a year. The increase starting at day 84 coincided with a decrease in the calcium sulfite hemihydrate (CSf) intensity and the reappearance of gypsum in the diffractograms. It is likely that this sulfite phase oxidized to a sulfate which permitted the formation of additional Et by reaction with available calcium aluminate and water. However, the supply of calcium aluminate must have been limited as both the CSf and Et peak heights decreased from day 142 to one year and some Gp remains. The decrease in CSf intensity occurred at about the same point in the L/S = 0.8 and 1.2 experiments, but here the Et intensity did not follow the previous trend. In the 0.8 experiment, Et intensity did not decrease and in the 1.2 experiment it did not decrease until about day 200. In the later stages of these experiments, loss of Et intensity correlated with gain in Cc intensity, possibly an indicator of excess solubilized calcium. This suggested that aluminate, but not calcium or sulfate, was the limiting factor in Et stability. No monosulfate (AFm) peaks were observed in the x-ray diffractograms.

  Codisposal Mix. The hydration behavior of the codisposal mix as monitored by XRD was essentially what would be expected from the reactions of its ash components (61% scrubber ash, 26% GPGA-4w and 13% bottom ash) with allowances for the reduction in intensities of some phases in the gasification ash due to dilution by the other ashes.

  Anhydrite and bassanite hydrated within an hour to give Gp. Et formed within the first day and gradually increased in abundance up to day 10. The Et intensity remained constant for 30 days and at about day 40 began to increase to eventually double its intensity. This increase coincided with the oxidation of CSf and the formation of Gp. Calcite is formed, and it shows the inverse relationship with Et intensities noted above. The zeolite phases and gaylussite were not observed in the CD Mix diffractograms, either because they were diluted below the level of detection of the method or, more likely, because they are not stable at the considerably lower pH's of the pore solutions compared to those of gasification ash (Table III). The principal bottom ash phases, melilite and pyroxene, showed little decrease in intensity over the entire experiment. Periclase, present in all three ashes, shows the late stage, slow, depletion noted above for gasification and scrubber ash. The peaks of other crystalline phases were either too weak or not sufficiently distinct in the diffractograms to be of value in monitoring reactions.

  A diffractogram of CD mix before and after curing for 28 d was given by McCarthy et al. [4]. It shows that Et and Cc were the only apparent crystalline reaction products.

  Fly Ash. As with scrubber ash, anhydrite reacted rapidly to form Gp. The formation of Et, coinciding with reductions in Gp and the C3A/Mw composite peak, suggests that the latter two phases are principal reactants of Et. After about 40 days, Et peaks become weaker and peaks of the calcium aluminate monosulfate hydrate phase (AFm) are detected. The low sulfate/sulfite content of the fly ash (Table I), compared to scrubber ash and CD Mix, may be responsible for the formation of AFm. Calcite forms and increases in abundance as the sulfate peaks become less intense. As one year of reaction approached, the overall crystallinity seemed to be decreasing.

  Manz [11] has found that the ash from this power plant is so cementitious that it can develop compressive strengths well in excess of 1000 psi. It was interesting to note that in the experiments described here, this fly ash had the lowest resistance to scratching of any of the six hydrated materials studied.

CONCLUSIONS

  All of the lignite-derived ashes studied, gasification ash, scrubber ash and fly ash, exhibited hydraulic (self-hardening) behavior. Based on a qualitative scratch resistance test, the gasification ashes produced the greatest hardness. The self-hardening behavior of these materials has been applied to production of monolithic waste forms made up of the abundant nonhazardous ash by-products and the small quantities of hazardous wastes produced at lignite gasification facilities [3,4].

  Except for the bottom ash, the pH's in ash-water mixtures having L/S = 2.4 exceeded 12, and in the case of the gasification ashes, exceeded 13. Pore solutions in hardening ash should have equal or higher pH's. The pH of the composite codisposal waste form is apparently controlled by the dominant scrubber ash component. The especially high pH's established by the "sluiced" gasification ash yielded at least three different zeolite reaction products detectable by x-ray diffraction. Calcite was a ubiquitous reaction product in all of the samples studied. It is apparently formed largely by carbonation of the high-pH high-Ca solutions.

It is proposed that carnegieite is the important crystalline phase in establishing the high-pH and reactivity in gasification ash. Its composition has been found to have an excess of Na and Al compared to the nominal composition of $NaAlSiO_4$ by Stevenson and McCarthy [5] who report a typical composition near $Na_{1.5}Al_{1.5}Si_{0.5}O_4$. This phase is apparently rapidly leached and this process yields a solution rich in NaOH and soluble species of Al. The Al could participate in reactions to form sulfate-carbonate ettringite in GPGA-3 and zeolites in the somewhat more alkaline GPGA-4w.

Ettringite is the principal crystalline reaction product observed for scrubber ash, codisposal mix and the fly ash and this reaction is probably responsible for the initial self-hardening of these materials. The XRD work indicated that the Et peaks typically reached a maximum in intensity after 6-7 weeks of hydration and that decreased thereafter. With the fly ash, monosulfate formation coincident with a decrease in ettringite abundance was observed, but no monosulfate was detected in the other ash hydration products.

For the experiments with the two ashes (GPGA-3 and scrubber ash) where liquid to solid ratio was varied, several trends were noted. The L/S ratio of 0.4 gave more rapid self-hardening than the mixtures with L/S twice and three times as great. Calcite abundance was greater at the larger L/S ratios, indicating greater Ca extraction from the ash relative to Al and sulfate that could combine with Ca to form zeolites or Et. A distinct interplay between decreasing Et and increasing Cc abundances was noted.

Because all of the ash materials contained noncrystalline solids (chiefly glass), this XRD study could have missed consolidation reactions that involved such solids as either reactants and products. The formation of calcium silicate hydrate or calcium aluminum silicate hydrate gels in the later stages of reaction is possible. While beyond the scope of the present study, studies of the contributions of noncrystalline solids constitutes a logical extension of the work reported here.

This study, although limited in scope to crystalline reactants and products, has given the background information necessary to define components of the reactive ashes to be included in ongoing examinations of the consolidation reactions in lignite coal ash.

ACKNOWLEDGMENTS

This research was supported by the Gas Research Institute, Chicago, IL, under Contracts 5082-253-0926, 5082-253-0771 and 5082-253-1283. Partial support was also provided by the sponsors of the Western Fly Ash Research, Development and Data Center. Chemical analyses of ashes were performed by D.J. Hassett. The fly ash samples and the C-618 analyses were provided by O.E. Manz.

REFERENCES

1. D.J. Hassett, G. J. McCarthy and K.R. Henke, in Fly Ash and Coal Conversion By-Products: Characterization, Utilization and Disposal II, edited by G.J. McCarthy, F.P. Glasser and D.M. Roy, Mat. Res. Soc. Symp. Proc. Vol. 65 (Materials Research Society, Pittsburgh, 1986), pp. 285-300.
2. G.J. McCarthy, D.J. Hassett, K.R. Henke, R.J. Stevenson, G.H. Groenewold, Characterization, Extraction and Reuse of Coal Gasification Solid Wastes. Vol. II. Leaching Behavior of Fixed-Bed Gasification Ash Derived From Northern Great Plains Lignite, (North Dakota Mining and Mineral Resources Research Institute, Grand Forks, ND) Final Report to the Gas Research Institute, October, 1986.
3. R.J. Stevenson, D.J. Hassett, G.J. McCarthy, O.E. Manz, Solid Waste Codisposal Screening Study, (North Dakota Mining and Mineral Resources

Research Institute, Grand Forks, ND) Draft Final Report to the Gas Research Institute, October, 1986.

4. G.J. McCarthy, D.J. Hassett, O.E. Manz, G.H. Groenewold, R.J. Stevenson, K.R. Henke and P. Kumarathasan, in Fly Ash and Coal Conversion By-Products: Characterization, Utilization and Disposal II, edited by G.J. McCarthy, F.P. Glasser and D.M. Roy, Mat. Res. Soc. Symp. Proc. Vol. 65 (Materials Research Society, Pittsburgh, 1986), pp. 310–310.

5. R.J. Stevenson and G.J. McCarthy, in Fly Ash and Coal Conversion By-Products: Characterization, Utilization and Disposal II, edited by G.J. McCarthy, F.P. Glasser and D.M. Roy, Mat. Res. Soc. Symp. Proc. Vol. 65 (Materials Research Society, Pittsburgh, 1986), pp. 77–90.

6. G.J. McCarthy and S.J. Steinwand, in Thirteenth Biennial Lignite Symposium: Technology and Utilization of Low-Rank Coal, edited by M.L. Jones, DOE/METC-86/6036 (1985).

7. G.J. McCarthy, O.E. Manz, D.M. Johansen, S.J. Steinwand and R.J. Stevenson, this volume.

8. K. Luke and F.P. Glasser, in Fly Ash and Coal Conversion By-Products: Characterization, Utilization and Disposal II, edited by G.J. McCarthy, F.P. Glasser and D.M. Roy, Mat. Res. Soc. Symp. Proc. Vol. 65 (Materials Research Society, Pittsburgh, 1986), pp. 173–180.

9. R.G. Garvey, North Dakota Equilibrium Geochemical Model (based on the WATEGM-SE program of C. Palmer, PhD thesis, University of Waterloo, 1983), IBM-PC Implementation, North Dakota State University, 1984.

10. R. von Ballmoos, Collection of Simulated XRD Powder Patterns for Zeolites, (Butterworth Scientific, Ltd., Surrey, UK,1984) 106 pp.

11. O.E. Manz, University of North Dakota, personal communication.

LEGAL NOTICE

ALUMINUM SULFATE HYDRATION RETARDERS FOR HIGH-CALCIUM FLY ASH
USED IN HIGHWAY CONSTRUCTION

M. TOHIDIAN* and JOAKIM G. LAGUROS**
*Standard Testing and Engineering, Oklahoma City, Oklahoma
**School of Civil Engineering and Environmental Science, The University of
Oklahoma, Norman Oklahoma 73019 USA

Received 20 November, 1986; refereed

ABSTRACT

The rapid hydration and setting associated with the use of high-calcium fly ash as an additive in soil and aggregate base stabilization in highway construction imposes certain limitations in regards to operational time and volume of work executed. Aluminum sulfate and its ammonium salt were evaluated as hydration reaction retarders. Mixtures of Ottawa sand and Class C high lime fly ash in a 1:1 weight ratio were used for the evaluations. These additives minimized the adverse effects of delayed compaction by recovering some of the compressive strength lost to the rapid hydration, although in all cases the density of the mixes decreased. The recovery of strength was related to the heat of hydration, wherein the peak temperature was reduced from 90°F to the range of 86-78°F at 2 hours; further temperature decreases were observed as reaction time increased. The availability of the sulfate ions, as manifested by the presence of ettringite, helps the hydration process continue, minimizes the adverse effects of delayed compaction and assists positively in the reduction of the void area of mixes and in stratlingite formation, which contributes to a strong crystalline framework.

INTRODUCTION

Fly ash containing free lime is an effective stabilizing agent in highway construction. Such high-lime fly ashes are used for improving the strength characteristics of substandard subgrade soil material [1] and for providing adequate cementation in coarse and fine granular base courses [2]. However, fly ash mixtures lose their workability and strength if not shaped and densified immediately [1,3], and become more time sensitive as the percentage of fly ash is increased [2]. These characteristics are reflected in the dual problems of loss of strength due to delayed compaction and rapid setting. In soil stabilization the amount of fly ash used is less than 20% [1], and thus only the effects of delayed compaction are of concern. Aggregate base course mixes, however, require high fly ash contents, in excess of 25%, and thus the maximization of strength gain in actuality is not attained because rapid setting interferes. Studies in delayed compaction are very few [5], but they all point to the fact that density and strength are, at times, substantially reduced.

Retarders have rarely been used with fly ash mixtures [5]. Some of the critical studies of retardation mechanisms [2,5-10] with other cementitious materials present the dominant conclusion that retardation is caused by the precipitation and adsorption of insoluble materials on surfaces. This study introduces aluminum sulfate chemicals which, as additives, retard the process of hydration and provide adequate operational time before the fly ash mixtures harden. In order to minimize the number of variables, the study was limited to mixtures of self-hardening Class C high-calcium fly ash and Ottawa sand.

Table I

Chemical (wt.%) and Physical Characteristics of Fly Ash

SiO_2	34.00
Al_2O_3	22.56
Fe_2O_3	7.29
$SiO_2+Al_2O_3+Fe_2O_3$	63.89
SO_3	2.75
CaO	26.88
MgO	5.06
Na_2O	1.46
Moisture Content, %	0.07
Loss of Ignition	0.45
Fineness	11.74
Fineness Variation, %	0.35
Specific Gravity	2.74
Soundness	0.43
Lime Pozz, 7 d (psi)	900
Cement Pozz, 28 d (% of control)	118.4
Water Requirement (% of Control)	98.0

*ASTM C 311 analysis conducted by the Bruce Williams
Laboratories for the Walter N. Handy Company.

MATERIALS

The high-calcium fly ash was collected from the Oklahoma Gas and Electric Company electrical generating station near Red Rock, Oklahoma. Its chemical and physical properties are given in Table I. According to ASTM C 618, it is classified as class C fly ash. Standard Ottawa sand, grade 30, was chosen because of its uniform particle size and consequent workability. The chemical additives used were aluminum sulfate and aluminum ammonium sulfate.

TESTING METHODOLOGY

Specimen Preparation

Mixtures of fly ash and Ottawa sand in a 50:50 weight ratio were prepared. Two Harvard miniature specimens (diameter 15/16 in.; height 2.816 in.; volume 1/454 ft^3) were compacted. The following were the test parameters:

Chemical retarders: none (control mix); aluminum sulfate; aluminum ammonium sulfate
Quantity of retarder: 0, 1, 3, 5 wt.%
Delay compaction time: 0, 30, 60, 120 minutes (min)
Curing period: 1, 7, 28 days (d).

Density-moisture relationships (proctors) for control mix (no chemicals), and for the various percentages of chemicals were determined according to AASHTO T 99-81. The chemicals were mixed with the water required for the optimum moisture contents and added to the sand/fly ash mixtures. The batch size was approximately 600 g. Specimens were extracted from the molds, wrapped in a polyethylene wrap and cured at near 100% relative humidity at room temperature (70°F).

Heat of Hydration

Samples of 100 g mixes were placed in small cylinders and stored in an insulated enclosure. A probe connected to a strip recorder was placed in the fresh mix and the temperature of the mixes for a period of 24 h was recorded continuously.

Compressive Strength

At the end of the specified curing periods the Harvard miniature samples were unwrapped, trimmed at both ends to obtain a smooth surface and tested for unconfined compressive strength (AASHTO designation T 208-78), both dry and after being immersed in a water tank for 24 h.

Scanning Electron Microscopy (SEM)

Following strength testing, fragments of each type of mix were removed and broken into approximately one-quarter inch in size. Since SEM was to be done after a considerable number of samples were collected, the fragments were put in separate pans and doused with large quantities of acetone in order to halt the hydration process by evaporating the free mix water. They were then dried at 110°C for a minimum of 4 h and stored in bottles. Samples for SEM were cemented to aluminum stubs and gold coated.

X-Ray Diffraction (XRD)

Acetone treated samples were ground and sieved to -200 mesh. A Siemens diffractometer unit operated under the the following conditions was used:

radiation and tube power: CuK-alpha; 35 kV, 18 mA;
scan conditions: 1 deg 2-theta/min; 200 and 1000 cts/sec scale factor.

RESULTS AND DISCUSSION

The results of the tests performed on the sand/fly ash /retarder mixtures, are given in Table II and discussed below.

Moisture Density

As the amount of retarder was increased, the required water to obtain optimum moisture increased from 8 to 13%, and resulted in the lowering of maximum dry density from 129 to 114 lbs/ft^3 (pcf).

Unconfined Compressive Strength

Table II shows the results of compressive strength measurements on specimens cured for 1, 7 and 28 d. The addition of small amounts of the retarders increased the compressive strength of mixtures over that obtained from control mixture (no retarder). The increase in strength was considerable for the 28 d cured samples. The optimum amount of retarder to give the maximum 28 d compressive strength appears to be about 1%.

Heat of Hydration

Results of heat of hydration measurements in mixes with aluminum ammonium sulfate retarder are shown in Figure 1. The results for the aluminum sulfate were very similar. The addition of the retarder chemicals lowered the overall temperatures of the setting sand/fly ash mixtures; the greater the additions, the greater the temperature reduction.

Table II

Summary of Test Data

Mix Type	Mix No.	Delay Time	Moisture Content	Dry Density	Compressive Strength, psi					
					1 day		7 day		28 day	
		min.	%	pcf	dry	wet	dry	wet	dry	wet
Control Mix	1	0	7.8	129.2	628	480	1330	1109	1420	1050
50% Sand +	2	30	7.4	102.2	NOT RECOVERABLE					
50% Fly Ash	3	60	0.0	0.0	COULD NOT BE MOLDED					
50% Sand +	5	0	8.4	121.7	813	606	1293	946	3104	2328
50% Fly Ash	6	30	8.4	115.8	458	392	1146	833	1811	1515
1% Alum	7	60	8.2	110.1	273	273	961	769	1131	1057
	8	120	7.7	100.7	96	52	200	222	429	273
50% Sand +	9	0	12.9	117.6	750	791	1183	1072	2323	961
50% Fly Ash +	10	30	12.4	115.9	706	776	998	850	1052	1002
3% Alum	11	60	12.1	112.7	624	606	916	717	983	980
	12	120	11.4	107.9	370	370	347	333	614	495
50% Sand +	13	0	13.9	109.4	281	251	443	433	814	769
50% Fly Ash +	14	30	13.0	103.2	200	307	348	333	495	492
5% Alum	15	60	12.8	101.6	170	244	318	281	440	488
	16	120	12.2	98.5	148	74	237	222	395	392
50% Sand +	29	0	9.3	125.9	813	850	2513	2291	2794	3104
50% Fly Ash +	30	30	9.1	121.2	739	591	1823	1633	2527	2661
1% Aluminum Ammonium Sulfate	31	60	8.8	116.7	436	480	1037	1097	1478	1574
	32	120	8.7	111.7	214	310	748	758	961	909
50% Sand +	33	0	11.6	120.0	472	488	776	1131	1404	1390
50% Fly Ash +	34	30	11.0	118.1	429	466	961	776	961	961
3% Aluminum Ammonium Sulfate	35	60	10.6	114.8	259	296	443	407	517	540
	36	120	10.4	111.2	85	133	345	394	407	414
50% Sand +	37	0	12.5	119.3	37	41	377	517	1293	1123
50% Fly Ash +	38	30	11.9	113.3	22	4	355	310	739	813
5% Aluminum Ammonium Sulfate	39	60	11.7	109.7	15	7	177	111	576	399
	40	120	11.3	109.0	11	0	140	81	281	299

Delayed Compaction

The 30 min delay in compaction of the sand/fly ash mixture without retarder caused the specimen to become very friable; it could not be extracted from the mold. There was also a considerable loss of density. Compaction of the mixture after a delay of more than 30 min was not possible because the mixture had already hardened. By adding a small dose of the retarder the mixture sand/fly ash could be molded even after a compaction delay of up to 2 h.

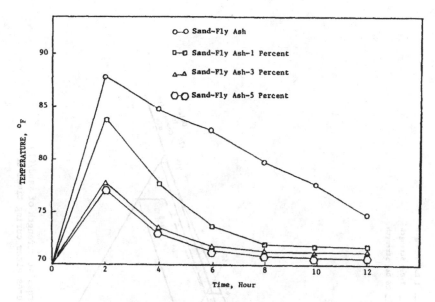

Figure 1. Heat of hydration of sand/fly ash/aluminum ammonium sulfate mixtures. (Results with aluminum sulfate were similar.)

Figure 2 is a plot of data from Table II for a 1% addition of aluminum ammonium sulfate as a function of delay time and for each time of curing. Delay in compaction of the sample resulted in a strength loss. Yet, with the addition of just 1% of retarder, up to one-third of the zero delay time strength could be retained even after a 2 h delay time. The compressive strength results in Table II also indicate that the amount of retarder added has a significant effect. Of the three amounts studied, the 1% additions resulted in the greatest overall retention of compressive strength for various compaction delay times.

X-Ray Diffraction

Crystalline phases were identified using the JCPDS Powder Diffraction File [12]. If three or more peaks of proper intensity were found, the phase was considered to be present. If fewer peaks were found, identification was considered tentative. Table III lists the mineral name, chemical formula, and strong peaks for all the crystalline phases identified in this study and Table IV gives the phases identified in the sand/fly ash mixtures cured for 1 d and 28 d, and after 0 h and 2 h of compaction delay.

Mix Components. As expected, the sand was pure quartz. The fly ash diffractogram indicated the presence of quartz, tricalcium aluminate, calcium oxide, magnesium oxide and anhydrite.

Sand/Fly Ash Mixtures. Diffractograms of the 50:50 mixture with no delay in compaction cured for 1 d and 28 d showed the presence of quartz, residual tricalcium aluminate or merwinite, and monosulfoaluminate. A trace amount of stratlingite appeared to be present in the diffractogram of the 28 d sample. A small amount of calcite formed by carbonation during drying and grinding was also present.

176

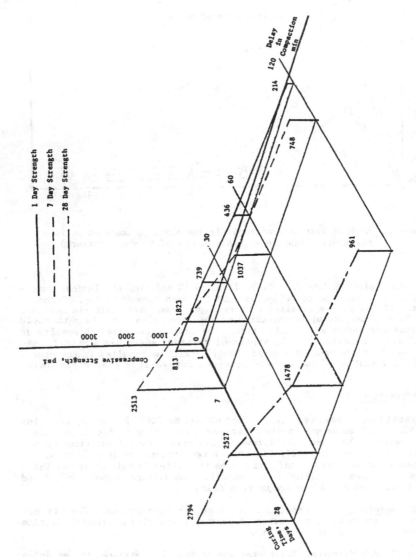

Figure 2. Effect of delay in compaction on compressive strength of sand/fly ash/1% aluminum sulfate retarder mixtures after curing times of 1, 7 and 28 d.

Table III

Crystalline Phases Identified by XRD

Mineral Name	Chemical	symbol	d-spacing Å	relative intensity percent		PDF Number
			Diffractogram			
Tricalcium aluminate	$3CaO \cdot Al_2O_3$	T	2.70 (100)	1.91(36)	1.56(27)	8-5
Quartz	SiO_2	Q	3.34(100)	4.26(35)	1.82(17)	5-0490
Anhydrite	$CaSO_4$	CS	3.49(100)	2.85(32)	2.33(22)	6-226
Lime	CaO	L	2.41(100)	1.70(45)	2.78(34)	4-777
Monosulfoaluminate	$3CaO \cdot Al_2O_3 \cdot CaSO_4 \cdot 13H_2O$	Ms	8.92(100)	2.87(70)	4.46(60)	11-179
Ettringite	$3CaO \cdot Al_2O_3 \cdot 3CaSO_4 \cdot 32H_2O$	E	9.73(100)	5.61(80)	3.88(80)	9-414
Periclase	MgO	Mg	2.11(100)	1.49(52)	1.22(12)	4-829
Merwinite	$3CaO \cdot MgO \cdot 2SiO_2$	Mr	2.66(100)	1.90(70)	1.53(70)	4-728
Stratlingite	$Ca_2Al_2SiO_7 \cdot 8H_2O$	S	12.5(100)	4.80(70)	6.27(40)	29-285
Calcite	$CaCO_3$	Cc	3.04(100)	2.29(18)	2.10(18)	5-0586

Table IV

Phases Identified by XRD in Sand/Fly Ash Mixtures

Curing Period (days)	Delay Period (hours)	Mixtures of Sand-Fly Ash		
		plain	aluminum sulfate	aluminum ammonium sulfate
1	0	Q T Ms Mg(tr) CS(tr)	Q T E Mg(tr) CS(tr)	Q T E Mg(tr)
28	0	Q T Ms Mg(tr)	Q T E S Mg(tr) CS(tr)	Q E T S Mg(tr)
1	2		Q T E Mg(tr)	Q T E Mr
28	2		Q T E S Mg(tr)	Q E S Mg(tr)

* (tr) = trace

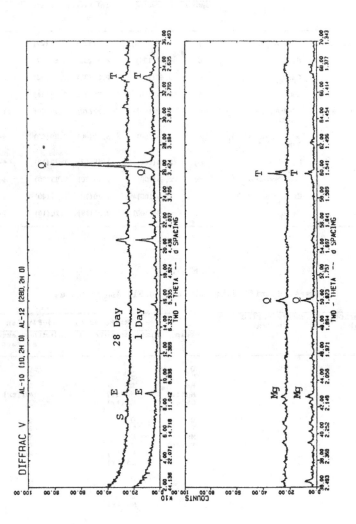

Figure 3. X-ray diffractograms of sand/fly ash/1% aluminum sulfate cured for 1 d and 28 d after a 2 h delay in compaction. Symbols for phases as in Table III.

Sand/Fly Ash/Retarder. In the no delay sample, the major crystalline compounds were quartz, ettringite, and possibly some monosulfoaluminate. Both 1 d and 28 d specimens appeared to contain some residual tricalcium aluminate or merwinite. The 28 d sample contained a calcium-aluminate-silicate hydrate, identified tentatively as stratlingite ($Ca_2Al_2SiO_7 \cdot 8H_2O$). There was no significant difference between the no delay and the 2 h delay samples. The major reaction products were ettringite and stratlingite. At 28 d, the ettringite had started to decompose to a lower hydrate monosulfoaluminate. Figure 3 shows diffractograms for 1 d and 28 d cured mixtures with 1% aluminum sulfate, and a 2 h delay in compaction. XRD results were similar for the two retarders (Table IV).

In general, stratlingite appears in all of the 28 d diffraction traces regardless of the treatment, and is an indicator of fly ash reaction to form cementitious products. The availability of sulfate ions has an influence on the crystalline compounds formed. Tricalcium aluminate, free CaO and calcium sulfate (anhydrite) apparently react to form monosulfoaluminate. Fly ash initially reacts to form the trisulfate ettringite ($(3CaO \cdot Al_2O_3 \cdot 3CaSO_4 \cdot 32H_2O)$), but as the available sulfate content decreases monosulfoaluminate ($3CaO \cdot Al_2O_3 \cdot CaSO_4 \cdot 12H_2O$) begins to form. When the sulfate retarder is added, additional ettringite forms and later conversion of ettringite to monosulfoaluminate is prevented.
Ettringite forms a protective coating on the fly ash grains. This coating would be expected to rupture due to the expansive nature of ettringite formation, but it is healed continually as long as sulfate ions are present. In sulfate containing samples, the hydration process is continued, thus creating a much stronger crystalline matrix for the silica and alumina which is present, apparently in part, in the sample in the form of stratlingite. The formation of ettringite directly from CaO, $Al_2(SO_4)_3$ and H_2O could not be completely ruled out. However, more in depth studies are required to verify even the probability of its formation.

Scanning Electron Microscopy

Figures 4-8 show a series of SEM photomicrographs of sand/fly ash/1% retarder mixtures cured for 1 d and 28 d with and without a delay in compaction. The 1 d cured/no delay specimen (Fig. 4) shows that most fly ash particles are intact and very few needle crystals have formed compared to the one day cured/ 2 h delay specimen (Fig. 5). The 28 d cured/no delay specimen (Fig. 6) shows heavy formation of needle and hexagonal plate type crystals. The 28 d cured/2 h delay specimen (Fig. 7) shows less formation of apparently densely crystalline material, more needle crystals and more voids compared to 28 d cured/no delay sample.

CONCLUSIONS

1. When 50:50 mixtures sand and fly ash were compacted with a delay in compaction of 30 minutes, they became very friable and lost bonding between particles. There was a considerable loss in dry density of samples with a delay in compaction. The sand/fly ash mixtures could not be compacted when compaction was delayed for 60 minutes.

2. By adding sulfate-based retarders to the sand/fly ash mixture, unconfined compressive strength for specimens cured for 28 d doubled. Compaction delays of up to 2 h still resulted in reasonable unconfined compressive strengths and dry densities that were substantial fractions of the no delay strengths. The optimum addition of retarder was about 1%.

180

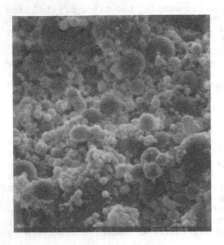

3000X   ⊢⊣ = 1 μm

Figure 4
SEM photomicrograph of a fracture
surface of sand/fly ash/1% aluminum
sulfate cured for 1 d with no delay
in compaction.

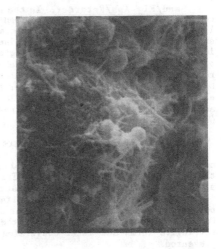

3000X   ⊢⊣ = 1 μm

Figure 5
SEM photomicrograph of a fracture
surface of sand/fly ash/1% aluminum
sulfate cured for 1 d with a 2 h
delay in compaction.

3000X   ⊢⊣ = 1 μm

Figure 6
SEM photomicrograph of a fracture
surface of sand/fly ash/1% aluminum
sulfate cured for 28 d with no delay
in compaction.

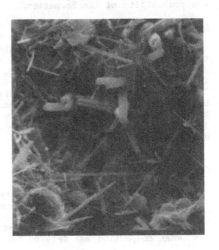

3000X   ⊢⊣ = 1 μm

Figure 7
SEM photomicrograph of a fracture
surface of sand/fly ash/1% aluminum
sulfate cured for 28 d with a 2 h
delay in compaction.

3. Addition of the retarders to the mixture of sand-fly ash lowered the peak temperature (seen at 2 h) of hydration. The higher the percent of retarder used, the lower the peak temperature.

4. XRD of sand/fly ash mixtures showed, that due to the low content of sulfate ions (as calcium sulfate) in the fly ash , monosulfoaluminate is formed in 1 d and 28 d cured samples. Formation of ettringite helped the hydration process to continue by creating a protective gel around fly ash grains. Due to the expansive nature of ettringite formation, rupture occurred but the ruptures healed continually as long as sulfate ions were present.

5. The aluminum sulfate additives apparently enhance the formation of stratlingite.

REFERENCES

1. J.G. Laguros and M.S. Keshawarz, in Fly Ash and Coal Conversion By-Products: Characterization, Utilization and Disposal II, edited by G.J. McCarthy, F.P. Glasser and D.M. Roy, Mat. Res. Soc. Symp. Proc. Vol. 65 (Materials Research Society, Pittsburgh, 1986) pp. 37-46.
2. A. Arman, and T.J. Dantin, The Effect of Admixtures on Layered Systems Constructed with Soil-Cement, Eng. Res. Bull. No. 86, (Louisiana State University Division of Engineering Research, Baton Rouge 1965).
3. J.G. Laguros, A Study of Aggregate-Fly Ash Mixes (Department of Civil Engineering, University of Oklahoma, Norman, 1981) 20 pp.
4. S.I. Thornton and D.C. Parker, Construction Procedures Using Self-Hardening Fly Ash, FHWA-AR-80, September 1980.
5. S.I. Thornton and D.G. Parker, Fly Ash as Fill and Base Material in Arkansas Highways, Highway Research Project 43, October 1975.
6. J.G. Laguros, and D.T., Davidson, Highway Research Board Record 36, 1963.
7. F.M. Lea, The Chemistry of Cement and Concrete, Third Edition, (Chemical Publishing Company, Inc., London, 1970).
8. V. Lorprayoon and D.R. Roosington, Cem. Concr. Res. 11, 267-277 (1981) .
9. Y. Watanbl, S. Suzuki and S. Nishi, Journal of Research, Onado Cement Company, Volume 11, 1969, p. 184-196.
10. J.F. Young, Transportation Research Record 564, 1-10 (1976).
11. J.L. Minnick, in Proceedings Fly Ash Utilization Symposium, Bureau of Mine Information Circular No. 8488, (U.S. Department of Interior, Washington, DC, 1970).
12. W.F. McClune, Editor-in-Chief, Powder Diffraction File (JCPDS-International Centre For Diffraction Data, Swarthmore, PA).

SOURCES OF SELF-HARDENING PROPERTIES IN FLY ASHES

R.C. JOSHI and D.T. LAM
Department of Civil Engineering, The University of Calgary, 2500 University
Drive N.W., Calgary, Alberta, Canada T2N 1N4.

Received 6 March, 1987; Communicated by G.J. McCarthy

SUMMARY

Laboratory investigations of the self-hardening properties of selected
subbituminous fly ashes have been conducted. Chemical analyses of the fly
ashes are given in Table I. The self-hardening value of the fly ashes was
determined by conducting unconfined compressive strength tests on compacted
samples of the moistened ashes. Various physical and chemical tests were performed to identify the reaction products, if any, of the hardened compacted
fly ash paste, and to delineate the source of self-hardening properties.
Results from x-ray diffraction analyses, scanning electron microscopic examination and differential thermal analyses indicated that the hydration products include calcium silicate and aluminate hydrates, and ettringite. Chemical and physical tests conducted to evaluate pertinent fly ash properties
included chemical analysis, water soluble fraction, dilute hydrochloric and
hydrofluoric acid soluble fractions, heat of solution on dissolving in dilute
hydrochloric acid, specific surface area, and electrical conductivity tests.
Results of chemical and physical tests suggest that calcium (in the form
of free calcium oxide), water soluble fraction, and specific surface area of
the fly ash play important roles in the self-hardening mechanism. Correlation
of compressive strength and various other data suggest that self-hardening
value of the fly ashes studied may be evaluated from a self-hardening number
which is the product of specific surface in cm^2/g and the weight fraction of
available free lime (Fig. 1).

Table I. Chemical Analyses of Fly Ashes

Constituents (Wt. %)	Forestburg	Sundance	Wabamun	Montana Colstrip	Kansas LaCygne	Kansas Montrose	Kemmerer
SiO_2	48.41	48.86	50.61	38.98	32.28	46.40	54.72
Al_2O_3	23.47	23.89	22.41	21.51	18.72	16.53	16.91
Fe_2O_3	4.95	3.40	4.50	3.70	5.06	19.65	7.21
CaO	16.99	15.30	16.83	25.16	31.25	9.06	8.11
MgO	0.53	0.49	1.72	4.73	3.99	0.48	2.36
SO_3	0.27	0.29	0.28	0.67	2.75	1.23	0.93
Na_2O	3.16	2.94	0.65	0.60	1.97	0.61	1.38
K_2O	1.06	0.69	0.97	0.35	0.39	1.58	1.27
TiO_2	< 1.0	< 1.0	< 1.0	< 1.0	< 1.0	< 1.0	< 1.0
LOI (950°C)	0.43	0.57	0.49	2.27	0.46	2.84	2.51
free CaO*	9.7	8.1	8.8	20.0	17.0	5.3	4.4

* Determined as per ASTM Specification C114-78.

184

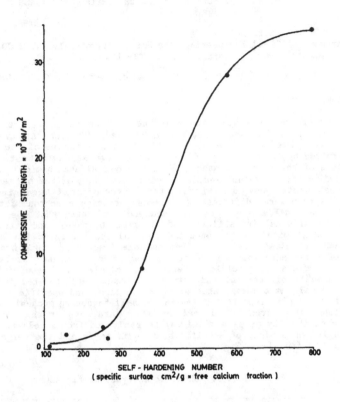

Figure 1. Compressive strength vs. self-hardening number.

MICROSTRUCTURE OF CEMENT BLENDS INCLUDING FLY ASH, SILICA FUME,
SLAG AND FILLERS

MICHELINE REGOURD
Microstructures Department, CERILH, 23 rue de Cronstadt, 75015 PARIS, FRANCE

Received 13 January, 1987; refereed

ABSTRACT

The hydration of a blended cement through hydraulic or pozzolanic reactions results in heterogeneous polyphase materials. Because portland cement clinker is the major component in most cement blends, the microstructural development of portland cement hydrates, including C-S-H and pore structures, is first discussed. Slag, fly ash, silica fume and limestone filler cements are then compared to portland cement with regards to C-S-H morphology and composition, aluminate crystallization, cement paste interfaces and pore size distribution.

INTRODUCTION

The microstructure of hydrated cements has been recently reviewed for the 8th International Congress on the Chemistry of Cement (Rio de Janeiro, 1986) in principal reports by Taylor [1], Diamond [2] and Uchikawa [3]. The microstructural description of a cement paste includes the characterization of the structural elements (particles, crystals, films, voids) and of their juxtaposition and mutual arrangement [2]. Progress has been made thanks to new procedures of sample examination under electron microscope (wet cell, cryostage, ion-beam thinned specimens) and to the combination of such electron techniques as TEM, HVEM, EPMA, SEM, STEM, AEM and image analysis [1,2].

This paper will present the microstructure of blended cements containing mineral additives, i.e. blast furnace slag, fly ash, silica fume, and limestone filler. Because in most of these cements the portland clinker is the major component, the microstructure of ordinary portland cement (OPC) pastes will be described first.

MICROSTRUCTURE OF PORTLAND CEMENT PASTES

Portland cement hydrates are C-S-H (calcium silicate hydrate) and CH (calcium hydroxide) generated from silicates (C_3S, tricalcium silicate (alite), and C_2S, dicalcium silicate (belite)), along with AFt (ettringite) and AFm (monosulfates) phases from aluminates (C_3A and C_4AF). Quantitative phase determinations using XRD, TG, DTA, DSC and chemical extraction methods allow the calculation of an approximate distribution of the structural elements. Table I shows that in volume proportion, C-S-H and voids are the most important components of a largely hydrated cement [4,5].

Microstructural Development

New results on the microstructural development during cement hydration have appeared since the Paris Congress (1980). Clearer pictures of the cement paste microstructure have been made possible through the following combination of techniques and careful preparation of specimens: high voltage transmission electron microscopy (HVEM) with a wet cell [6]; scanning transmission electron microscopy (STEM) of ion-beam thinned sections [7]; scanning electron microscopy (SEM) with either backscattered electron images (BEI) of polished sections [8] or secondary electron images of matched surfaces [9] and using a cryostage to avoid dehydration [10]; and image analysis [11].

Table I

Calculated Phase Analysis of Portland Cement Pastes[a] (% by volume)

Cement paste sample	C-S-H	CH	AFm	CaCO$_3$	Clinker	Porosity
1 year, 11% RH [4]	37	11	15	1	5	31[b]
3 months, saturated [5]	40	12	16	1	8	24

a. W/C = 0.5.
b. micropores <5 nm = 0.13; capillary pores = 0.18.

Three stages of cement hydration (at 20°C) have been considered by Jennings, Dalgleish and Pratt [12], and Scrivener [13]: the early period, up to 3 hours (h); the middle period, between 3 and 20 to 30 h; and the late period. The microstructural development, as illustrated in Figure 1 [13], can be described as follows:

Early Period. During the first stage of hydration, including the pre-induction and induction periods, a gel rich in aluminate forms at the surface of clinker grains, and stubby rods of AFt phase nucleate on the outer surface of gel or in the solution.

Middle Period. This corresponds to setting and hardening, with a rapid formation of C-S-H and CH. The outer C-S-H, which can possibly nucleate on AFt rods, entirely covers the clinker grains, leaving a space of about 1 µm between the anhydrous grains and the hydrated shells. This gap is filled with highly concentrated solutions and possibly with colloidal material. At the end of this middle period, an inner C-S-H develops inside the shells. According to Scrivener [13], cement grains smaller than 3 µm are completely dissolved, producing only outer C-S-H and hollow shells. According to Diamond [2], the hollow shells are related to high W/C ratios. However, the existence of hydrated shells confirms the first description of "Hadley grains" observed on fracture surfaces [14]. Calcium hydroxide crystallizes in hexagonal plates from the solution between cement grains.

AFt in long needles (1-2 µm, sometimes 10 µm) is formed mainly outside the shells, but it can also be observed inside the shells. As the hydration progresses, the shells become less permeable to the outer solution, and inside the shells, the surface ions in lower concentrations give rise to AFm.

Late Period. In the late period, C-S-H grows slowly by three different means:

a. The outer C-S-H gives thick layers of fibres, Type I according to the classification of Diamond [15].
b. The inner C-S-H fills the space inside shells, probably as Type III.
c. The late C-S-H, which occupies the space left by the interface of C$_3$S moving inward [12], would be Type IV C-S-H.

These three types of C-S-H have been observed on a BEI of a 23-year old cement paste [13]. Outer and inner C-S-H are formed through solution. Late forming C-S-H occurs by a topochemical mechanism. CH appears as large crystals, very often superimposed. It may also crystallize as small particles inside shells. AFm phase is observed inside shells as formed directly from aluminate phases and solution. Outside the shells, AFm may be produced from the decomposition of AFt.

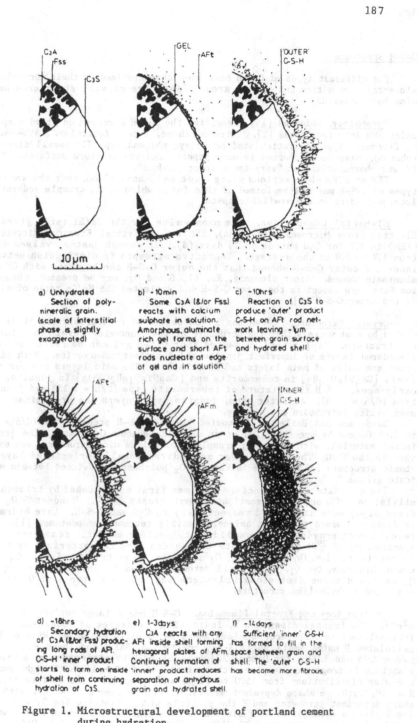

a) Unhydrated
Section of poly-
nineralic grain.
(scale of interstitial
phase is slightly
exaggerated)

b) ~10min
Some C_3A (&/or Fss)
reacts with calcium
sulphate in solution.
Amorphous, aluminate
rich gel forms on the
surface and short AFt
rods nucleate at edge
of gel and in solution.

c) ~10hrs
Reaction of C_3S to
produce 'outer' product
C-S-H on AFt rod net-
work leaving ~1μm
between grain surface
and hydrated shell.

d) ~18hrs
Secondary hydration
of C_3A (&/or Fss) producing long rods of AFt.
C-S-H 'inner' product
starts to form on inside
of shell from continuing
hydration of C_3S.

e) 1-3days
C_3A reacts with any
AFt inside shell forming
hexagonal plates of AFm.
Continuing formation of
'inner' product reduces
separation of anhydrous
grain and hydrated shell.

f) ~14days
Sufficient 'inner' C-S-H
has formed to fill in the
space between grain and
shell. The 'outer' C-S-H
has become more fibrous.

Figure 1. Microstructural development of portland cement
during hydration.

C-S-H Structure

The different types of C-S-H have been characterized by their morphology, elemental composition and surface area. Tentative crystal structures have also been offered.

Morphology. Diamond [15] classified the C-S-H observed on dried samples under SEM into four types [2]. Three of these, Type I (acicular, lath-shaped or fibrous), Type II (reticulated or honeycomb) and Type III (small discs or spheres), have been reported in many papers studying fracture surfaces. Type IV was characterized as "very dense inner product".

TEM or STEM observations using a wet cell have shown that the various types of C-S-H may all be formed of thin foils which roll, crumple and interlock, depending on the available space.

Elemental Composition. The mean value for the Ca/Si ratio given by Electron Probe Microanalysis (EPMA), SEM and Analytical Electron Microscopy (AEM) is 1.7 for C_3S pastes at 28 days (d). In cement pastes, values vary from 1.7 to 2.0 on the average. Tentative attempts to distinguish between inner and outer C-S-H showed that the outer C-S-H can be mixed with CH and aluminate phases. Minor elements, Na, Al, S and Mg, may be present. Magnesium ions were absent in the outer C-S-H and reflected the Mg/Ca ratio of alite in the inner C-S-H [1].

Crystal Structure. C-S-H has a low degree of crystallinity. It is admitted by most workers that the structure at the "nanometer level" is a layered structure. Taylor [16] suggested that C-S-H gel can be represented by disordered layers of imperfect jennite and imperfect tobermorite. Both silicates are built of main layers and interlayers. The main layers are respectively $[Ca_4Si_6O_{18}H_2]$ in tobermorite and $[Ca_8Si_6O_{18}H_2(OH)_8]$ in jennite. At early ages, C-S-H is a mixture of tobermorite type (C/S = 1.25) and jennite type (C/S = 2.25). At later ages, jennite type layers are predominant and some silica tetrahedra are missing.

Henderson and Bailey [17] studied various C-S-H preparations (C/S = 0.8 to 1.9) formed by precipitation at an organic aqueous interface. The interfacial material, with typically crumpled, foil-like microstructure, was examined in the TEM. The foils appear as a "directionally orientated layered atomic structure related to a backbone CaO_x polyhedra sandwiched between silicate groups."

The silicate anion structure has been first established by trimethylsilylation (TMS) and gel permeation chromatography (GPC). Monomer SiO_4 and dimer Si_2O_7 were attributed respectively to C_3S and C-S-H. Late hydrated products contained polymeric anions, possibly tetramer or pentamer [1]. New results were provided by the solid state nuclear magnetic resonance (NMR) spectroscopy of ^{29}Si [18]. Silicate tetrahedra were characterized by their connectivity, i.e. Q_0 as monomer, Q_1 as dimer, Q_2 as trimer. In the C_3S and cement hydrates, Q_0, Q_1, Q_2 were all detected in C-S-H. The absence of Q_3 and Q_4 excluded three dimensional clusters but the coexistence of Q_1 and Q_2 suggested a chain-like structure.

Surface Area and Fractal Dimension. C-S-H has a large surface area, 200 m^2g^{-1}. The fractal dimension D, indicating the degree of roughness of the internal surface structure, has been measured by different ways. Winslow [19] calculated D using small angle X-ray scattering (SAXS) and found values between 2.95 and 3.09. Van Damme [20] determined D by adsorption of a single adsorbate (nitrogen) on samples with variable size. Samples with eight particle size distributions from 1500 to 10 μm were prepared. The surface area is $S = R^D$, with a shape dependent parameter. The volume is $V = R^3$, with a shape dependent parameter and R the particle size. Plotting log Sv = S/V vs. log R gives a straight line of slope D−3. The fractal dimensions D of a white portland cement paste are 2.90 (W/C = 0.6), 2.93 (W/C = 0.4), 2.95 (W/C = 0.2)

Table II

Alite Porosity (volume fraction) [24]

	Hydration time (days)			
	1	3	10	28
Degree of hydration	0.22	0.30	0.50	0.83
Impregnated porosity (by microscopy)	0.50	0.42	0.26	0.25
Open porosity (by microscopy)	0.44	0.38	0.15	0.12
Estimated total porosity	0.61	0.58	0.55	0.49
Estimated porosity >50 nm in width	0.51	0.44	0.34	0.27

and 2.97 (W/C = 0.2 + 0.8% superplasticizer). Thus lower W/C is associated with higher D, corresponding to a rougher surface.

Porosity

As shown in Table I, a cement paste is a porous material. Cavities may contain "pore solution" or be air filled spaces [2]. In his principal report at the Rio Congress, Feldman [21] discussed both pore structure determinations by such procedures as mercury intrusion, gas adsorption and desorption and electron microscopy with image analysis, and the effects of sample preparation methods. It is known that drying evaporates the pore solution and shrinkage results in cracks that alter the microstructure. New procedures have been offered. They concern either a water solvent replacement followed by evaporation of the fluid [22], or wet cells for microscopy. Table II reports results of porosity determined by microscopy and image analysis [24] for an alite at different times of hydration. The impregnated porosity can be related to the coarse porosity.

Air-filled spaces are distributed in two families: large, almost spherical voids; and "geometrically complex" capillary pores between hydrates [2]. When a dried sample is rehumidified, capillary pores become resaturated but air voids do not. Very fine pores, so-called "gel pores", are features of the pore structure of such hydrates as C-S-H. Porosity is still in discussion and will not be developed here as there is a separate session on pore structure of hydrated cement at this symposium.

MICROSTRUCTURE OF BLENDED CEMENT PASTES

Mineral additives to portland cement include blast furnace slag, fly ash, silica fume and limestone filler. Their characterization and activation have been presented in two principal reports at the Rio Congress by Regourd [25] and Uchikawa [3]. Hydration and properties of blended cement pastes have also been discussed during two international conferences at Montebello, Canada, in 1983 and Madrid, Spain, in 1986.

Blast Furnace Slag Cement

Blast furnace slags are hydraulic materials. Although they usually react more slowly than portland cement, they produce the same hydrates, C-S-H and sulfoaluminates. Tanaka, Totani and Saito [26] gave a schematic hydrated microstructure of a glassy slag in portland cement (Fig. 2). The slag surface is first covered by portland cement hydrates, then attacked by Ca^{2+} ions from the supersaturated solution, producing the "inner hydrate". The dissolution of Ca^{2+} and Al^{3+} ions from slag leaves a skeleton hydrated layer, which

190

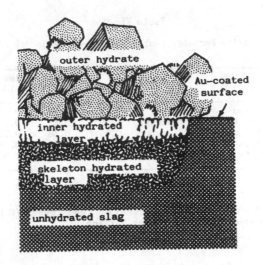

Figure 2. Schematic microstructure of hydrated glassy
           slag in portland cement paste [26].

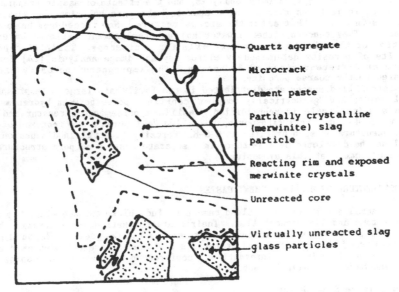

Figure 3. Sketch of an electron micrograph of reactive merwinitic slag
           compared to other glassy grains [27].

transforms gradually into "inner hydrate" by the supply of Ca^{2+} ions. Under
the microscope, the inner hydrate appears crystallized and dense. The skeleton hydrate is a porous solid containing equant grains of Type III C-S-H. The

elemental distribution shows that Ca/Si increases and Al decreases from the skeleton to the outer hydrate. There is a restricted mobility of Mg^{2+} ions, which are found at higher concentrations in the skeleton hydrate than in anhydrous slag and inner hydrate.

Frearson and Uren [27] showed that merwinitic slag grains containing merwinite dendrites are more reactive than fully glassy particles (Fig. 3), thus confirming results of Demoulian et al. [28] that the glass around merwinite dendrites is lower in Mg but richer in Al and more reactive than the glass of fully glassy particles.

Hydrates observed under SEM are described as: (i) C-S-H, either amorphous, honeycomb Type II or in thin plates -- all forms have a lower Ca/Si than OPC (C/S = 1.3) and contain Al, Mg and Fe; (ii) AFt as thin needles and AFm as hexagonal plates; (iii) $C_4AH_x-C_4MH_x$ and $Mg_6Al_2CO_3(OH)_{12}\cdot4H_2O$ (hydrotalcite) [29].

Harrisson, Winter and Taylor [30] suggested the possible presence of a magnesium silicate hydrate other than C-S-H that contains Mg, Al, Si and Ca.

Fly Ash Cements

Fly ash is a heterogeneous material varying from one source to another [2,3,25]. As classified by ASTM C 618, fly ashes with low calcium contents are pozzolanic materials, while fly ashes with high calcium contents are hydraulic.

When a pozzolanic fly ash cement is mixed with water, the first hydrates formed are the same as those from portland cement clinker. These hydrates deposit on fly ash spheres as a duplex film composed of amorphous CH and C-S-H gel [2]. The densification of this duplex film gives rise to hydrated shells (Fig. 4). The pozzolanic reaction of fly ash is characterized by etching of glass at the sphere surface. Gaps or Hadley grains have been observed at early ages. Inert crystals, e.g. mullite, quartz and magnetite, are observed when the superficial glassy layer has reacted. AFt, AFm and late aluminates are also formed as in portland cement pastes.

The C-S-H from fly ash occupies the space previously occupied by the pore solution and CH [30]. The pozzolanic reaction is illustrated by BEI of CH distribution in Figure 5 [31]. The C-S-H, apparently more amorphous than the C-S-H from OPC, also contains more Al. Its Ca/Si ratio is between 1 and 1.50. Stratlingite, C_2ASH_8, has also been detected.

Silica Fume Cements

Silica fume, a by-product of electrometallurgy [25], is used in cements, mortars and concretes as a filler. In the presence of superplasticizer, it produces a microstructure completely different from that of portland cement pastes [2]. The matrix appears homogeneous and dense with no gaps and no large crystals of CH. The high pozzolanic reactivity of the silica fume is easily followed by XRD measurement of the decrease in CH liberated by the portland cement silicates [32]. The C-S-H is close to C-S-H Type III (Fig. 6). Fibrous Type I C-S-H and honeycomb Type II are not observed. The Ca/Si ratios of C-S-H are inversely proportional to the amount of silica fume. As an example, at 28 d and in the same conditions, the C/S ratio and the amount of alkalies in C-S-H are as follows: OPC, 1.8 Ca/Si and 0.5% K_2O; and OPC + 30% silica fume + 1% superplasticizer, 0.9 Ca/Si and 1.3% K_2O.

The same C-S-H is observed in mortars and concretes with a strong bond between the matrix and sand grains or aggregates and a transgranular fracture (Fig. 7).

Blended Cements with Silica Fume

Silica fume has been used as an early reactive filler in blended cements with either fly ash or slowly reactive slag or inert material such as quartz

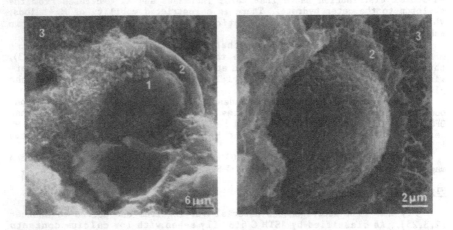

Figure 4. Microstructure of fly ash cement paste. (1 = duplex film; 2 = hydrated shell; 3 = OPC C-S-H.)

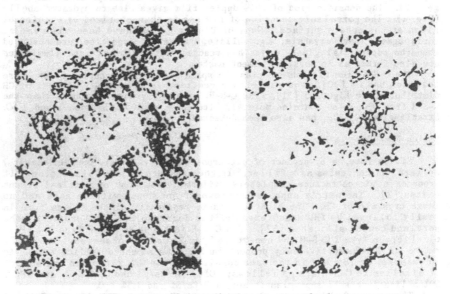

Figure 5. BEI showing CH distribution in a portland cement paste (left) and in a fly ash cement at 90 d (right) [31].

[33]. With slightly or slowly reactive slag or pozzolanic blends, an improvement in mechanical strength and microporosity is clearly observed at 28 d; 5% of silica fume replacement lead to a compressive strength increase of 15% to 22%. At three months both characteristics show small gains as compared with

 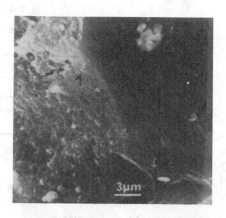

Figure 6
Apparently amorphous C-S-H in a
silica fume cement paste.

Figure 7
Cement paste bonded strongly to
aggregate in the presence of silica
fume.

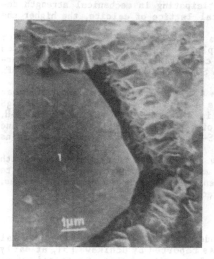 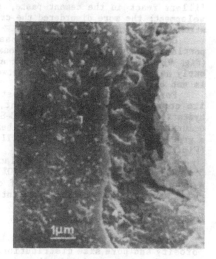

Figure 8.   Comparison between paste with 30% slag (left), and 25 slag + 5%
silica fume (right), showing the gap filled in by silica fume
hydrate [33].

blended cements with no silica fume. The maximum benefit is obtained with
quartz, with an increase in mechanical strength of 20% at three months. The
filler effect is shown clearly in Figure 8, where a gap observed between a
slag grain and an hydrated slag at 28 d in a cement paste with 30% granulated
slag is filled in with amorphous C-S-H in the presence of silica fume. The

Table III

Compressive Strengths of Cement Pastes[a] (MPa)

Type of cement	28 d	3 months
OPC	105	122
OPC + 25% disordered calcite	87	102
OPC + 25% well crystallized calcite	81	91
OPC + 25% quartz (inert)	74	88

a. W/C = 0.3.

difference in C-S-H microstructure is also reflected in the Ca/Si ratios, 1.55 and 1.20 respectively.

Cements with Limestone Filler

Blended cements with limestone fillers are produced in Europe, particularly in France [25]. The content of filler averages 15%. Fillers used at cement plants are composed mainly of calcite. This calcite appears more or less disordered, as analyzed by XRD [25] or IR spectroscopy [34]. Limestone fillers react in the cement paste, participating in mechanical strength development; the more disordered the crystal lattice of calcite, the higher the strength developed (Table III).

Limestone fillers react in pastes of either aluminous cement [34] or portland cement [35] to form hexagonal plates of calcium monocarboaluminates (Fig. 9). In portland cements, AFt and carboaluminate phases coexist, apparently as a result of competition between SO_4^{2-} and CO_3^{2-} for Al ions. The AFm is not observed.

For a given W/C ratio, a limestone filler cement contains more CH than its corresponding portland cement, as a result of a higher rate of C_3S hydration. The microstructure of C-S-H is slightly different from OPC C-S-H, occurring as shorter and thicker fibers of Type I C-S-H. CH microcrystals and a grainy C-S-H, which may be Type III C-S-H, are observed around filler grains (Fig. 10) [36].

The use of 5% silica fume replacing 5% limestone filler in a cement with 30% additive results in a gain of 20% in mechanical strength, corresponding to a more amorphous C-S-H, absence of orientated CH crystals on filler grains, and a strong filler-cement paste interface (Fig. 11).

Pore Structure of Blended Cements

Compared to portland cements, blended cements show differences in total porosity and pore size distribution. As reported by Uchikawa [3], at early ages a slag cement exhibits the same total porosity as that of portland cement (Fig. 12). At later ages when the slag has reacted, the pores of 3 to 5 nm size become larger [37]. In fly ash cements, the slow pozzolanic reaction reflects a higher total porosity at ages up to six months, which has been related to both gaps around fly ash particles and pores inside particles. Even at early ages, silica fume cement blends have, at the same total porosity, more fine pores than portland cement. This has been considered as the result of filler effect and initiation of the pozzolanic reactions of the very fine silica particles.

From porosity measurements with the mercury intrusion process and after repeated intrusions following mercury removal by distillation, Feldman [21] concluded that in blended cements the pore structure is characterized by rel-

Figure 9
Hexagonal crystals of calcium
monocarboaluminate in a filler
cement paste [35].

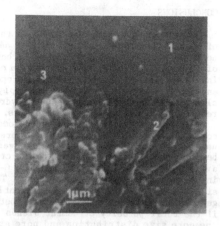

Figure 10
Filler-cement paste interface,
with 1 = filler, 2 = CH crystals
and 3 = grainy C-S-H [36].

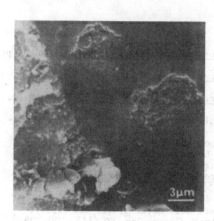

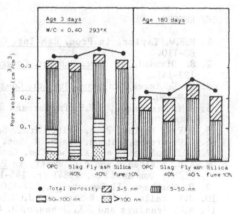

Figure 11
Filler-cement paste interface
in the presence of silica fume
[35].

Figure 11
Pore volume and pore size distribution of blended cements [3].

atively large, but discontinuous and thin-walled, pores. As this characteristic was the same for slag and fly ash cements, Feldman related the pore
structure to the CH content. Continuous capillary pores in portland cement
pastes should be due to deficient CH/C-S-H interfaces. In cement blends, the
amount of CH is lower; when it is less than 7%, the same pore size distribution is obtained for slag and fly ash cements.

CONCLUSIONS

Great progress has been made in the last few years on the microstructure of cement pastes that include mineral additives. The microstructural development of portland cement hydrates is better understood based on the formation of various C-S-H that precipitate on each side of hydrated shells as outer, inner and late products, and on the occurrence of AFt and AFm phases outside and inside shells. These results apply to blended cements in which portland clinker is the first component to hydrate. When mineral additives start to react, they provide their own hydrates, namely C-S-H and aluminates.

The microstructure of hardened cements includes hydrate morphology, C-S-H composition, solid interfaces and pore size distribution. TEM observations using a wet cell have shown that the various C-S-H of portland cements may all be formed of fine foils which roll, crumple and interlock, depending on the available space. In blended cements, C-S-H appears less crystalline and lower in calcium than in portland cements. The largest differences are observed in silica fume blends, where an apparently amorphous C-S-H constitutes a homogeneous paste with strong interfaces between solid particles and matrix. Even if the total porosity of cement blends is similar to that of portland cements, the pore size distribution and pore structure are different. When mineral additives react, there is a trend toward the finest pores, and discontinuous pores substitute for continuous capillary pores, in relation with the increase in CH.

These new data should be used for monitoring rheological properties, performance characteristics and durability of concretes in various aggressive environments.

REFERENCES

1. H.F.W. Taylor, in Proc. 8th Int. Congr. Chem. Cem., Rio, 1986, Vol. I, pp. 82-110.
2. S. Diamond, in Proc. 8th Int. Congr. Chem. Cem., Rio, 1986, Vol. I, pp. 122-147.
3. H. Uchikawa, in Proc. 8th Int. Congr. Chem. Cem., Rio, 1986, Vol. I, pp. 249-280.
4. A.M. Harrisson, H.F.W. Taylor and N.B. Winter, Cem. Concr. Res. 15, 775-780 (1985).
5. H.F.W. Taylor, Proc. Brit. Ceram. Soc. 35, 65-82 (1984).
6. G.W. Groves, Philos. Trans. R. Soc. London A 310, 79-83 (1983).
7. B.J. Dalgleish and K. Ibe, Cem. Concr. Res. 11, 729-739 (1981).
8. K.L. Scrivener and P.L. Pratt, in Proc. 6th. Int. Conf. Cem. Microsc., (ICMA, Duncanville, 1984) pp. 145-155.
9. B. Marchese, Il Cemento 80, 107-120 (1983).
10. J.E. Bailey and H.R Stewart, J. Mater. Sci. Lett. 3, 411-414 (1984).
11. H.M. Jennings and S.K. Johnson, J. Amer. Ceram. Soc. 69 (11), 790-795 (1986).
12. H.M. Jennings, B.J. Dalgleish, P.L Pratt, J. Amer. Ceram. Soc. 64, 567-572 (1981).
13. K.L. Scrivener, PhD thesis, Imperial College, University of London (1984).
14. D.N. Hadley, PhD thesis, Purdue University (1972).
15. S. Diamond, in Hydraulic Cement Pastes (Cement and Concrete Association, Wexham Springs, 1976) pp. 2-30.
16. H.F.W. Taylor, J. Amer. Ceram. Soc. 69 (6), 464-467 (1986).
17. E. Henderson and J.E Bailey, in Proc. 8th Int. Congr. Chem. Cem., Rio, 1986, Vol. III, pp. 376-381.
18. N.J. Clayden, C.M. Dobson, G.M. Groves and S.A Rodger, in Proc. 8th Int. Congr. Chem. Cem., Rio, 1986, Vol. III, pp. 51-56.
19. D.N. Winslow, Cem. Concr. Res. 15, 817-824 (1985).
20. H. Van Damme, (Personal Communication) (1986).
21. R.F. Feldman, in Proc. 8th Int. Congr. Chem. Cem., Rio, 1986, Vol. I, 336-356.

22. L.J. Parrott, Philos. Trans. R. Soc. London, A 310, 155-166 (1983).
23. H.M. Jennings, J. Grimes and P.L. Brown, U.S Patent Application (1986).
24. L.J. Parrott, R.G. Patel, D.C Killoh and H.M. Jennings, J. Amer. Ceram. Soc. 67, 233-237 (1984).
25. M. Regourd, in Proc. 8th Int. Congr. Chem. Cem., Rio, 1986, Vol. I, pp. 200-229.
26. H. Tanaka, Y. Totani and Y. Saito, in Proc. 1st Int. Conf. on the Use of Fly Ash, Silica Fume and Other Mineral By-Products in Concrete, edited by V.M. Malholtra, ACI SP 79 (American Concrete Institute, Detroit, 1983) Vol. 2, pp. 963-977.
27. J.P.H. Frearson and J.M. Uren, in Proc Int. Conf. Fly Ash, Silica Fume, Slag and Natural Pozzolanics in Concrete, edited by V.M. Malhotra, ACI SP 91-69 American Concrete Institute, Detroit, 1986) Vol. 2, pp. 1401-1421.
28. E. Demoulian, P. Gourdin, F. Hawthorn and C. Vernet, in Proc. 7th Int. Congr. Chem. Cem., Paris, 1980 Vol. II, pp. 89-94.
29. G. Mascolo and O. Marine, in Proc. 7th Int. Congr. Chem. Cem., Paris, 1980 Vol. II, pp. 23-29.
30. A.M. Harrisson, W.B. Winter and H.F.W. Taylor, in Proc. 8th Int. Congr. Chem. Cem., Rio, 1986, Vol. IV, pp. 170-175.
31. K.L. Scrivener, K.D. Baldie, Y. Halse, P.L. Pratt. in Very High Strength Cement-Based Materials, edited by J.F. Young, Mat. Res. Soc. Symp. Proc. Vol. 42 (Materials Research Society, Pittsburgh, 1985) pp. 39-43.
32. M. Regourd, in Condensed Silica Fume, edited by P.C Aitcin (Sherbrooke University, 1983) pp. 20-24.
33. M. Regourd, B. Montureux and H. Hornain, in Proc. 1st Int. Conf. on the Use of Fly Ash, Silica Fume and Other Mineral By-Products in Concrete, edited by V.M. Malholtra, ACI SP 79 (American Concrete Institute, Detroit, 1983) Vol. 2, pp. 847-865.
34. A. Bachiorrini, Thesis, University of Lyon (1985).
35. B. Montureux, H. Hornain and M. Regourd, Int. Coll. Liaison, Pate de ciment – Materiaux Associes (Toulouse) A 64 (1982).
36. P. Gegout, H. Hornain, B. Thuret, B. Montureux, J. Volant and M. Regourd, in Proc. 8th Int. Congr. Chem. Cem., Rio, 1986, Vol. IV, pp. 197-203.
37. D.M. Roy and K.M. Parker, in Proc. 1st Int. Conf. on the Use of Fly Ash, Silica Fume and Other Mineral By-Products in Concrete, edited by V.M. Malholtra, ACI SP 79 (American Concrete Institute, Detroit, 1983) Vol. I, pp. 397-411.

MICROSTRUCTURE AND MICROCHEMISTRY OF SLAG CEMENT PASTES

A.M. HARRISSON*, N.B. WINTER* and H.F.W. TAYLOR**
*Blue Circle Industries PLC, Research Division, London Road, Greenhithe, Kent,
DA9 9JQ, England
**Department of Chemistry, University of Aberdeen, Meston Walk, Old Aberdeen,
AB9 2UE, Scotland

Received 24 October, 1986; refereed

ABSTRACT

Pastes of a portland cement (60%) blended with a granulated blastfurnace
slag (40%) were examined, principally by SEM with EDX analysis. Reaction rims
around slag particles and relicts of fully reacted slag particles had compositions compatible with mixtures, in varying proportions, of C-S-H having
Si/Ca -0.62 and a phase of the hydrotalcite family having Al/Mg -0.38. Calculations taking into account relevant densities and water contents indicated
that replacement of the slag by its in situ hydration products entails little
or no change in the numbers of Mg and O atoms per unit volume, but that substantial proportions of the Ca, Si and Al are released and an equivalent
amount of H gained. In other respects, the microstructures qualitatively
resembled those of pure portland cement pastes of similar ages, but less CH
was formed and the C-S-H not formed in situ from the slag had a Si/Ca ratio of
0.56, higher than that of 0.50 to 0.53 found in the absence of slag. None of
the individual phases in the slag cement pastes showed significant compositional variation with time in the 28 day to 14 month period studied. The
relative amounts of Ca, Si and Al expelled from the slag are such that, in
order to form C-S-H and AFm phase, more Ca is required. It is obtained partly
at the expense of CH formation, and partly through increase in the Si/Ca ratio
of the C-S-H formed from the clinker phases. Mass balance, volume composition
and bound water content were calculated for the 14 month old paste and compared with corresponding results for the pure portland cement.

INTRODUCTION

Previous studies have shown that the principal hydration products of slag
cements are essentially the same as those of portland cements, though less
calcium hydroxide is formed even after allowing for the dilution of the cement
clinker by the slag [1-6]. The C-S-H surrounding the clinker grains has a
lower Ca/Si ratio than that formed in portland cement pastes [6-8]. Reaction
rims around the slag grains have been observed; they broadly resemble C-S-H in
composition, but are richer in Al_2O_3 and MgO [5,6,9]. Hydrotalcite
($Mg_6Al_2(OH)_{16}(CO_3)\cdot 4H_2O$), or possibly a carbonate-free phase of closely
similar structure, has sometimes been reported [10,11]. An attempt was made
to estimate the amounts and compositions of all the phases in some cement
pastes, and to test the results by making mass balance calculations for each
element and by calculating some derived quantities, e.g. bound water, that
could be compared with experimental data [8,12]. With slag cements, poor
agreements were reported for Mg balance and between calculated and observed
contents of non-evaporable water.

We have recently reported preliminary results of a study of some slag
cement pastes by scanning electron microscopy (SEM) with X-ray microanalysis
[13]. In the present paper, these results are extended and more fully interpreted. Calculations of mass balance and of derived quantities were carried out, with more success than in the previous work. Results of comparison
studies of pure portland cement pastes are also described.

Table I

Chemical Analyses (wt.%) and Other Data for the Starting Materials

	Na2O	MgO	Al2O3	SiO2	SO3	K2O	CaO	TiO2	Mn2O3	Fe2O3	LOI	Total
Cement	0.2	1.2	5.4	20.7	2.6	0.46	4.2	0.3	0.1	3.1	1.6	100.0[a]
Slag	0.8	8.1	10.4	33.6	0.02	0.44	1.9	0.6	0.6	1.0	2.2	99.8[b]

a. Also 0.2% P_2O_5, 0.5% free CaO (included in other components), 0.4% insoluble residue, 3130 kg/m^3 density, 330 m^2/kg specific surface area, and 11.6% >45 μm.
b. Also 1.2% S^{2-}, 0.01% P_2O_5, LOI determined in N_2, total corrected for O = S, 2880 kg/m^3 density, 477 m^2/kg specific surface area, 3.5% >45 μm.

MATERIALS, PREPARATION OF PASTES, X-RAY DIFFRACTION AND THERMOGRAVIMETRY

Table I gives analyses and other data for the portland cement and granulated slag used. We have reported elsewhere [14] on the quantitative phase composition of the clinker and the compositions of its individual phases; the alite (C_3S) content of the cement, from a modified Bogue calculation using the actual compositions of the clinker phases, was 62%. X-ray diffraction (XRD) of the slag showed it to consist almost entirely of glass.

A blended cement (60% portland cement, 40% slag) was prepared, and pastes of it, and of the pure portland cement, were made by hand mixing at W/S = 0.5 and curing at 25°C for 28 days (d), 6 months and 14 months. After the first 2 d, curing was in water in sealed polyethylene containers. Hydration was stopped at the two earlier times by immersion in methanol; in agreement with an observation by Day [15], this was later found to lead to formation of CO_2 and consequent partial carbonation of the C-S-H during subsequent thermogravimetric analysis (TG), and was abandoned in favor of evacuation using a rotary pump. XRD of the 14 month old portland cement paste showed C-S-H, CH, AFm and AFt phases, and a little unreacted belite. The AFm phases had layer thicknesses of 0.79 nm and 0.88 nm. The results for the 14 month old slag cement paste were similar, but the CH peaks were relatively weaker and no AFt phase was detected. CH contents were determined by TG, in CO_2-free N_2 at a heating rate of 10°C/min. Carbon dioxide contents, determined chemically by an acid decomposition method [16], were 0.7% to 1.5% on the ignited weights.

SCANNING ELECTRON MICROSCOPY AND X-RAY MICROANALYSIS

Scanning Electron Microscopy

Polished sections of all the pastes were made, coated with 30 nm of carbon, measured using an oscillating quartz crystal, and examined by SEM. Backscattered electron images, using a Robinson detector, and secondary electron images were obtained. The pure portland cement pastes showed the normal features of usually polymineralic clinker grains in which the clinker phases had in varying degrees been replaced by hydration products, together with regions of CH and of less well defined material presumed to have formed in space initially occupied either by water or by the smaller cement grains. This latter material will be described as "undifferentiated product." The slag cement pastes showed similar features; but at any given age, CH was notably less abundant, and angular particles of unreacted slag were prominent.

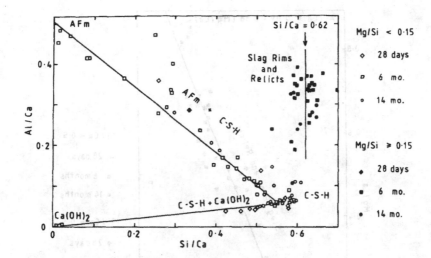

Figure 1.   Results of X-ray microanalyses for slag cement pastes: Al/Ca
ratios plotted vs. Si/Ca ratios.

In the 28 d old paste, little or no sign of reaction was detectable by electron microscopy, but the C-S-H composition differed from that found in the
absence of slag, showing that reaction had begun; this is described later. In
the older pastes, reaction rims around the larger slag particles were clearly
visible, up to 10 μm wide at 6 months and up to 15 μm wide at 14 months.
Fully reacted relicts of smaller slag particles were also seen.

X-ray Microanalysis

X-ray microanalysis was carried out in the SEM, which was fitted with an
energy dispersive X-ray system. Spots for analysis were chosen by examining
the secondary or backscattered electron images, and do not represent a random
selection of the entire material. An accelerating voltage of 15 kV was chosen
to provide the best compromise between spatial resolution and adequate exci-
tation of the FeK-alpha peak; the beam current, measured in a Faraday cup on
the edge of the specimen holder, was 0.7 nA. X-rays were collected for 100 s
at each point analyzed; the reference standards were generally specimens of
natural minerals; matrix corrections were made by the ZAF method.

In Figure 1, Al/Ca ratios are plotted against Si/Ca ratios for the slag
cement pastes. Analyses giving totals above 85% were excluded as arising
wholly or partly from anhydrous phases, but all others are included. Six
groups of spots may be defined as follows:

1. CH, with Si/Ca and Al/Ca both <0.05;
2. C-S-H, with Si/Ca >0.5 and Al/Ca <0.10;
3. AFm phases, with Si/Ca <0.05 and Al/Ca >0.4;
4. mixtures of C-S-H with AFm phases;
5. mixtures of C-S-H with CH;
6. slag rims and relicts, with Si/Ca >0.5 and Al/Ca >0.15.

The analyses of this last group, unlike almost all the others, had relatively
high Mg/Ca ratios (up to 0.86). The compositional limits for C-S-H (group 2)

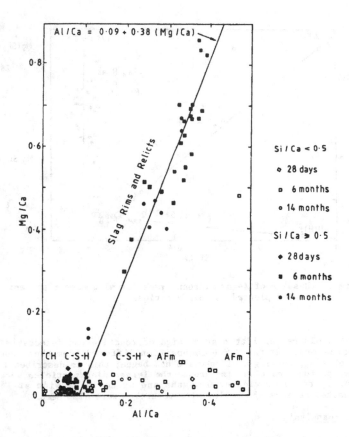

Figure 2.   Results of X-ray microanalyses of slag cement
pastes: Mg/Ca plotted vs. Al/Ca ratios.

given above are somewhat arbitrary. Most of the spots in this group were
located in alite or belite rims or relicts, and most of those in groups 3 and
4 were in the undifferentiated product. Analyses of the slag rims or relicts
could not be obtained at 28 days, but there were no significant changes in
compositions of individual phases with time in the period studied.

In Figure 2, Mg/Ca ratios are plotted against Al/Ca ratios for the same
set of analyses, and Table II includes mean results for the C-S-H and the slag
rims and relicts in each case for all three ages combined. The compositions
of the slag rims and relicts could not be correlated with any positional parameter, such as the distance from the boundary with unreacted slag, nor with
the analysis totals, and they could not be attributed to mixture of hydrated
material with unreacted slag.

X-ray microanalyses were also obtained for the pure portland cement
pastes. Plots of Al/Ca ratio against Si/Ca ratio were broadly similar to that
in Figure 1 for the slag cement pastes, apart from the absence of points for
the slag rims or relicts, but the C-S-H compositions differed and showed a
possibly significant tendency to change with time. The mean composition for

Table II

Results of SEM Microanalyses -- Mean Atomic Ratios and Anhydrous Oxide Sums

Sample	N		Na	Mg	Al	Si	S	K	Ca	Ti	Mn	Fe	Oxide Sum
C-S-H[a]	22	$\bar{x}$	0.01	0.03	0.06	0.56	0.02	0.00	1	0.00	0.00	0.01	74
		s	0.01	0.02	0.01	0.03	0.01	0.00		0.00	0.00	0.01	4
C-S-H[b]	23	$\bar{x}$	nd	0.03	0.04	0.53	0.02	0.01	1	0.01	0.00	0.01	78
		s	-	0.01	0.01	0.02	0.01	0.00		0.00	0.00	0.01	2
Slag Rims and Relicts[c]	30	$\bar{x}$	0.04	0.60	0.32	0.62	0.02	0.01	1	0.02	0.01	0.00	70
		s	0.03	0.14	0.05	0.03	0.01	0.01		0.01	0.01	0.01	4
C-S-H[d]	-		0.04	0.00	0.09	0.62	0.02	0.01	1	0.00	0.00	0.00	-
Hydrotalcite[e]	-		0.00	0.60	0.23	0.00	0.00	0.00	0	0.02	0.01	0.00	-

a. Clinker rims and relicts in slag cement pastes of all three ages.
b. Clinker rims and relicts in 14-month old paste of portland cement.
c. Slag cement pastes aged 6 and 14 months.
d.,e. Components of (c) according to interpretation given in the text.

the C-S-H in the 14-month old portland cement paste is included in Table II. The mean Si/Ca ratios for both the 28-day and 6-month old pastes were 0.50.

SLAG RIMS AND RELICTS

The compositions found in the individual spot analyses (Figs. 1,2) are approximately represented by the following: Si/Ca = 0.62, and Al/Ca = 0.09 + 0.38 (Mg/Ca). The simplest explanation of this observation is that the material is a mixture of two phases, which are mixed on a scale somewhat below 1 um, too small to be resolved. A few analyses were made at 6 kV, but resolution was still not achieved. One phase is C-S-H with Si/Ca ~0.62 and Al/Ca ~0.09; these ratios differ from those found for the C-S-H in the alite and belite rims or relicts. The other phase is a magnesium aluminum hydroxide with Al/Mg ~0.38. Based on some assumptions about the distributions of the minor elements, compositions of the two phases are included in Table II. The alkalies are included arbitrarily with the C-S-H, but were probably present at least in part in the pore solution and deposited in or on the solid phases on drying.

In the saturated state, the C-S-H has a H_2O/Ca ratio of 2.3 and a density of ~1950 kg/m^3 [17]. From its composition, the magnesium aluminum hydroxide is likely to be a member of the hydrotalcite family. Its constitution was assumed, on the basis of the known crystal structures of these phases, to be $[Mg_{0.70}Al_{0.27}Ti_{0.02}Mn_{0.01}(OH)_2](OH)_{0.3}(H_2O)_{0.70}$, where square brackets enclose the contents of the brucite type principal layer. The composition is near that of hydrotalcite, $[Mg_{0.75}Al_{0.25}(OH)_2](CO_3)_{0.125}(H_2O)_{0.5}$, and even closer to that of its hydroxyl analog, meixnerite [18], in which the 0.125 CO_3 in the above formula is replaced by 0.25 OH. The X-ray density of meixnerite is 1950 kg/m^3.

Table III

Atoms per $1000/N_0$ cm^3 in Slag and Its In Situ Hydration Product

	Na	Mg	Al	Si	S	K	Ca	Ti	Mn	Fe	O	H
Slag	0.4	5.8	5.9	16.1	1.1	0.3	21.5	0.2	0.2	0.4	68.9	0.0
Product	0.4	5.6	3.0	5.8	0.2	0.1	9.4	0.2	0.1	0.0	68.1	72.8

The weight percentage of Mg in the rim or relict material, in the partly dehydrated condition in which the latter exists under the conditions of the analysis, is often higher than that of the unreacted slag. Mg X-ray images showed a generally higher content; they did not resolve it into distinct components. Regourd et al. [5] found slag rim material to have Ca/Si ratios of 0.9 to 1.3, Ca/Al ratios of 9.2 to 9.6, and Ca/Mg ratios of 1.4 to 3.2. The Ca/Mg ratios are similar to those found in the present work, but the Ca/Si ratios are lower, and the Ca/Al ratios much less variable. We were unable to confirm the observation of Tanaka et al. [9] that several compositionally distinct layers occur within the rims.

From the proportions, compositions and densities of the two phases, and the composition and density of the unreacted slag, one may calculate the numbers of atoms of each element in any given volume of the latter and of the in situ product that constitutes the rim and relict material. It was assumed that no pores existed in the latter, other than the micropores implied by the density and water content assumed for the C-S-H component.

Table III gives the results of these calculations. It will be seen that, when the slag is replaced by its in situ product, there is little or no significant change in the numbers of Mg or O atoms in a given volume, but that substantial proportions of the Ca, Si and Al atoms must be expelled and an equivalent amount of H taken up. A similar situation exists for the in situ hydration of alite and belite [19]. The results are subject to a possible small error associated with uncertainties regarding the oxidation state of the S and any changes in this quantity that occur on hydration; the oxidation state was taken to be −2 in both the slag and the product.

INTERACTIONS BETWEEN SLAG AND CLINKER

The Ca, Al and Si released from the slag contribute to the formation of C-S-H and AFm phase in the initially water filled space, but are relatively deficient in Ca (Fig. 3). The additional Ca that is required is provided from two sources. Firstly, the CH content is reduced. Table IV gives the CH contents, determined by TG, for the slag cement and comparison portland cement pastes. Lowering of CH content beyond that attributable to dilution of the portland cement by the slag is barely detectable at 28 days, but is considerable at 6 or 14 months. Secondly, the Si/Ca ratio of the C-S-H formed from the clinker phases is increased from 0.50 to 0.53 in the absence of slag, to 0.56 if the latter is present.

MASS BALANCE, BOUND WATER CONTENT AND VOLUME COMPOSITION

Taylor [20] has developed a procedure for estimating the amounts and compositions of all the phases in a portland cement paste, testing the results for mass balance for each element, and calculating the bound water content, volume composition and other quantities. This procedure was applied to the 14-month old portland cement paste described in the present paper.

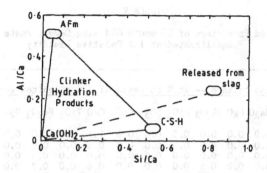

Figure 3. Mean atom ratios of the principal elements released from the slag, compared with those of C-S-H, AFm phase and calcium hydroxide.

Table IV

Calcium Hydroxide Contents

Sample	Basis	28 days	6 months	14 months
Portland Cement	% ignited weight	19	23	23
Slag Cement	% ignited weight	11	10	9
Slag Cement	% ignited weight of clinker	18	17	16

Making appropriate modifications to allow for the slag and its in situ hydration products (see Appendix 1), it was also applied to the 14-month old paste of the slag cement. Based on earlier results for the same slag [8], it was assumed that 22 g of slag per 100 g ignited weight of blended cement had reacted. Other data were obtained by methods described elsewhere [20].

Samples were equilibrated over saturated aqueous LiCl at room temperature (11% RH), heated to constant weight in a vacuum furnace at 105°C and p(H₂O) = 6 torr, and then heated to constant weight in a stream of CO_2-free N_2 at 1100°C. Results are given in Table V. The observed content of volatiles retained at 11% RH after correction for CO_2 (1.2%) was 24.5%, which agrees with the predicted bound water content of 24.5%. The observed content of volatiles retained at 105°C, similarly corrected, was 17.6%, which agrees with the predicted content of non-evaporable water (17.3%). All these quantities are referred to the ignited weight. The mass balance is good for all the elements; as with the portland cement, the iron from the ferrite phase appears to enter an iron hydroxide or calcium ferrite gel. The poor agreements in the previous work [8] probably arose from neglect of the hydrotalcite type phase. The amount of the latter is too small to be detected by XRD, in view of its probably low degree of crystallinity; in any case, it would be difficult to detect this phase in the presence of a larger amount of AFm phase of similar layer thickness.

TABLE V

Calculated Parameters of 14-month Old Slag Cement Paste (40% slag)
Equilibrated at 11% Relative Humidity

Mass balance -- wt.% on basis of 100 g ignited weight

	Na_2O	MgO	Al_2O_3	SiO_2	SO_3	K_2O	CaO	TiO_2	Fe_2O_3	H_2O	CO_2	Total
Alite	0.0	0.0	0.0	0.5	0.0	0.0	1.4	0.0	0.0	0.0	0.0	1.9
Belite	0.0	0.0	0.0	0.3	0.0	0.0	0.6	0.0	0.0	0.0	0.0	1.0
Aluminate	0.0	0.0	0.0	0.0	0.0	0.0	0.0	0.0	0.0	0.0	0.0	0.0
Ferrite	0.0	0.0	0.3	0.0	0.0	0.0	0.6	0.0	0.2	0.0	0.0	1.2
Slag	0.1	1.5	2.0	6.2	0.0	0.1	7.7	0.1	0.2	0.0	0.0	18.0[a]
CH	0.0	0.0	0.0	0.0	0.0	0.0	7.1	0.0	0.0	2.3	0.0	9.4
CaCO3	0.0	0.0	0.0	0.0	0.0	0.0	1.5	0.0	0.0	0.0	1.2	2.7
C-S-H(c)	0.1	0.6	1.5	16.2	0.8	0.2	26.9	0.0	0.4	12.1	0.0	58.8
AFm	0.0	0.1	2.7	0.3	0.7	0.1	5.9	0.1	0.2	5.9	0.0	16.0
AFt	0.0	0.0	0.0	0.0	0.0	0.0	0.0	0.0	0.0	0.1	0.0	0.1
C-S-H(s)	0.1	0.0	0.3	2.6	0.1	0.0	3.9	0.0	0.0	1.8	0.0	8.9
Hydrotalcite	0.0	1.7	0.8	0.0	0.0	0.0	0.0	0.0	0.1	2.3	0.0	5.0[b]
Other	-0.1	0.1	0.0	0.0	0.0	0.0	0.4	0.2	1.3	0.1	0.0	2.8[c]
Total	0.3	4.0	7.6	26.1	1.6	0.4	56.2	0.4	2.3	24.5	1.2	125.8[d]

Volume Percentages

Alite	0.7	CH	5.2	C-S-H(s)	4.7
Belite	0.4	CaCO3	1.2	Hydrotalcite	3.1
Aluminate	0.0	C-S-H(c)	31.4	Other	1.4
Ferrite	0.4	AFm	9.8	Pores	33.8
Slag	7.7	AFt	0.1		

	% by volume		g/100 g ignited wt.	
Capillary Porosity	21.7	Gel Water (sat'd paste)	16.2
"Total" Porosity	44.4	Non-Evaporable Water	17.3

Notes:
C-S-H(s) = component of in situ product from slag.
C-S-H(c) = all other C-S-H.
Other components included in totals: [a]0.3%, [b]0.1%, [c]0.7%, [d]1.2%.
Discrepancies in row and column totals arise from rounding.

DISCUSSION AND CONCLUSIONS

The results show that a satisfactory account can be given of the broad
microstructures of slag cement pastes, and of the compositions of the constituent phases. It is also possible to estimate the quantitative phase composition and to predict values of such independently determinable quantities as
water contents and porosities for various drying conditions, though the results in this respect are tentative due to the lack of a sufficiently reliable
method for determining unreacted slag.

It is of interest to compare the calculated volumetric phase composition (Table V) with that of the comparable portland cement paste [20]. From

the standpoint of physical properties, C-S-H, AFm phase, and the hydrotalcite-like phase are probably to a first approximation identical, as all have layer structures, and to a considerable extent appear to be intergrown on a sub-micron scale. The total percentage by volume of these phases is 51% for the portland cement paste and 49% for the slag cement paste. The portland cement paste contained 13% of CH by volume; it may be fair to compare this with the total of 13% for CH and unreacted slag for the slag cement paste, because both phases occur as relatively massive and nonporous particles. These similar-ities may account for the fact that the physical properties of the mature pastes, for a given W/S ratio, are not grossly different.

The porosities calculated in the present work do not depend on the degree of continuity of the pore system, and in this respect they differ from the results obtained by some, if not all of the experimental methods. The calcu-lated 11% RH porosity is higher for the slag cement (34%) than for the port-land cement (29%). This supports the view that the generally lower permeabili-ties of mature slag cement pastes are due, not to lower porosity, but to a lower degree of continuity in the pore system, which in turn may be due to the lower content of calcium hydroxide.

The extent to which the CH content at a given age, referred to the ig-nited weight of clinker, is lowered by addition of slag appears to vary con-siderably, depending on both the slag and the cement clinker used. The re-ductions in the present case are relatively small, considerably larger ones having been observed with other combinations of clinker and slag [21]. This applies especially to the results at 28 days.

REFERENCES

1. C. Cesareni and G. Frigione, in Proc. 5th Int. Symp. Chem. Cem. , Tokyo, 1968, 4, 237-247 (1969).
2. R. Kondo and S. Ohsawa, in Proc. 5th Int. Symp. Chem. Cem., Tokyo, 1968, 1, 255-269 (1969).
3. M. Daimon, in Proc. 7th Int. Congr. Chem. Cem , Paris, 1980, 1, III-2/1 to III-2/9 (1980).
4. M. Regourd, in Proc. 7th Int. Congr. Chem. Cem , Paris, 1980, 1, III-2/10 to III-2/26 (1980).
5. M. Regourd, B. Mortureux, E. Gautier, H. Hornain and J. Volant, in Proc. 7th Int. Congr. Chem. Cem , Paris, 1980, 2, III-105 to III-111 (1980).
6. H. Uchikawa, in Proc. 8th Int. Congr. Chem. Cem., Rio, 1986, 1, 249-280 (1986).
7. M. Regourd, B. Mortureux and H. Hornain, in Proc. 1st Int. Conf. on the Use of Fly Ash, Silica Fume and Other Mineral By-Products in Concrete, Montebello, 1983 (ACI SP 79, 1983) 2, pp. 847-865.
8. H.F.W. Taylor, K. Mohan and G.K. Moir, J. Amer. Ceram. Soc. 68, 685 (1985).
9. H. Tanaka, Y. Totani and Y. Saito, in Proc. 1st Int. Conf. on the Use of Fly Ash, Silica Fume and Other Mineral By-Products in Concrete, Montebello, 1983 (ACI SP 79, 1983) 2, pp. 963-977.
10. K. Kuhle and U. Ludwig, Sprechsaal Keram., Glas, Email, Silik. 105, 421 (1972).
11. G. Mascolo, A. Nastro and V. Sabatelli, Cemento 74, 45 (1977).
12. H.F.W. Taylor, K. Mohan and G.K. Moir, J. Amer. Ceram. Soc. 68, 680 (1985).
13. A. Harrisson, N.B. Winter and H.F.W. Taylor, in Proc. 8th Int. Congr. Chem. Cem., Rio, 1986, 4, 170-175 (1986).
14. A.M. Harrisson, H.F.W. Taylor and N.B. Winter, Cem. Concr. Res. 15, 775-780 (1985).
15. R.L. Day, Cem. Concr. Res. 11, 341 (1981).
16. F.E. Jones, J. Soc. Chem. Ind. 51, 29 (1940).
17. H.F.W. Taylor, J. Amer. Ceram. Soc. 69, 464 (1986).

18. S. Koritnig and P. Susse, Tschermaks Mineral. Petrogr. Mitt. <u>22</u>, 79 (1975).
19. H.F.W. Taylor, Materials Science Monographs, 28A (Reactivity of Solids, Part A), 39 (1985).
20. H.F.W. Taylor, in <u>Microstructural Development During Hydration of Cement</u>, edited by L.J. Struble and P.W. Brown, Mat. Res. Soc. Symp. Proc. Vol. <u>87</u> (Materials Research Society, Pittsburgh, 1987).
21. G.K. Moir, in <u>Proc. of the 1985 Beijing International Symposium on Cement and Concrete</u>, (China Building Industry Press, 1986) <u>1</u>, pp. 366-379.

APPENDIX: Calculation of Mass Balance Table and Other Quantities

The procedure described in Appendix 1 of Ref. 20 for a portland cement paste was modified as follows. The compositions and amounts of the hydrotalcite type phase and the C-S-H formed in situ from the slag were calculated from the data in Table II, which are based on the results of the SEM microanalyses; it was assumed that the hydrotalcite type phase contained all the Mg from the slag that had reacted. The amount of each element thus accounted for, together with that in unreacted slag, was subtracted from the bulk analysis in Step 2 of the calculation. For the saturated and 11% RH conditions, the hydrotalcite type phase was assumed to contain 0.89 g of H_2O per g of (MgO + Al_2O_3 + TiO_2 + Mn_2O_3 + Fe_2O_3) assigned to it, and to have a density of 1950 kg/m^3. For the 105°C condition, the factor 0.89 was replaced by 0.45, and the density was assumed to be 2200 kg/m^3.

PORE STRUCTURE DEVELOPMENT IN PORTLAND CEMENT/FLY ASH BLENDS

DAVID J. COOK*, HUU T. CAO* and EVERETT P. COAN**
*National Building Technology Centre, Chatswood, NSW 2067, Australia.
**University of New South Wales, Kensington, NSW 2033, Australia.

Received 24 October, 1986; refereed

ABSTRACT

Pore structure development in portland cement/fly ash blends was investigated using mercury porosimetry and methanol exchange techniques. The progress of hydration was monitored using compressive strength tests. The specimens were made using four water-cement ratios and were hydrated over a one-year period in lime-saturated water. Mercury porosimetry results indicated that the blended cement pastes generally had higher total porosity than plain cement pastes. The major contribution to this increase in porosity was in the form of smaller pore sizes. With reactive fly ash at 20% replacement, the pore structure of mature paste consists mainly of pores nominally smaller than 0.05 µm in diameter. Diffusion parameters obtained from the methanol exchange results were found to be inversely related to the volume of large pores (nominally >0.05 µm) and also to the volume of small pores (nominally <0.05 um). The effects of the physical and chemical properties of cements and fly ashes on pore structure development are discussed.

INTRODUCTION

The durability of concrete is essentially controlled by the rate of penetration of aggressive gases and liquids. If its permeability is low, the concrete should have a good service life. The pore structure of the cement paste has a great influence on the permeability of concrete [1-3]. The objective of the work reported in this paper was to investigate pore structure development of hardened blended cement pastes considering water-cement ratios (W/S), and duration of hydration, and to relate diffusion properties to the pore structure.

EXPERIMENTAL PROCEDURE

Two ordinary portland cements (C#1 and C#2) and two fly ashes (F#1 and F#2) were used for this investigation. Their compositions and particle size analyses are given in Table I and Figure 1. Blended cement pastes containing 20% and 40% fly ash by weight were cast at W/S ratios of 0.42, 0.56, 0.69 and 0.77 by weight. The pastes were molded in plastic vials, vibrated, sealed and then rotated during setting to prevent bleeding and segregation. After 24 hours, the specimens were demolded, cut into two 4 g discs and stored in lime-saturated water at 23°C until ready for porosity and diffusion testing.

For measurement of pore size distribution, a mercury intrusion apparatus capable of exerting pressures up to 355 MPa was used. Hydration of the samples was stopped by immersion in methanol for one day (d) and drying in vacuum for one day. The contact angle was assumed to be 130.

A methanol exchange technique, reported in detail elsewhere [4], was employed to obtain the diffusion parameters of the blended cement pastes.

Mortar cube strength tests were performed in accordance with ASTM C 109-84.

Table I
Chemical and Phase Compositions of Raw Materials (wt.%)

		Cements		Fly ashes	
		C#1	C#2	F#1	F#2
CaO		64.2	64.5	5.2	1.3
SiO_2		19.9	21.7	52.6	61.2
Al_2O_3		5.7	4.7	27.2	25.6
Fe_2O_3		3.4	3.2	4.1	3.3
MgO		1.0	0.9	2.4	0.9
K_2O		0.62	0.29	1.20	1.9
SO_3		2.6	2.7	0.16	0.1
TiO_2		0.3	0.3	1.4	1.0
L.O.I.		1.1	1.1	0.9	3.8
S.S.A.	(m^2/kg)	333	348	605	321
C_3S		60	53		
C_2S		12	22		
C_3A		10	7		
C_4AF		10	10		

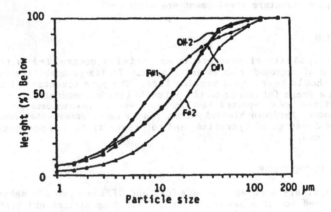

Figure 1. Particle size distibutions of raw materials.

RESULTS

Strength Development

The strength development of the blended cement mortars is shown in Figure 2. It is evident that fly ash F#1 showed better pozzolanic reactivity than fly ash F#2. At 20% replacement, blends using F#1 continued to gain strength after 50 d of lime-saturated water curing. The strength curves of blends using F#2 flattened off after the same period.

It should be noted that the percentage of strength gained in the blended cement samples with respect to control samples was higher when fly ash was used with cement C#1 than with C#2.

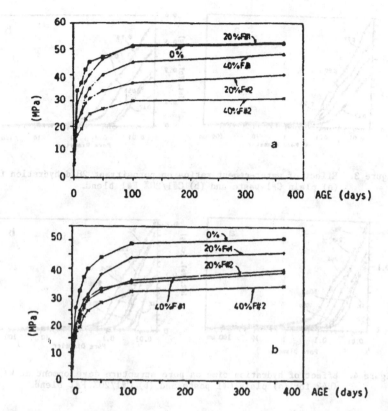

Figure 2. Mortar cube strength for (a) cement C#1 blends and (b) C#2 blends.

Pore Size Distribution Development

The pore structure information from the porosimetry tests was obtained for pores down to a nominal 3.6 nm radius. The results are illustrated in Figures 3 through 9.

From these figures, it can be seen that:

1. As the W/S ratio increased, the total measured porosity increased. The major contribution to the increase was the proportion of larger pores (Fig. 3), as observed by Mehta and Manmohan [5].

2. The total measured porosity reduced with time. The volumes of larger pore sizes reduced significantly in the first 28 d of hydration (Figs. 4 and 5).

3. The addition of fly ash increased the total measured porosity of the pastes (Figs. 6 and 7).

4. An increase in fly ash replacement percentage did not significantly change the pore size distribution at early ages (Fig. 6).

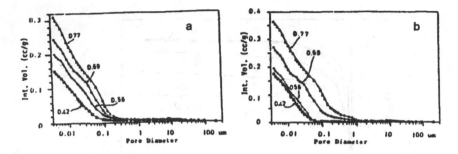

Figure 3.   Effect of water-cement ratios on porosity at 28 d hydration for
(a) plain C#1 paste and (b) C#1/20% F#1 blend.

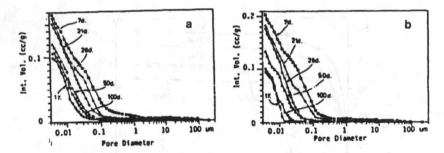

Figure 4.   Effect of hydration time on pore structure development at W/S =
0.42 for (a) plain C#1 paste and (b) C#1/20% F#1 blend.

5. The specific fly ash used clearly affected the pore distribution development in the pastes. Fly ash F#1 provided a higher rate of pore reduction
(particularly a larger pore size reduction) than F#2 (Fig. 8).

6. Cement C#2 paste at early ages had a higher larger pore size volume (and
total measured porosity) than C#1 (Fig. 9).

7. Cement C#2 blends had greater larger pore volumes than plain paste (Fig.
7). In C#1 blends, the larger pore volumes were approximately the same
as that of plain paste (Fig. 6).

Diffusion Properties

The diffusion parameters were measured by the methanol exchange method.
A typical exchange curve is illustrated in Fig. 10. It was observed that the
shape of the curves was not uniform and there were changes of slope after the
half-saturation point. The exchange process has been attributed to the counter diffusion of methanol and it has been found to be very sensitive to the
pore structure [4,6]. In order to characterize the observed changes in slope
of the exchange curve, the parameter T' was introduced and is defined as the
duration from the half-exchange point to saturation. The results of the
methanol exchange experiments are tabulated in Table II.

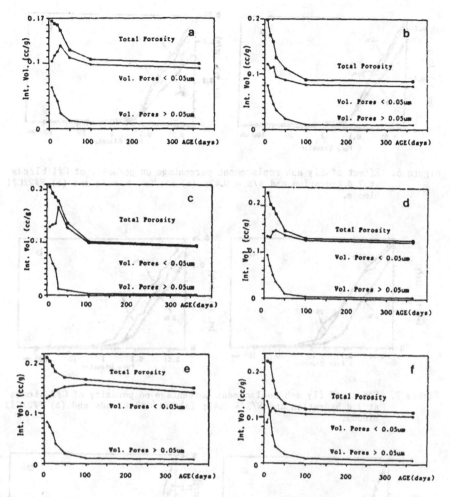

Figure 5. Pore structure development at water-solids ratio 0.42 for (a) plain
C#1 paste, (b) plain C#2 paste, (c) C#1/20% F#1 blend, (d) C#1/20%
F#2 blend, (e) C#1/40% F#1 blend, and (f) C#2/20% F#1 blend.

From these results, the following observations can be made:

1. $T_{1/2}$ increased with age of hydration.

2. $T_{1/2}$ decreased with higher W/S and fly ash replacement ratios.

3. Fly ash F#1 blends showed higher values of $T_{1/2}$ than blends using F#2.

4. Cement C#1 blends showed higher $T_{1/2}$ values than C#2 blends for the same
type of fly ash.

5. At early ages and 20% fly ash replacement, there was no apparent effect
on T'.

214

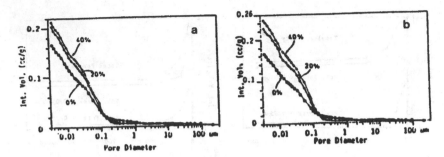

Figure 6. Effect of fly ash replacement percentage on porosity of C#1 blends at 7 d hydration and W/S = 0.42; (a) F#1/C#1 blends and (b) F#2/C#1 blends.

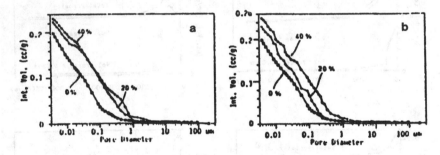

Figure 7. Effect of fly ash replacement percentage on porosity of C#2 blends at 7 d hydration and W/S = 0.42; (a) F#1/C#2 blends and (b) F#2/C#2 blends.

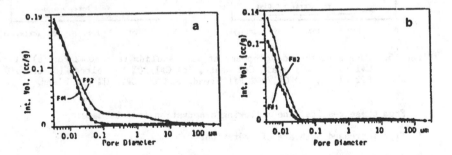

Figure 8. Effect of fly ash type on pore distribution development in cement C#1/20% fly ash replacement blends, W/S = 0.42 at (a) 28 d hydration and (b) 100 d hydration.

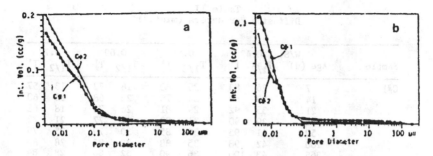

Figure 9. Pore size distribution in plain cement paste with W/S = 0.42 at (a) 7 d hydration and (b) 100 d hydration.

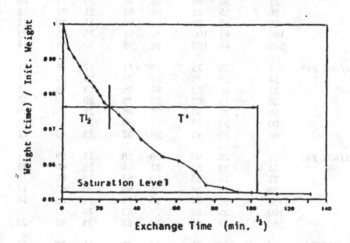

Figure 10. Methanol exchange curve of C#1/F#1, 20% replacement, with W/S = 0.42 at 7 d hydration.

Relationship Between Porosity and Diffusion Parameters

The capillary pores in hydrated cement pastes are often conventionally subdivided into large capillaries (10 – 0.05 μm) and small capillaries (<0.05 μm). The water in the large pores behaves as bulk water while there are moderate to strong surface tension forces generated in the small capillaries [7]. The counter diffusion process, in which pore water is exchanged by methanol, is influenced by these surface tension forces. As observed from experimental data, the weight changes of hardened blended cement pastes due to methanol exchange went through a number of slope variations, particularly during the later stages of the experiment. It is assumed that this effect is due to the changes in pore water behavior caused by the surface tension forces.

In the course of this investigation, it was assumed that $T_{1/2}$ is related to the volume of large capillary pores and T' to the volume of small capillary pores.

Table II
Diffusion Parameters ($min^{1/2}$)

Sample	Age (d)	W/S: 0.42		0.56		0.69		0.77	
		$T_{1/2}$	T'	$T_{1/2}$	T'	$T_{1/2}$	T'	$T_{1/2}$	T'
C#1	7	26	83	20	80	18	72	13	57
	14	30	80	24	76	22	68	15	60
	21	35	92	29	81	26	74	18	62
	28	39	90	31	83	28	77	21	72
	50	41	95	33	85	30	79	25	75
	100	42	98	35	88	32	85	28	80
	365	43	100	36	90	32	86	28	84
C#1/20% F#1	7	24	80	19	82	17	72	12	61
	14	26	84	23	84	19	72	14	60
	21	30	86	25	85	20	72	16	60
	28	36	82	26	87	21	76	19	61
	50	38	92	27	89	24	80	22	74
	100	43	100	30	91	26	82	23	80
	365	44	102	36	95	30	86	24	85
C#1/20% F#2	7	24	75	19	75	17	66	11	57
	14	27	75	22	75	18	66	14	55
	21	28	82	24	75	19	66	15	60
	28	28	82	25	75	20	70	18	62
	50	30	85	26	88	21	74	19	67
	100	33	93	27	88	22	75	21	75
	365	35	95	28	92	23	83	22	78
C#1/40% F#1	7	20	60	18	58	15	59	11	57
	14	23	62	21	64	18	60	13	60
	21	26	63	23	67	19	60	15	64
	28	28	65	24	69	20	65	17	66
	50	31	74	25	70	21	65	18	67
	100	36	82	27	83	24	72	20	73
	365	37	97	29	85	25	82	22	75
C#1/40% F#2	7	19	61	16	59	14	50	11	49
	14	23	63	19	61	16	50	12	51
	21	25	64	19	63	17	53	13	52
	28	27	66	20	64	18	55	15	55
	50	29	67	22	66	20	63	16	60
	100	30	70	25	71	23	65	17	60
	365	32	73	26	72	24	72	17	65

Table II (continued)

Sample	Age (d)	W/S: 0.42 $T_{1/2}$	T'	0.56 $T_{1/2}$	T'	0.69 $T_{1/2}$	T'	0.77 $T_{1/2}$	T'
C#2	7	24	80	17	70	15	70	14	65
	14	29	78	20	70	19	75	15	67
	21	33	90	23	80	21	80	18	72
	28	37	93	26	83	24	82	22	80
	50	40	95	28	85	25	84	23	82
	100	41	97	31	87	27	85	24	83
	365	43	100	32	90	29	86	25	84
C#2/20% F#1	7	20	81	14	69	12	70	10	67
	14	23	78	18	68	16	70	12	65
	21	25	83	23	69	20	72	15	65
	28	27	85	24	73	21	72	16	73
	50	30	86	27	79	22	77	17	73
	100	33	89	28	82	26	80	20	76
	365	39	91	30	85	29	80	22	79
C#2/20% F#2	7	18	65	15	60	11	54	10	52
	14	20	65	18	62	13	57	11	54
	21	23	70	21	69	17	57	13	60
	28	25	75	23	69	19	69	15	65
	50	29	80	26	76	21	71	16	70
	100	32	85	28	77	24	75	17	73
	365	36	88	30	80	27	80	19	76
C#2/40% F#1	7	14	58	11	55	10	52	9	50
	14	17	58	14	56	13	57	10	53
	21	22	60	17	60	16	59	12	53
	28	25	67	22	61	19	63	14	56
	50	28	75	23	69	21	65	15	60
	100	29	78	25	73	23	69	19	61
	365	34	86	28	77	26	72	20	69
C#2/40% F#2	7	14	48	11	40	10	40	9	35
	14	17	49	15	45	14	41	10	40
	21	21	49	17	49	15	45	12	45
	28	23	57	20	55	18	47	13	47
	50	26	64	22	63	21	59	15	55
	100	27	67	23	66	22	62	18	60
	365	34	73	24	72	24	68	19	64

The relationship between pore volumes and diffusion parameters is shown in Figure 11. The volume of large pores (>0.05 μm) is plotted against $T_{1/2}$ (Fig. 11a) and the volume of small pores (<0.05 μm) is plotted against T' (Fig. 11b). It can be seen that there is a reasonable inverse linear relationship between $T_{1/2}$ and the large pore volumes of plain cement pastes, which seems to be independent of cement type. For blended cement pastes, the relationship is quite linear in the first 28 d and appears to depend on the mix proportions. After 28 d, there was no apparent relationship between large pore volumes and $T_{1/2}$.

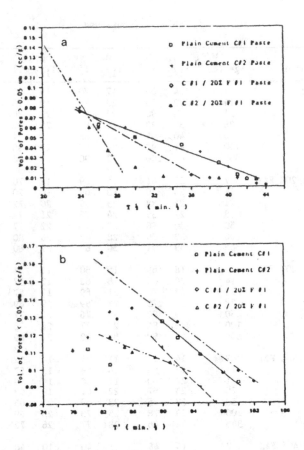

Figure 11. Pore volume versus diffusion parameters for (a) large pore volume (>0.05 μm) vs. $T_{1/2}$ and (b) small pore volume (<0.05 μm) vs. T'.

Figure 11b indicates that if the results from the first 14 d are ignored, there is an inverse linear relationship between the volumes of small pores and T'. It is evident that these relationships depend on materials used and mix proportions.

DISCUSSION

The pore structure development in cement pastes tends to follow a pattern of significant reduction in the volume of large pores during the initial stage of hydration. This reduction in large pores has been attributed to the preference of hydration products to fill the space of least resistance [6] and also to the increase in the volume of small pores measured by mercury intrusion (Fig. 5). The apparent increase of the small pore volumes is most likely caused by the formation of hydration products around the large pore necks.

This in turn causes higher intrusion volumes to be assigned to smaller pore diameters.

In blended cement pastes, the total measured porosity was generally larger than that of equivalent plain cement pastes. This increased porosity consisted of slightly greater large pore volumes and a significant increase in the volume of small pores. In the case of C#2 blends, there was a substantial increase in the volume of large pores at early ages.

The results indicated that there was a unique interaction between fly ash and ordinary portland cement, i.e. fly ash reacted differently with different cements. This phenomenon was reflected in the strength and porosity development patterns. For both types of fly ash, a higher rate of strength gain and a higher rate of pore size diminution were obtained when blended with C#1 (which has higher C_3S and C_3A contents than C#2). The uniqueness of the interaction and the importance of choosing a cement with a high early rate of CH formation in order to exploit the fly ash reactivity have been noted in the literature [8,9].

Fly ash F#1, with a finer particle size distribution, was expected to possess a higher pozzolanic reactivity than F#2. This effect was confirmed by the strength and also pore structure development results. In fact, the strength obtained for the 40% replacement mixes using F#1 was higher than that of the 20% replacement F#2 mixes.

At 20% replacement, F#1 blends were characterized by a high rate of large pore reduction in the first 28 d (Fig. 5c). This pore reduction was much slower in F#2 blends. The long term pore structure of the reactive F#1/C#1 blends consisted mainly of pores nominally less than 0.05 μm in diameter. The results also indicated that the volume of small pores in plain cement pastes remained almost constant after approximately 50 d while that in blended cement pastes continued to decrease with age.

At 40% replacement, the pore refinement process seemed to involve only the large pore volumes (Fig. 5e). There was no apparent reduction in the volume of small pores. A substantial porosity was retained in these pastes even after one year of curing. It should be noted that even at 40% replacement, the volume of large pores in blended cement paste was approximately the same as that of plain cement paste after 100 d.

The methanol exchange experiments have been found to be very sensitive to the pore structure of the hardened pastes [4,6,10,11]. The sensitivity of the diffusion parameters $T_{1/2}$ and T' with respect to changes in pore volume found in this work is assumed to be due to the changes in continuity of pores in the hardened pastes. If the pore structure is continuous (i.e. high degree of interconnection between pores), a small variation in the pore volumes (narrowing of pore necks) would be reflected by a significant change in methanol exchange time. However, if the pore structure is partly discontinuous, the change in methanol diffusion due to pore volume change would be small. In the case of a completely discontinuous pore structure, the diffusion parameters are expected to be essentially independent of the pore volume.

Figure 11a shows that after approximately 28 d, there is no apparent relationship between $T_{1/2}$ and the volume of large pores in blended cement pastes. In plain cement pastes, an inverse linear relationship between $T_{1/2}$ and the large pore volume persisted up to 1 year. This implies that after approximately 28 d of hydration, the large pores in blended cement pastes were isolated pores, while a high degree of continuity of large pores remained in plain cement pastes. Figure 11b indicates that the small pore structure in blended cement pastes was less continuous than that of plain cement pastes (slopes were steeper in plain cement pastes).

The continuous nature of pores in plain cement pastes has been associated with the high CH content [2]. In fly ash blends, the CH content has been found to reach a maximum value after 7-14 d and decrease thereafter as a consequence of the pozzolanic reaction [8]. The response of $T_{1/2}$ to the volume of large pores appears to follow the same pattern.

In general, the permeability of cement paste is affected by the volume of large pores [5] and the continuity of pore structure [2]. The permeability of mature blended pastes is expected to be low due to the discontinuous nature of the pores and limited large pore size volume. It should be noted that the beneficial effect of fly ash replacement could only be realized after approximately 28 d of curing. This is due to the high volume of large continuous pores in the blended cement pastes at early ages.

The "optimum" replacement percentage of fly ash F#1, taking strength and pore structure into account, appeared to be between 20% and 40% when used with cement C#1, and less than 10% when C#2 was used.

CONCLUSIONS

1. The total porosity of fly ash blended pastes measured by mercury porosimetry is higher than that of portland cement pastes, due to a greater volume of small pores (<0.05 μm).

2. The pore structure of blended cement pastes depends on the specific cement used. Reactive fly ash can provide a mature pore structure which consists of pores <0.05 μm in nominal diameter.

3. At 40% replacement, the pore refinement process by pozzolanic reaction is limited.

4. The pore structure of fly ash blends is relatively discontinuous after approximately 28 d of curing. The continuous nature of pores in ordinary portland cement pastes persists with age.

ACKNOWLEDGEMENT

The finacial assistance of the Cement and Concrete Association of Australia in support of this project is gratefully acknowledged.

REFERENCES

1. P.K. Mehta, in Performance of Concrete in Marine Environment, edited by V.M. Malhotra (ACI SP 65, Detroit 1980) pp. 1-20.
2. R.F. Feldman, in Proc. 1st Int. Conf. on the Use of Fly Ash, Silica Fume and Other Mineral By-Products in Concrete, Montebello, 1983 (ACI. SP 79, 1983) 1, pp. 415-434.
3. D.C. Hughes, Mag. Concr. Res. 37, (133) 227-233 (1985).
4. L.J. Parrott, Mater. Constr. (Paris) 17, (98) 131-137 (1984).
5. P.K. Mehta and D. Manmohan, in Proc. 7th Int. Congr. Chem. Cem., (Paris, 1980) 3, VII 1-5.
6. L.J. Parrott, R.G. Patel, D.C. Killoh and H.M. Jennings, J. Acer. Ceram. Soc. 67 (4), 233-237 (1984).
7. S. Mindess, J.F. Young, Concrete (Prentice-Hall, 1981) p. 99.
8. Y. Halse, P.L. Pratt, J.A. Dalziel and W.A. Gutteridge, Cem. Concr. Res. 14, 491-498 (1984).
9. J.A. Dalziel, Technical Report 555 (Cement and Concrete Association, Wexham Springs, 1983).
10. R.G. Patel, L.J. Parott, J.A. Martin and D.C. Killoh, Cem. Concr. Res. 15, 343-356 (1985).
11. L.J. Parrott, Cem. Concr. Res. 13, 18-22 (1983).

RESTRICTED HYDRATION OF MASS-CURED CONCRETE CONTAINING FLY ASH

R.H. MILLS* and N. BUENFELD**
*Department of Civil Engineering, University of Toronto, 35 St. George St.,
Toronto, Ontario, Canada M5S 1A4
**Department of Civil Engineering, Imperial College, Imperial College Road,
London, SW7 2BU, United Kingdom

Received 1 December, 1986; refereed

ABSTRACT

Cement paste and concrete specimens containing three different mixtures
of portland cement (PC) and high-lime fly ash (FA) were subjected to various
curing conditions, and the strengths, non-evaporable water, and porosities
were compared with control mixes containing portland cement only. Strength
and porosity data indicated that the cementing action of the mixtures was, in
all cases, inferior to portland cement. For each mixture the strength loss
resulting from imperfect curing, i.e. sealed or exposed to 50% relative
humidity, was greater than for portland cement. In the case of concrete
drying from one surface, strength differences were found, but these were not
as clearly defined as those obtained in the paste specimens.

INTRODUCTION

Laboratory concrete is invariably cured by immersion in water or exposure to 100% relative humidity. Site concrete receives little water curing
except, perhaps, near the surface. The current ACI standard [1] allows the
use of sealing materials to prevent evaporation of water from the surface, but
does not address the effect of self-desiccation due to chemical shrinkage [2].
Shalon and Ravina [3] reported substantial differences between water cured and
sealed concrete even when the water/cement ratio was as high as 0.92. The ACI
Standard [1] recognises that portland-pozzolan cements require 50% more curing
than portland cements.

In an earlier paper [4], micrographs of cement pastes sealed against gain
or loss of moisture showed that the pastes were deficient in hydration
products in comparison with water cured specimens.

EXPERIMENTAL

Materials

Table I shows the composition of the portland cement (PC) and fly ash
(FA) used. The fly ash had a pozzolanic activity value of 16.4 MPa at 7 days
(d), and a portland cement pozzolanic activity value at 104%, tested according
to ASTM C 618-83.

Cement Paste Specimens

Because an important aspect of the tests concerned the capacity of mixing
water to support hydration, it was important to avoid bleeding. This was done
by dosing the high water/cement (w_o) mixes with inert rock flour (RF) derived
from crystalline quartz. All mixes were made with 491 kg/m^3 initial water
content. Mix proportions used are given in Table II.

Thirty specimens were cast from each mix in plastic containers, 50 mm x
28 mm diameter, each with a tightly fitting lid. After 24 hours (h), 10
specimens were placed in a water bath with the lids removed (W), 10 were left
sealed (S) and 10 were exposed to the laboratory atmosphere (D). After 30 and

Table I
Composition of Portland Cement and Fly Ash

	Na_2O	MgO	Al_2O_3	SiO_3	SO_3	K_2O	CaO	Fe_2O_3
PC	0.4	2.9	4.7	21.3	4.0	0.4	62.9	2.1
FA	7.6	3.9	19.2	35.0	4.9	0.8	20.4	6.0

Table II
Cement Paste Mixes

Mix	Reference	w_o	Mass (kg/m^3) PC	FA	Water	RF
100	30	0.30	1638	–	491	–
100	45	0.45	1091	–	491	454
100	60	0.60	818	–	491	680
85	30	0.30	1391	246	491	–
85	45	0.45	927	164	491	423
85	60	0.60	696	123	491	657
70	30	0.30	1145	491	491	–
70	45	0.45	764	327	491	392
70	60	0.60	573	245	491	634
55	30	0.30	900	736	491	–
55	45	0.45	600	491	491	361
55	60	0.60	450	368	491	610

Density values (kg/m^3): PC = 3220; FA = 2620; RF = 2670.

90 d, specimens for test were stripped, the ends were ground flat and specimens were vacuum soaked to constant weight before crushing.

Concrete Specimens

Concrete mixes having the proportions shown in Table III were cast to a depth of 0.7 m in 0.85 m x 0.1 diameter plastic tubes. The tops were sealed for the first 24 hours. At 1 day, the tube above the level of the concrete was either (a) filled with water, (b) filled with hot wax to a depth of 20 mm to prevent gain or loss of moisture or (c) filled with silica gel that was regenerated at frequent intervals to cause rapid drying from the surface.

After 112 days the specimens were cut into 20 mm thick discs and vacuum saturated before testing. Specimens were loaded on the diameter. The splitting tension was calculated from the load using

$$\sigma_t = 2P/\pi Dt \tag{1}$$

where σ_t is tensile strength (MPa), P is load (MN), D is diameter (m) and t is thickness (m).

Table III
Concrete Mix Proportions

PC/FA	PC	FA	Water	Sand	Pea Gravel (-9.6 mm)
			(kg/m^3)		
100/10	511	–	230	843	777
85/15	434	77	230	825	777
70/30	356	155	230	807	777
55/45	281	230	230	788	777

Table IV
Linear Regression Coefficients in Equations
of the Form $f_c' = A + B (w_n)$

Sample		A	B	Correlation Coefficient (R^2)
All	$w_o = 0.3$	−11.2	721	0.93
All	$w_o = 0.45$	−22.5	541	0.87
All	$w_o = 0.6$	−15.7	356	0.91

Determination of Non-evaporable and Evaporable Water

The evaporable water was determined by drying to constant weight at 110°C, and the non-evaporable water (w_n) by igniting the oven-dried specimens at 1050°C. The porosity was calculated on the assumption that evaporable water filled all the void space.

RESULTS

Compression Tests on Paste Specimens

Compressive strengths versus FA content are summarized in Figure 1. It is seen that strength is reduced with the proportion of fly ash for all water/cement ratios and all curing conditions.

Variation of Compressive Strength with Non-Evaporable Water

Pooled results from specimens subjected to the 3 curing conditions are shown in Figure 2. The corresponding factors for the linear regression equation are given in Table IV.

There appears to be a fair correlation between strength and non-evaporable water for each water/cement ratio regardless of PC/FA proportions or curing regime.

Variation of Compressive Strength with Porosity

Figure 3 shows the pooled results of compressive strength f_c' versus porosity. The regression equation is

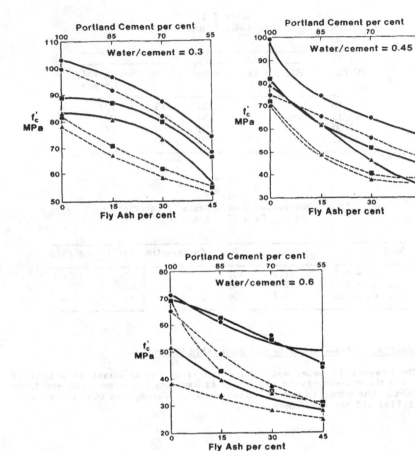

Figure 1. Compressive strengths of paste specimens, showing specimens cured
for 90 days (solid lines), 35 days (broken lines), in water
(circle), sealed against gain or loss of moisture (square) and
allowed to dry from one face (triangle).

$$f'_c = 17.66 \; \varepsilon^{-1.46} \tag{2}$$

where ε is the ratio of the volume of evaporable water to volume of (mixing
water + PC + FA). The correlation coefficient (R^2) is 0.64. This is a poor
correlation; but it appears, from a study of the data, that better correlations may be obtained by treating data from each PC/FA combination and each
curing regime separately. The results of this analysis are summarised in
Table V, where it is seen that good correlations were obtained for the water
cured specimens. The correlations for sealed and dried specimens were only
fair, possibly due to variations in the degree of saturation at the time of
test.

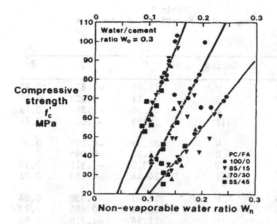

Figure 2. Variation of compressive strength with non-evaporable water, with each line fitted to pooled data from 4 values of PC/FA and three curing specimens.

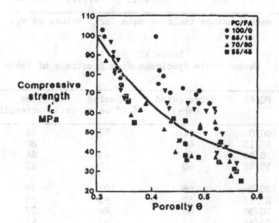

Figure 3. Variation of compressive strength of pastes with porosity, with points from pooled data for 4 values of PC/FA and 3 curing regimes.

The results of imperfect curing may be illustrated by evaluating each of the equations of Table V for porosity values (ε) 0.4 and 0.5. These values were chosen within the range of interpolation for each group of data. The results are given in Table VI.

The reductions of strength resulting from imperfect curing shown in Table VI are less than expected and insufficient to cause serious concern. Of more importance, perhaps, is the clear evidence of lower binding strength in all categories of curing, which results from partial substitution of this particular fly ash for portland cement.

Table V
Regression Equations of the Form

PC/FA	Curing	P	Q	R^2
100/10	W	35.99	−0.879	0.95
	S	44.11	−0.623	0.83
	D	20.36	−1.367	0.79
	All	29.89	−1.02	0.71
85/15	W	25.89	−0.197	0.96
	S	26.20	−0.045	0.72
	D	153.0	−1.511	0.93
	All	24.44	−1.183	0.71
70/30	W	18.92	−0.338	0.84
	S	13.62	−0.605	0.72
	D	11.24	−1.719	0.94
	All	20.00	−1.329	0.64
55/45	W	15.63	−0.463	0.90
	S	16.27	−0.294	0.77
	D	13.31	−1.557	0.64
	All	14.55	−1.476	0.74

Note: each equation includes data for 3 values of w_o.

Table VI
Strengths of Cement Paste Specimens from Equations of Table V

ϵ	PC/FA	Water cured (MPa)	Sealed	Dried
			(% of water cured strength)	
0.4	100/0	81	97	88
	85/15	69	90	80
	70/30	64	92	84
	55/45	60	89	93
0.5	100/0	66	103	79
	85/15	52	93	75
	70/30	48	86	77
	55/45	43	93	91

Concrete Specimens

Bar charts showing splitting strengths of samples taken at 20 mm intervals are shown in Figure 4. The station numbers identify the 20 mm slice tested, Station 1 being the top slice at the exposed surface, and slice 35 being the end 0.7 m remote from the exposed surface.

Specimens in which the top surface was in contact with water were appreciably stronger than specimens with sealed surface or those subjected to drying with silica gel. Contrary to the starting hypothesis, there is no critical distance from the surface beyond which the material is insensitive to curing.

227

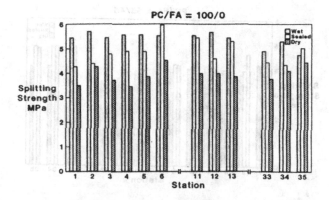

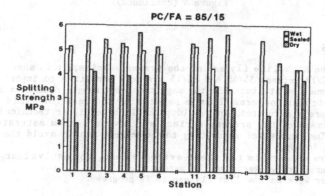

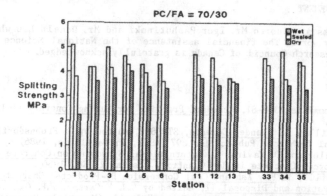

Figure 4. Bar charts showing splitting strengths at 20 mm intervals between
surface exposed to treatment (station 1) and sealed end (station
35) for PC/FA = 100/0, 85/15, 70/30 and 55/45.

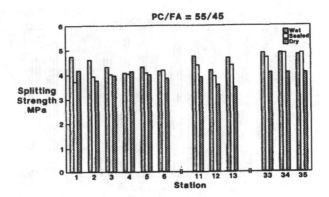

Figure 4 (continued).

CONCLUSIONS

For the high-lime fly ash of the present series, it is shown that mixtures of PC/FA between 55/45 and 85/15 are more sensitive to imperfect curing than portland cement, but that the magnitude of strength reduction should not cause particular concern. This is especially the case for field concrete, which is rarely saturated, and would probably have higher technical strength than specimens of the present series, which were vacuum saturated before testing. The purpose of saturating the specimens was to avoid the secondary effects of drying.

Of more concern is the clear evidence that, for equivalent porosity, PC/FA mixtures are weaker than the parent PC.

Further work on low lime fly ash is in progress.

ACKNOWLEDGEMENT

Thanks are due to Mr. Igor Pashutinski and Mr. Donald Lau who did the laboratory work. The financial assistance of the National Science and Engineering Research Council of Canada is gratefully acknowledged.

REFERENCES

1. ACI Committee 308–81, <u>Standard Practice for Curing Concrete</u> (ACI, Detroit, 1986).
2. R.H. Mills, in <u>Blended Cements</u>, STP 897, edited by G. Frohnsdorff ASTM Special Technical Publication 897 (ASTM, Philadelphia, 1986) pp. 49–61.
3. R. Shalon and D. Ravina, in <u>International Symposium on Concrete and Reinforced Concrete in Hot Countries</u>, Haifa (RILEM, 1960).
4. <u>R.H. Mills, in Fly Ash and Coal Conversion By-Products: Characterization, Utilization and Disposal II</u>, edited by G.J. McCarthy, F.P. Glasser and D.M. Roy, Mat. Res. Soc. Symp. Proc. Vol. <u>65</u>, (Materials Research Society, Pittsburgh, 1986) pp. 207–217.

EFFECT OF FLY ASH INCORPORATION ON RHEOLOGY OF CEMENT PASTES

M. RATTANUSSORN*, D.M. ROY** and R.I.A. MALEK
Materials Research Laboratory, The Pennsylvania State University, University
Park, PA 16802
*Current address: The Siam Cement Co., Ltd., Bangkok, Thailand
**Also affiliated with the Department of Materials Science and Engineering

Received 12 November, 1986; refereed

ABSTRACT

The predominant spherical shape of fly ash particles combined with mainly
glassy composition and texture of its surfaces have a special effect on
rheology of cement pastes containing fly ash. The early ages rheological
behavior of cement pastes (ASTM Type I) incorporating 30% low-calcium fly ash
was monitored by measuring viscosity of the fresh pastes prior to initial
hardening and stiffening (up to ~2 hours) as a function of time. The viscosities were determined using a co-axial rotoviscometer (HAAKE). The effects of
fly ash content, water to cement ratio, and presence and concentration of
superplasticizer, were evaluated. In addition, the dispersivity of fly ash
spheres was evaluated by determining the zeta-potential of fly ash suspensions
in water using a microelectrophoresis technique and the results were correlated to the chemical composition of fly ash as well as the viscosities of fresh
pastes.

INTRODUCTION

The rheological properties of fresh concrete and cement paste have been
intensively investigated by many researchers [1-8]. From these studies, it
has been established that the rheology of fresh Portland cement pastes and
concrete is comparable, both theoretically and practically. Furthermore, it
has been recognized that incorporation of finely divided particles such as fly
ash generally improves workability by reducing the size and volume of voids in
fresh concrete mixtures.

Ivanov and Zacharieva [9] studied the effect of two fly ashes having
different specific surface areas on the rheology of cement pastes having 0.45
to 0.7 water to cement (w/c) ratios. The fly ash contents ranged from 0 to
100% of the solids by weight. They found that both yield stress and viscosity
increased with increasing fly ash content in the paste. The higher specific
surface area fly ash (Blaine air permeability = 515 m^2/kg) had much more
effect in increasing both yield stress and viscosity at the same proportion
(by mass) of fly ash replacement and w/c ratio. However, in terms of water to
solid ratio (w/s) both yield stress and viscosity decreased with increasing
fly ash content. The lower specific surface area fly ash (Blaine = 465 m^2/kg)
was more effective in reducing both parameters. The coarser fly ash, in this
case, was considered to be more effective because the specific surface area is
closer to that of cement (340 m^2/kg). Nagataki et al. [10] found that addition of fly ash was advantageous for improving fluidity of cement paste,
especially in the case of low w/s ratio.

The purpose of the present paper is to investigate the effect of fly ash
on the rheological properties of fresh cement pastes, and the effects of the
amount of fly ash, w/c ratio, and the presence and concentration of superplasticizer, on fluidity of pastes composed of fly ash/cement blends.

Table I

Chemical Analyses and Other Characteristics of Materials Used

	Portland Cement (I-09)	Fly Ash (B-86)	Fly Ash (B-87)
Chemical Analyses			
SiO_2	19.80	48.90	37.10
Al_2O_3	5.74	27.40	12.30
Fe_2O_3	2.17	11.00	38.90
CaO	62.98	2.70	4.45
MgO	3.13	0.93	0.78
SO_3	2.55	0.40	1.77
LOI	1.40	3.60	1.54
Na_2O	0.19	0.23	0.64
K_2O	0.78	2.45	1.64
Total Alkalis as Na_2O	0.78	1.85	1.72
TiO_2	0.27	1.30	0.60
P_2O_5	0.21	0.38	1.55
Other Characteristics			
BET*, m^2/g	0.83	0.89	0.40
Blaine*, m^2/kg	330	325	213
Density**, kg/m^3	3,200	2,340	2,970

*Specific Surface Area.   **Kerosene Density.

EXPERIMENTAL APPROACH

Starting Materials

An ASTM Type I cement and two low-calcium fly ashes were the basic in-gredients used in the present investigation. Details of chemical analyses as well as some physical characteristics are listed in Table I. A formaldehyde-based sulfonated naphthalene condensate was used as a superplasticizer (PSU Code No. A-76).

Rheological Studies

Background. Roy and Asaga [11] found that the behavior of certain pastes could be Bingham type, antithixotropic, thixotropic, dilatant or Newtonian, depending upon the experimental conditions (i.e., admixture presence and con-centration, water to cement ratio, shear rate, and the time after mixing with water). However, for the low w/c pastes, the behavior can be approximated to that of Bingham fluids. The Bingham equation is:

$$\tau = \tau_o + \eta_p \dot{\gamma}$$

where τ = shear stress, τ_0 = yield stress, η_p = plastic viscosity, $\dot{\gamma}$ = shear rate. This rheological properties of cement pastes may be charac-terized by viscosity and yield stress.

A typical flow curve of cement paste obtained from the rotational viscom-eter is illustrated in Figure 1. In the present investigation, the yield stress value (τ_0) is determined by extrapolation of the straight line portion near the starting point of the increasing curve.

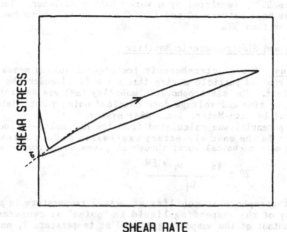

Figure 1. Typical flow curve of cement paste.

For a non-Newtonian material like cement paste, in contrast to a Newtonian material, the ratio of shear stress to shear rate is not constant but depends on the shear rate at which the measurement is made. Hence, the term "apparent viscosity" (η_a) is sometimes used, and is defined as the viscosity of a Newtonian material that would behave in the same way as the non-Newtonian medium at the particular shear rate. The intermediate shear rates of 44, 88, and 132 s^{-1} were chosen for calculation of apparent viscosity because at these rates, there was a comparatively small degree of structure development, and the rates were suitable for study of fluidity [12].

Equipment and Technique. The pastes were measured with a rotating coaxial cylinder viscometer (HAAKE Rotovisco Model RV3) using a standard measuring head MK 500 (maximum torque is 0.49 N·cm) and serrated sensor system MV IIP. Details of the sensor system are discussed elsewhere [19].

The maximum measurable stress is 376 Pa. The system can measure a viscosity up to about 10^5 cp (1 cp = 1×10^{-3} Pa·s). In operation, the rotor speed was increased from 0 to 650 rpm. The viscometer output was recorded by a Hewlett Packard Model 7040 X-Y recorder. The flow curves were obtained and apparent viscosities at various shear rates were calculated.

Preparation of Samples. The pastes were prepared by mechanical mixing in accordance with the procedure of ASTM C 305, section 5 which can be briefly described as follows:

1. Add the sample to water and allow 30 seconds (sec) for the absorption of the water.
2. Start the mixer and mix at a slow speed (140±5 rpm) for 30 sec.
3. Stop the mixer for 15 sec to scrape down any paste that may be on the side of the bowl.
4. Mix at a medium speed (285±10 rpm) and mix for 1 minute (min).

The dry sample weight was 650 g. The rheological measurement was started at 6.5 min after the sample was mixed with water. This is within the early hydration period where rapid initial dissolution of alkali sulfates and aluminates, initial hydration of C$_3$S and formation of AFt occur. The measurement

temperature was 25°C, regulated by a water bath circulator. The temperature
of the room was kept between 20 and 27.5°C and the relative humidity of the
room was greater than 50%.

Zeta Potential and Electrophoretic Mobility

Background. The electrophoresis technique involves measurement of the
velocity of a charged particle moving through a fluid under the influence of
an electric field. The electrophoretic mobility (EM) was determined directly
from the tracking time and voltage data obtained using a Zeta-Meter Model ZM-
77 (manufactured by Zeta-Meter, Inc., USA) system.

The zeta potential was calculated from the Helmholtz-Smoluchowski equation (2), which is the most elementary expression of Zeta Potential, and is
sufficient for most technical work, though it gives an approximate value:

$$ZP = \frac{4\pi \cdot \nu_T \cdot EM}{D_T} \tag{2}$$

where EM = electrophoretic mobility at actual temperature in $\mu m \cdot s^{-1}/V \cdot cm^{-1}$;
ν_T = viscosity of the suspending liquid in "poise" at temperature T; D_T =
dielectric constant of the suspending liquid at temperature T, and ZP = potential in electrostatic units.

When a voltage is imposed across the electrophoretic cell, frictional
heat is generated by moving ions, therefore, the temperature of the liquid
rises continuously. If the rate of temperature rise is too high, thermal
turbulence of the suspended particles makes the measurement inaccurate. The
rate of temperature rise depends on the specific conductance and the applied
voltage. The specific conductance of the sample was measured prior to zeta
potential determination, and a maximum recommended voltage was applied to
minimize the temperature rise at a particular specific conductance value.
However, in some cases, the applied voltage was further reduced to maintain a
particle velocity of less than 64 μm/s (1 full scale division is 160 μm) which
would avoid operator error. The temperature of suspension was recorded before
and after each zeta potential determination, and the average temperature was
used to find the corresponding viscosity and dielectric constant of the
suspending liquid.

Technique. A small amount of the unfractionated fly ash sample (0.3 g)
was immersed in 100 mL deionized water. The suspension was dispersed using a
medium-speed magnetic stirrer for 15 min, and transferred to the measurement
cell. The time required for a particle to move a fixed distance in response
to specific field gradient was measured. The electrophoretic movements of 30
particles were determined and the average was reported.

RESULTS AND DISCUSSION

Zeta-Potential

The zeta potentials (ZP) of fly ash samples in water with varied concentrations of superplasticizer (A-76) were determined, with results as presented in Figure 2. ZP measurement of a mixture of cement and fly ash particles
at three different superplasticizer concentrations are also presented in the
figure.

It can be seen that the zeta potential of B-86 fly ash increases from
-11.9 to -17 mV when the amount of superplasticizer increases from 0 to 0.1%.
Any further increase of the amount of superplasticizer added was found to have
little effect on the zeta potential. Consequently, from this and previous
studies it is considered that 0.1% of superplasticizer is sufficient to effectively disperse this type of fly ash particle. Malek et al. [13] found that

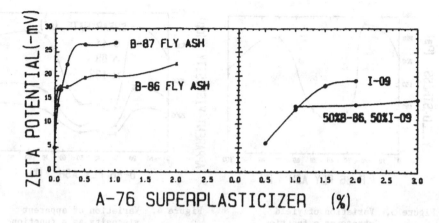

Figure 2. Variation of zeta potential of fly ashes, cement and fly ash/cement mixture at various dosages of superplasticizer.

slag containing cement pastes could be effectively dispersed using 0.5-1% of superplasticizer. The same trend was observed with a sample of B-87 fly ash, with the zeta potential changing from -4.9 to -27 mV when the amount of super-plasticizer was increased from 0 to 0.5% (Figure 2). Beyond this limit (0.5%) further addition of superplasticizer was found to cause no appreciable change. Therefore, the addition of a sulfonated naphthalene formaldehyde condensate superplasticizer (A-76) had a lesser effect on the higher specific surface area (Blaine = 325 m^2/kg) fly ash (B-86) investigated. This phenomenon may be related to not only the adsorption mechanism but also the surface properties and leaching characteristics of the fly ashes which are in turn related to their composition [14,15].

Rheological Studies

Effect of Fly Ash on Rheological Properties of Portland Cement Paste.
The variations of yield stress and apparent viscosity as a function of the fly ash content in the cement paste, w/s of 0.3 are illustrated in Figures 3 and 4. It can be seen that the yield stress has its lowest value when the fly ash content is in the range of 30-50 wt% (37-58 vol.%). The apparent viscosities measured at different shear rates, namely, 44, 88, and 132 s^{-1} show similar trends.

It appears that fly ash plays a dual role which affects the rheological properties of fresh cement pastes:

1. A predominantly physical effect consists of two parts: (a) the predom-inant spherical shape of fly ash particles combined with glassy texture of its surface which permits effective particle packing, reducing water demand without decreasing the workability of the mix, (b) at the same time the lower density of fly ash relative to that of cement results in higher volume percent for a given weight of fly ash to cement, causing a higher water demand (or lower workability).
2. A chemical effect results from difference in hydration rates of cement and fly ash particles. The retardation effect of fly ash, either by chemisorption of Ca^{2+} (Grutzeck et al. [16]) or through retarding the hydration of C$_3$A and C$_4$AF (Cabrera and Plowman [17]), results in de-

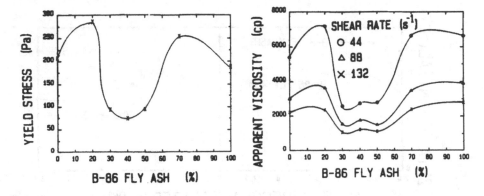

Figure 3. Variation of yield
          stress as a function
          of fly ash content.

Figure 4. Variation of apparent
          viscosity as a function
          of fly ash content.

creased viscosity (and yield stress) of fly ash cement pastes. A second
chemical effect is the better dispersion properties of the fly ashes due
to their negative zeta potentials, which may dominate when 37-58 vol.% of
fly ash is present.

Therefore, it is hypothesized that (i) at low fly ash replacement levels
(<30 wt%), the retardation of hydration and decrease in water demand are the
dominant factors in decreasing viscosity (and yield stress) with increasing
level of fly ash replacement; (ii) at high fly ash replacement levels (>50
wt.%) the high volume proportion of solid to liquid in the fly ash cement
pastes is the predominant factor in increasing viscosity (and yield stress)
with increased level of fly ash replacement; and (iii) at 30-50 wt% levels
(approximately equal volume properties of cement and fly ash) the pastes
appear to acquire the optimum condition where the different factors help
achieve minimum viscosity (and yield stress).

Effect of Water Content on Rheological Properties of Fly Ash Cement and
Portland Cement Pastes. The variations of yield stress as a function of
changing water to solid ratio are illustrated in Figure 5. It can be seen
that at the lower w/s ratio (0.3) fly ash helped reduce the yield stress
substantially. At w/s = 0.35, fly ash still helps reduce the yield stress but
to a lesser extent. At high water to solid ratios (w/s = 0.4 and 0.5) the fly
ash had little influence on the yield stress (the yield stress is close to
that of pure cement paste). This trend applies also to the apparent viscosity
(Figures 6 and 7). It appears that the results could be explained in terms of
the characteristics of interaction of cement with water [3]. This behavior
could be ascribed to the immobilization of a part of free water and bound
water due to the marked deposition of hydrating cement particles. The effects
in lowering viscosity and yield stress with fly ash substitution are even more
striking when considered as volume proportions of liquid and solid.
     The effect of varied superplasticizer dosages on the rheological proper-
ties of fly ash cement and cement pastes was also studied. It was found that
the largest effect of the superplasticizer is on the cement component of fly
ash cement pastes, and that fly ash cement pastes have less need for
superplasticizer.

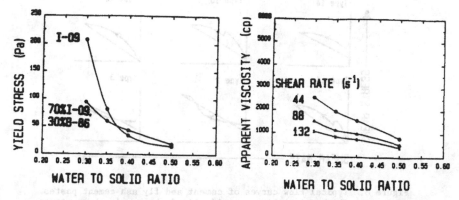

Figure 5. Yield stress--w/s ratio dependence.

Figure 6. Change of apparent viscosity of cement pastes containing 30% fly ash as a function of w/s ratio.

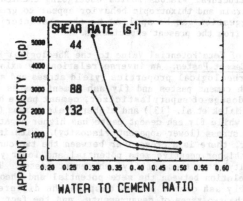

Figure 7. Change of apparent viscosity of cement pastes as a function of w/c ratio.

Flow Curve Characteristics. By using the same mixing method and specific time allowance (6.5 min) before measurement of cement and fly ash cement pastes, and a 15 min cycle time, the flow curves obtained of the pastes with and without superplasticizer, at different water to solid ratios, could be categorized by three main types: Type 1 (a,b,c,d,), Type 2, and Type 3 as illustrated in Figure 8.

It is noticeable that a low apparent viscosity paste tends to have a Type 1 flow curve and a higher apparent viscosity tends to have a Type 2 or Type 3 flow curve. This may be due to the fact that at higher viscosity, the paste hydrates and sets faster, accompanied by some hydrated particle linkage (Roy and Asaga [5]). In addition, for a high w/s ratio paste or very high dosage of superplasticizer the yield stress value is very low or close to zero (Flow curves Type 1b, 1c, and 1d).

236

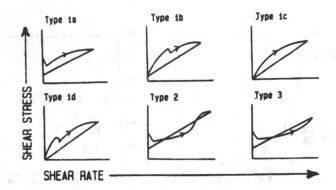

Figure 8. Typical flow curves of cement and fly ash cement pastes.
The straight lines are idealized changes in shear stress
as the shear rate is gradually decreased.

Nagataki and Kawano [10] suggested that cement paste and mortar should be
considered as structural viscous bodies undergoing chemical reactions, where
hysteresis of strain and thixotropic behavior appear to greatly affect fluidity. Their suggestion can be applied to the characteristics of the flow
curves obtained from the present experiments.

Comparison of Zeta Potential Value to the Rheological Properties of Fly
Ash Cement and Cement Pastes. An inverse relation may exist between the zeta
potential and rheological properties (yield stress and apparent viscosity).
Results for both cement pastes and fly ash cement pastes exhibit comparable
trends: at low dosage of superplasticizer, cement paste has lower negative
zeta potential (Malek et al. [13]) and higher yield stress (also higher apparent viscosity), while fly ash cement paste has higher negative zeta potential
and lower yield stress (lower apparent viscosity). With increasing dosage of
superplasticizer, there is a cross-over between the two curves in that cement
paste possesses higher negative zeta potential and lower yield stress (lower
apparent viscosity) than those of fly ash cement pastes. This indicates that
there is some relation between the zeta potential and rheological properties
of cement and fly ash cement pastes despite the different water to solid
ratios used in the two types of measurements, and the fact that only the fine
particles are tracked in the zeta potential measurement, while the larger
particles have much influence on the fluidity when measured with a viscometer.
It is observed that with increasing superplasticizer concentration, the
zeta potential values of cement containing 50% fly ash are lower than those of
cement alone or fly ash alone (Figure 2). These results agree with those
obtained by Nagataki et al. [18]. They attributed this to peculiar adsorption
of Ca^{2+} ions released accompanying hydration of cement.

CONCLUSIONS

From the above findings, the following conclusions were reached:

1. Low-calcium fly ash with a narrow particle size distribution has a tendency to improve the rheological properties of cement pastes, i.e., its
   substitution for cement in the range of 30-50% w/w fly ash replacement
   reduces both the yield stress and apparent viscosity to half of the
   values observed with cement alone.

2. Superplasticizer does not improve the fluidity of fly ash cement paste as much as it does for pure cement paste, and therefore is less needed with fly ash-substituted cement.
3. Water to solid ratio markedly influences the properties of cement and fly ash cement pastes. With increasing w/s ratio, yield stress and apparent viscosity of cement paste decrease more rapidly (non-linear) than those of fly ash cement paste (linear).

ACKNOWLEDGEMENT

This research was partially funded by the National Science Foundation (Grant No. DMR-8418160).

REFERENCES

1. M. Ish-Shalom and S.A. Greenberg, in Proceedings of the 4th International Symposium on the Chemistry of Cement (1960), pp. 731-748.
2. T.C. Powers, The Properties of Fresh Concretes (John Wiley and Sons, Inc., New York, 1968).
3. Y. Ivanov and E. Stanoeva, in Proceedings of the VIIth International Congress on Rheology, (Sweden, 1976) pp. 360-361.
4. C.R. Dimond and G.H. Tattersall, in Proceedings of a Conference held at University of Sheffield (1976), pp. 118-133.
5. D.M. Roy and K. Asaga, Cem. Concr. Res. 9, 731-739 (1979).
6. K. Asaga and D.M. Roy, Cem. Concr. Res. 10, 287-295 (1980).
7. R. Lapasin, A. Papo, and S. Rajgelj, Cem. Concr. Res. 13, 349-356 (1983).
8. G.H. Tattersall and P.F.G. Banfill, The Rheology of Fresh Concrete, (Pitman Books Ltd., London, 1983).
9. Y. Ivanov and S. Zacharieva, in Proceedings of the 7th International Congress on the Chemistry of Cement, Vol. III (Paris, 1980), pp. VI-103 - VI-107.
10. S. Nagataki and S. Kawano, in Proceedings of the 7th International Congress on the Chemistry of Cement, Vol. VIII (Paris, 1980), pp. VI-120 - VI-124.
11. D.M. Roy and K. Asaga, Cem. Concr. Res. 10, 387-394 (1980).
12. Y. Ivanov and E. Stanoeva, Silicates Industriels 9, 199-203 (1979).
13. R.I.A. Malek, M. Silsbee, and D.M. Roy, in Proceedings 8th International Congress Chemistry of Cement, Vol. IV (Brazil, 1986), pp. 270-275.
14. R.I.A. Malek and D.M. Roy, in Fly Ash and Coal Conversion By-Products: Characterization, Utilization and Disposal I, edited by G.J. McCarthy and R.J. Lauf, Mat. Res. Soc. Proc. Vol. 43 (Materials Research Society, Pittsburgh, 1985) pp. 41-50.
15. R.I.A. Malek, P.H. Licastro, and D.M. Roy, in Fly Ash and Coal Conversion By-Products: Characterization, Utilization and Disposal II, edited by G.J. McCarthy, F.P. Glasser and D.M. Roy, Mat. Res. Soc. Symp. Proc. Vol. 65, (Materials Research Society, Pittsburgh, 1986) pp. 269-284.
16. M.W. Grutzeck, Wei Fajun, and D.M. Roy, in Fly Ash and Coal Conversion By-Products: Characterization, Utilization and Disposal I, edited by G.J. McCarthy and R.J. Lauf, Mat. Res. Soc. Symp. Proc. Vol. 43 (Materials Research Society, Pittsburgh, 1985) pp. 65-72.
17. J.G. Cabrera and C. Plowman, in 7th International Congress on the Chemistry of Cement, Vol. III (Paris, 1980), pp. IV-84 - IV-92.
18. S. Nagataki, E. Sakai, and T. Takeuchi, in Research Sessions of First International Conference on the Use of Fly Ash, Silica Fume, Slag, and Other Mineral By-Products in Concrete (Canada, 1983), 13 pp.
19. M. Rattanussorn, M.Sc. Thesis in Ceramic Science, The Pennsylvania State University, University Park, PA 16802 (1986).

THE DIFFUSION OF CHLORIDE IONS IN FLY ASH/CEMENT PASTES AND MORTARS

R.I.A. MALEK, D.M. ROY* and P.H. LICASTRO
Materials Research Laboratory, The Pennsylvania State University, University Park, PA 16802
*Also affiliated with the Department of Materials Science and Engineering

Received 12 November, 1986; refereed

ABSTRACT

Fly ashes having three distinctly different levels of calcium, designated low-calcium (Class F), intermediate-calcium (Class F/C), and high-calcium (Class C) comprised the basic material for the present study. Pastes and mortars were made out of the three types of fly ashes and one type of cement (Type I) at various levels of replacement as well as different water/solid ratios (w/c). Chloride ion diffusion was measured by applying an electrical potential across cured cylindrical samples and measuring the amount of current passed in a certain period of time (proportional to amount of Cl^- passed in this time). Other supportive measurements were made, e.g., porosity and pore size distribution, water permeability and surface area. The Cl^- ion diffusivities were correlated with the chemical composition of fly ash (FA), mix proportioning, and water permeabilities of the tested hardened pastes or mortars.

INTRODUCTION

The diffusion of Cl^- ions has been found to be strongly influenced by the type of cement, and type and proportions of blending materials [1]. Blending cements (with 30% fly ash or 65% slag) results in lower diffusion rates than neat cement pastes [2,3].

Chloride diffusion follows Fick's law [4-9] and is lower in blast furnace slag cements than in portland cements. The reduction in diffusion rate in fly ash cements is significant [4,6,7,10] though variable, while sulfate-resistant portland cements may have higher diffusion rates. Silica fume has been shown to decrease the diffusivity [4,11], even at relatively high w/c ratios [4,7, 11]. In an attempt to predict time-Cl^- penetration depth plots in concrete, Pereira and Hegedus [12] developed a diffusion-reaction-equilibration model in which Langmuir adsorption and Fickian diffusion were coupled. Uchikawa et al. [13] showed that Na^+ diffusion was markedly decreased in slag and fly ash-containing pastes, which they related to the zeta-potential.

Pore solution compositions may provide a key to transport inhibition mechanisms and enable comparisons among the effectiveness of various materials. It has been observed [14-16] that the chloride content of pore solutions of fly ash-containing and slag-containing cement pastes is diminished by up to 80% over a 3-month period, compared with 40% reduction in a neat portland cement paste [17,18].

This study examines the effect of various factors on chloride ion diffusivity in cement/fly ash blends.

MATERIALS AND TECHNIQUE

Materials

An ASTM Type I cement (with 12.3% C_3A content) and two different classes of fly ashes were used in this investigation. Oxide compositions are presented in Table I. Two Class F fly ashes were used; the Baldwin ash having higher

Table I

Oxide Composition of the Cement and Fly Ashes Used

Material	Type I Cement	Baldwin Class F_1#	Crawford Class F_2#	Rockport Class C
MRL Designation	I-15	G-05	G-06	G-07
SiO_2	20.50	53.00	56.90	39.50
Al_2O_3	6.00	18.60	15.90	16.90
Fe_2O_3	2.10	15.60	5.50	6.40
CaO	62.80	6.37	9.12	24.80
MgO	2.90	1.26	4.01	6.30
Na_2O	----	0.94	5.02	1.44
K_2O	----	2.08	1.06	0.53
Total Alkali [as Na_2O]	0.90	2.31	5.72	1.79
SO_3	3.70	2.02	0.65	1.99
L.O.I.	1.20	1.68	0.54	0.22
Totals	100.10%	101.55%*	98.70%*	98.08%*
Fineness (m^2/Kg) [air permeability]	395	----	----	----
325 Sieve Res.	----	21.10	15.80%	12.60%
C_3A content	12.30%	----	----	----

*Contain 1.3–2% physically adsorbed moisture.

#F_1 = Class F ash with high Fe_2O_3 and low CaO and alkali contents.
  F_2 = Class F ash with low Fe_2O_3 and high CaO and alkali contents.

Table II

Mineralogies of the Fly Ashes

Class C (High-Calcium) Ash:	quartz, anhydrite, CaO, C_3A, hematite, magnetite, melilite, merwinite, periclase, and aluminosilicate glass
Class F* Ash:	quartz, mullite, hematite, Al- and Mg-substituted ferrite spinel or magnetite, anhydrite, CaO, with dominant silica-rich aluminosilicate glass

*Representative of F_1 and F_2 fly ashes (see differences in chemical composition in Table I).

Fe_2O_3 and lower CaO and alkali contents than the Crawford ash. The mineralogical compositions of the fly ashes as determined by x-ray diffraction (XRD) are given in Table II.

Mixture Formulations

The range of formulations tested is shown in Table III and schematically presented in Figure 1. The basic mixture design was oriented toward studying

Table III

Mixture Formulations

Mix Designation[a]	Cement (%)	Water (%)	Sand (%)	Fly Ash (%)			Remarks
				F_1	F_2	C	
86–01	23.43	12.12[b]	64.45	---	---	---	Ref.mortar
86–02	74.07	25.93	---	---	---	---	Ref.paste
86–03	64.52	35.48	---	---	---	---	Ref.paste
86–04	18.44	12.44[b]	63.36	---	5.76	---	Mortar[c]
86–05	16.15	11.76[b]	63.44	---	8.65	---	Mortar[e]
86–06	56.44	25.92	---	---	17.64	---	(c)
86–07	48.23	25.93	---	---	25.84	---	(e)
86–08	49.16	35.48	---	---	15.36	---	(c)
86–09	42.02	35.49	---	---	22.51	---	(e)
86–10	18.71	12.28[b]	64.33	---	---	4.68	Mortar[d]
86–11	16.43	11.97[b]	64.55	---	---	7.05	Mortar[f]
86–12	59.26	25.93	---	---	---	14.81	(d)
86–13	51.85	25.93	---	---	---	22.22	(f)
86–14	51.62	35.48	---	---	---	12.90	(d)
86–15	45.16	35.48	---	---	---	19.36	(f)
86–16	18.35	12.85[b]	63.07	5.73	---	---	Mortar[c]
86–17	16.00	12.57[b]	62.86	8.57	---	---	Mortar[e]
86–18	56.44	25.92	---	17.64	---	---	(c)
86–19	48.23	25.93	---	25.84	---	---	(e)
86–20	49.16	35.48	---	15.36	---	---	(c)
86–21	42.01	35.49	---	22.50	---	---	(e)

a. Each formulation was prepared in duplicate for curing at two different temperatures; 23°C and 38°C.
b. Adjusted to a flow table value of 110±5%.
c. 25 g fly ash for each 80 g cement.
d. 20 g fly ash for each 80 g cement.
e. 35 g fly ash for each 70 g cement.
f. 30 g fly ash for each 70 g cement.

the effect of the different variables on chloride ion transport (permeability) under an applied electric field, namely:

•Effect of w./(c+FA) ratio: two basic values were used, 0.35 and 0.55.

•Effect of including sand (mortars): standard mortars were prepared having cementitious materials:sand in the ratio 1:2.75. An ASTM C 109 Ottawa sand was used.

•Effect of various fly ash replacement levels: 25 and 35% replacement levels were used for Class F fly ashes; 20 and 30% for Class C ash.

•Effect of curing temperature: samples were subjected to two curing temperatures, 23°C and 38°C and >95% relative humidity. Samples were cured for 28 days (d) prior to testing.

| W/(C+F) ratio | Fly Ash Replacement | MRL Mix Designation[a] | | | |
		Reference Mixtures	Class F_2 Ash	Class C Ash	Class F_1 Ash
mortar [c]	0%	86-01			
0.35	0%	86-02			
0.55	0%	86-03			
mortar [c]	25% F[b] 20% C[b]		86-04	86-10	86-16
	35% F 30% C		86-05	86-11	86-17
0.35	25% F 20% C		86-06	86-12	86-18
	35% F 30% C		86-07	86-13	86-19
0.55	25% F 20% C		86-08	86-14	86-20
	35% F 30% C		86-09	86-15	86-21

a. All samples have been cured at two temperatures, 23°C and 38°C; saturated humidity.

b. F = Class F fly ash(see Table I for differences in oxide composition of F_1 and F_2 fly ashes); C = Class C fly ash.

c. Standard mortar: 1:2.75 (cement + fly ash):aggregate (sand); w/ (C+F) ratio was adjusted to a flow table value of 110 ± 5%.

Figure 1. Matrix of variables in the tested formulations.

Techniques

Chloride Permeability. Accelerated chloride ion diffusion tests, under application of an external electric field, were performed using a diffusion cell similar to that reported in Federal Highway Administration report number FHWA/RD-81/119 (August 1981). The samples (cylinders of 9.53 cm diameter x 5.08 cm height) were molded and their flat surfaces were polished with sandpaper to expose fresh surfaces prior to Cl^- permeability measurements. An epoxy coating was applied to the outer cylinder surface; the cylinders were then vacuum saturated overnight with previously boiled deionized water. The flat surfaces of the saturated samples were pressed against two 300 mL plastic compartments provided with two copper screens (serving as electrodes) so that the screens were in contact with the cement surfaces. Two rubber shims together with sealant application ensured sealing the cement sample against the two plastic compartments. One of the compartments was filled with 3% by

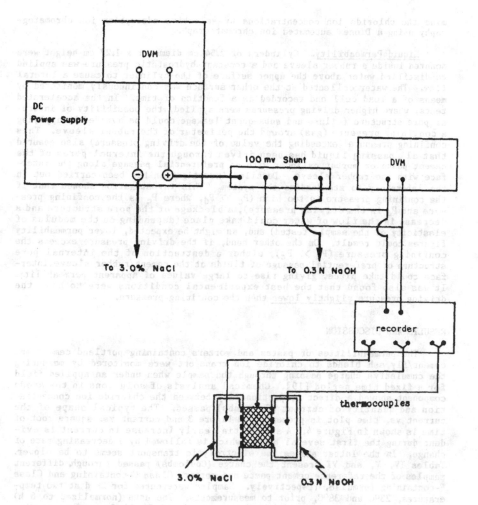

Figure 2. Block diagram of test set-up.

weight reagent grade sodium chloride solution and its copper screen was connected to the negative pole of a power supply. The other compartment was filled with 0.3 N sodium hydroxide (reagent grade) solution, with its screen connected to the positive pole of a power supply. A block diagram of the test set-up is presented in Figure 2. A 60 volt D.C. field was applied and the current and temperature (through thermocouples) were monitored continuously. Temperature eventually rose due to passage of current, and a maximum of 90°C was chosen for terminating each experiment. A period of six hours (h) was chosen for the duration of each experiment, unless it was terminated earlier due to excessive temperature rise. At the end of several selected runs (having appreciable differences in quantities of current passed), the solutions in the anode compartment were subjected to chemical analysis to determine the chloride ion concentrations by selective adsorption ion chromatography using a Dionex automated ion chromatograph.

Liquid Permeability. Cylinders of 2.54 cm diameter x 1.27 cm height were mounted inside a rubber sleeve and a constant hydrostatic pressure was applied to distilled water above the upper surface of the cylinder to cause a lateral flow. The water collected at the other surface was continuously monitored by means of a load cell and recorded as a function of time. In the accelerated tests, where higher driving pressures were applied, the possibility of internal pore structure failure and subsequent leakage could be avoided by applying a confining pressure (gas) around the perimeter of the rubber sleeve. This confining pressure (exceeding the value of the driving pressure) also ensured that all passing liquid has been driven through the internal pores of the cement paste or mortar cylinder with no preferential passage along the interface with the rubber sleeve. Detailed investigation has been carried out in the laboratory to standardize the system. This investigation showed that if the confining pressure is too high ($P_c \gg P_d$, where P_c is the confining pressure and P_d is the driving pressure), a blockage of the pore structure and a decrease in the flow of water could take place (depending on the modulus of elasticity of the sample tested) and, as might be expected, lower permeability figures could result. On the other hand, if the driving pressure exceeds the confining pressure ($P_d \gg P_c$), either a destruction of the internal pore structure or preferential passage of fluids at the cement/rubber sleeve interface could take place, giving rise to large values of apparent permeability. It was also found that the best experimental conditions were to have the driving pressure slightly lower than the confining pressure.

RESULTS AND DISCUSSION

The permeabilities of pastes and mortars containing portland cement and cement/fly ash blends to chloride ion transport were monitored by measuring the cumulative charge passing through the sample when under an applied field for a fixed time period [19]. Chemical analysis of solutions in the anode component showed a direct proportionality between the chloride ion concentration and quantity of current (Coulombs) passed. The typical shape of the current vs. time plot is presented in Figure 3 and current vs. square root of time is shown in Figure 4. A substantial early increase in current is evident during the first several hours, which is followed by a decreasing rate of change. In the latter stages the electrolytic transport seems to be slower. Tables IV, V, and VI present the charge (Coulombs) passed through different samples of the reference cement paste and mortar, Class C-containing and Class F-containing formulae, respectively. Samples were cured for 28 d at two temperatures, 23°C and 38°C, prior to measurements. The data (normalized to 6 h) are presented diagrammatically in Figures 5, 6 and 7(a,b) for reference mixtures, Class C-containing and Class F-containing mixtures, respectively. It was intended that all charge values (Coulombs) presented in Tables IV, V, and VI be based on 6 h duration of the electro-osmosis experiment. In several cases, however, it was found necessary to terminate the experiments at earlier times due to the very high passage of charge and subsequent excessive heat generation. In those cases, the time at which the run was terminated is indicated in parentheses. According to the experimental setup, the charge passed in those latter cases was estimated to exceed 10^5 Coulombs in one hour.

From the data given in Tables II, V, and VI and presented in Figures 5, 6 and 7, several interesting points emerge:

1. The most interesting outcome of this investigation is the benefit from the use of these fly ashes in hindering chloride ion penetration in cement mortars and pastes. In general, it was found that in mortars and pastes containing fly ashes from both curing temperatures, the total

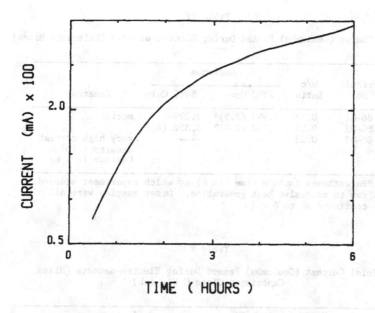

Figure 3. Typical current vs. time plot for sample no. 86-15
cured for 28 d at 38°C.

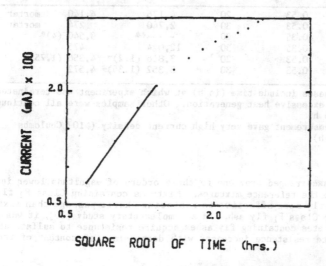

Figure 4. Current as a function of the square root of time for
sample 86-15 (28 d, 38°C).

Table IV

Total Charge (Coulombs) Passed During Electro-osmosis (Reference Mixes)

Formula No.	w/c Ratio	Coulombs		Remarks
		23°C Cure	38°C Cure	
86-01	0.53	2,999 (2.3)*	2,279	mortar
86-02	0.35	6,300 (2.8)*	8,352 (4)*	
86-03	0.53	------	------	very high current density (>10^5 Coulomb in 1 h)

*Parentheses include time (in h) at which experiment stopped due to excessive heat generation. Other samples were all continued up to 6 h.

Table V

Total Current (Coulombs) Passed During Electro-osmosis (Mixes Containing Class C Fly Ash)

Formula No.	Nominal w/(c+FA) Ratio	FA Replacement %	Coulombs		Remarks
			23°C	38°C	
86-10	0.53	20	4,176	4,140	mortar
86-11	0.53	30	2,740	274	mortar
86-12	0.35	20	-----**	9,540 (4)*	
86-13	0.35	30	12,024	475	
86-14	0.53	20	3,816 (1.2)*	4,356 (1.75)*	
86-15	0.53	30	3,352 (1.35)*	4,572	

*Parentheses include time (in h) at which experiment was terminated due to excessive heat generation. Other samples were all continued up to 6 h.
**23°C measurement gave very high current density (>10^5 Coulombs in one h).

charge passed ranged from one to three orders of magnitude lower in comparison to the reference mixtures. Mixtures containing Class F_2 fly ash are slightly more effective in resisting charge passage than mixtures containing Class F_1 fly ash. In a complementary study [20], it was found that concretes containing fly ashes acquire resistance to sulfate attack. The sulfate resistance increases with decreasing CaO content of the fly ash.

2. Increasing the percentage of fly ash suppresses the charge passage by one order of magnitude for either of the curing temperatures.

3. Temperature effects on chloride permeability are more emphasized in fly ash containing mixes. Thus, higher curing temperatures were found to decrease the charge densities passing through fly ash containing mixes by

Table VIa

Total Charge (Coulombs) Passed During Electro-osmosis (Mixes
Containing Class F_1 Fly Ash)

Mix No.	Nominal w/(c+FA) Ratio	FA Replacement %	Coulombs		Remarks
			23°C	38°C	
86–16	0.53	25	4,932	371	mortar
86–17	0.53	35	2,149	11	mortar
86–18	0.35	25	9,936 (5)*	630	
86–19	0.35	35	2,210	53	
86–20	0.53	25	3,708 (1.6)*	3,852	
86–21	0.53	35	1,670 (2.5)*	392	

*Parentheses include time (in h) at which experiment stopped due
to excessive heat generation. Other samples were all continued
up to 6 h.

Table VIb

Total Charge (Coulombs) Passed During Electro-osmosis (Mixes
Containing Class F_2 Fly Ash)

Formula No.	Nominal w/(c+FA) Ratio	FA Replacement %	Coulombs		Remarks
			23°C	38°C	
86–04	0.53	25	796	51(5)*⌐	mortar
86–05	0.53	35	23	102	mortar
86–06	0.35	25	5,076	976	
86–07	0.35	35	544	25	
86–08	0.53	25	5,328 (2.7)*	900	
86–09	0.53	35	1,242 (2)*⌐	109	

*Parentheses include time (in h) at which experiment stopped due to
excessive heat generation. Other samples were all continued up
to 6 h.
⌐Experiment stopped due to accidental power shutdown.

one order of magnitude. This is probably due to the more advanced degree
of re-action of the fly ash and greater proportion of C-S-H.

4. Tests of reference and fly ash containing mortars showed lower chloride
permeabilities than the corresponding pastes. A simple mathematical
normalization indicated that sand had negligible contribution to the process of chloride permeability which is mainly controlled by the composition of the cementitious matrix.

5. Water permeability values did not show a strong correlation with Cl⁻ permeability figures which may be a reflection of the different transport
mechanisms of the two systems.

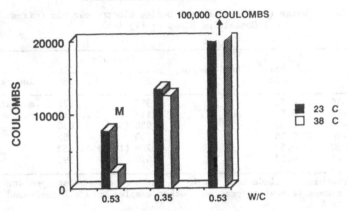

Figure 5. Comparison of various charges passed through reference mixtures. For the mixtures with w/c ratio = 0.53, the current passed exceeded 10^5 Coulombs (M = mortar).

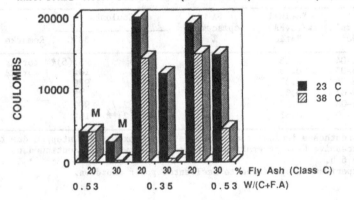

Figure 6. Comparison of various charges passed through mixtures containing Class C fly ash. For the mixture with 20% fly ash, w/c = 0.35 cured at 23°C the current passed exceeded 10^5 Coulombs (M = mortar).

CONCLUSIONS

Accelerated electro-osmosis experiments were carried out to measure the capability of chloride ions to penetrate cement pastes and mortars made either from cement (Type I) or cement/fly ash blends. It has been found that, by increasing the water/cementitious materials ratio the resistance to Cl⁻ diffusion decreases. Sand was found to impose no effect on the permeability to chloride ions.

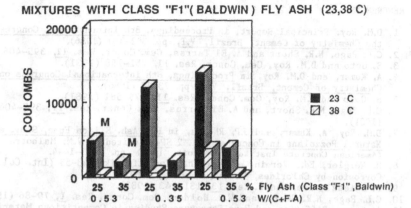

Figure 7a. Comparison of various charges passed through mixtures with Class F_1 fly ash (M = mortar).

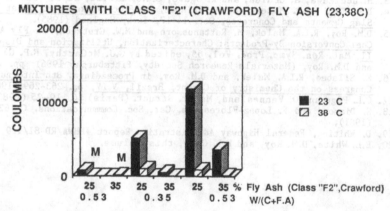

Figure 7b. Comparison of various charges passed through mixtures containing Class F_2 fly ash (M = mortar).

Class F fly ash is effective in increasing resistance of its blends with cement to chloride ion migration at any replacement level. Class C fly ash blends seem to provide high resistance to Cl⁻ transport only at higher replacement levels. Increasing curing temperature seems to suppress the chloride ion permeability, due to a higher degree of reaction of the fly ash.

ACKNOWLEDGEMENTS

This work was sponsored by the American Fly Ash Company. The authors thank Richard M. Majko, Vice President, for invaluable discussions.

REFERENCES

1. D.M. Roy, Principal Report, in Proceedings, 8th International Congress on the Chemistry of Cement, Brazil, V.I, pp. 362-380 (1986).
2. C.L. Page, N.R. Short and A. El Tarras, Cem. Concr. Res. 11, 395-406 (1981).
3. S. Goto and D.M. Roy, Cem. Concr. Res. 11, 751-758 (1981).
4. A. Kumar, and D.M. Roy, in Proceedings, 8th International Congress on the Chemistry of Cement, Brazil, V.5, pp. 73-79 (1986).
5. S. Goto and D.M. Roy, Cem. Concr. Res. 11, 575-581 (1981).
6. C.L. Page, N.R. Short, and A. El Tarras, Cem. Concr. Res. 11, 395-406 (1981).
7. D.M. Roy, A. Kumar, and J.P. Rhodes, in Fly Ash, Silica Fume, Slag, and Natural Pozzolans in Concrete, Vol. 2 SP-91 edited by V.M. Malhotra (American Concrete Institute, Detroit, 1986) pp. 1423-1444.
8. H. Weigler, Ed., Verein Oesterreich. Zementfabrik., 43-53 (Int. Colloq. Corrosion by Chlorides, Vienna, Austria, 1983).
9. P. Shiessl, Cem. Concr. Res. 13, 131-143 (1983).
10. C.L. Page, N.R. Short, and W.R. Holden, Cem. Concr. Res. 1, 79-86 (1986).
11. A. Kumar, Diffusion and Pore Structure Studies in Cementitious Materials, PhD Thesis in Solid State Science, The Pennsylvania State University, University Park, PA 16802 (1985).
12. C.J. Pereira and L.L. Hegedus, in I. Chem. E. Symp. Series #87, pp. 427-438 (1984).
13. H. Uchikawa, S. Uchida, and K. Ogawa, C.A.J. Review, 56-59 (1984).
14. D.M. Roy, R.I.A. Malek, and M. Silsbee, in Proc. Workshop on Blast Furnace Slag Cements and Concretes, Oct. 14-17, York, England (1985).
15. D.M. Roy, R.I.A. Malek, M. Rattanussorn and M.W. Grutzeck, in Fly Ash and Coal Conversion By-Products: Characterization, Utilization and Disposal II, Mat. Res. Symp. Proc. Vol. 65, edited by G.J. McCarthy, F.P. Glasser, and D.M. Roy, (Materials Research Society, Pittsburgh, 1986) pp. 219-226.
16. M. Silsbee, R.I.A. Malek, and D.M. Roy, in Proceedings, 8th International Congress on the Chemistry of Cement, Brazil, V.IV, pp. 263-269 (1986).
17. C.L. Page and O. Vennesland, Mater. Struct. (Paris) 16, 19-25 (1983).
18. S. Diamond and F. Lopez-Flores, Am. Cer. Soc. Communications, C-162-C-164 (1981).
19. D. Whiting, Federal Highway Administration Report #FHWA/RD-81/119 (1981).
20. E.L. White, D.M. Roy, and P.D. Cady, this volume.

MODELING THE EFFECTS OF FLY ASH CHARACTERISTICS AND MIXTURE PROPORTIONS
ON STRENGTH AND DURABILITY OF CONCRETES

ELIZABETH L. WHITE[a,b], DELLA M. ROY[c], and PHILIP D. CADY[a]
Materials Research Laboratory, The Pennsylvania State University, University
Park, PA 16802
a. also affiliated with the Department of Civil Engineering
b. also affiliated with Environmental Resources Research Institute
c. also affiliated with the Department of Materials Science and Engineering

Received 10 November, 1986; refereed

ABSTRACT

Factor analyses and cluster analyses were the modeling tools used to
relate the chemical and physical characteristics of fly ash and cement to the
strength, sulfate resistance, and freeze-thaw durability of fly ash-modified
concrete. A Type I Portland cement was mixed with base load and upset load
condition fly ashes from three different power plants in each of five regions
in the United States. Based on the interactions between the reactive constituents of the cement and fly ash, common factor loadings were identified.
Cement loaded onto the early strength factor; fly ash loaded onto the later
strength factor. In some subgroups the quantity of mixing liquid loaded separately as representative of the high water/cement ratio, which masked the
reactive interactions between the fly ash and cement. In other subgroups the
inter-relationships between sulfate resistance and strength with fly ash/cem-
ment fineness, CaO content, and alkali content were represented in the factor
analysis as well as in the numerical analysis models.

INTRODUCTION

To assess the performance (strength, sulfate resistance, and durability
characteristics) of fly ash in concrete, both a factor analysis model and a
cluster analysis model were proposed (Fig. 1) to specifically address the
chemical and physical properties of the fly ash and cement [1]. Fly ashes were
collected during base load and upset operating conditions from 16 plants
within five different regions in the United States. During base load and
upset conditions the coal source was constant, except for one case where the
clay (partings) content and composition of the coal source controlled the fly
ash properties and thus the properties of the resultant fly ash/concrete.
These ashes were used for preparing three sets of concrete mixture formulations in which several factors were varied. Mixtures were formulated with (a)
no additives, (b) a water reducer, and (c) a water reducer and an air entraining agent. Within each of the above subsets, comparisons were made at a
second level between each pure cement concrete mixture at 404, 351, 297 and
244 kg per cubic meter of cement (kg/m^3) (= 680, 590, 500 and 410 lbs of
cement per cubic yard (lbs/yd^3)). At a third level, the 351 kg/m^3 pure cement
concrete mixtures were compared with a 15 weight percent (wt.%) mixture of 53
kg (90 lbs) of fly ash with 297 kg (500 lbs) of cement, and a 30 wt.% mixture
of 107 kg (180 lbs) of fly ash with 243 kg (410 lbs) of cement. One hundred
and seventy-seven separate mixtures were investigated.

Fly ash is usually classified as a mineral admixture which will modify
the properties of the concrete in which it is used. Research from this study
and from other studies [2,3] indicates that fly ashes which are high in silica and alumina, light in color, high in fineness, low in carbon content, and
high in pozzolanic activity are high quality fly ashes; while high calcium fly
ashes are more cementitious and develop earlier strength. In general, a delay
in the strength development is observed with low calcium fly ashes such that
the optimal fly ash effect on the cured concrete is not usually seen until

Table I

Chemical Compositions of Fly Ash and Cement Samples

REGION 1 / REGION 2 / REGION 3

	1BL.1	1UP.1	1BL.2	1UP.2	1BL.3	1UP.3	CEM.1	2BL.1	2UP.1	2BL.2	2UP.2	2BL.3	2UP.3	CEM.2	3BL.1	3UP.1	3BL.2*
SiO_2	53.2	51.9	48.4	46.5	47.8	45.9	20.5	54.5	62.8	54.6	52.4	54.0	57.5	20.0	40.4	39.4	46.9
Al_2O_3	26.0	25.1	25.9	27.0	22.7	22.3	6.0	25.8	20.3	18.1	17.4	21.7	24.1	6.5	12.5	12.3	20.9
Fe_2O_3	8.0	12.8	16.1	15.3	21.9	22.2	2.1	11.4	10.9	24.4	25.6	14.5	12.0	4.0	7.4	8.7	5.3
CaO	3.57	1.71	1.73	1.79	2.18	2.18	62.8	1.50	1.34	2.65	3.37	1.08	1.10	65.1	24.0	26.6	15.0
MgO	0.97	0.89	0.75	0.69	0.92	0.93	2.9	1.30	1.15	0.88	0.92	0.92	0.94	3.3	5.3	5.3	4.7
SO_3	0.59	0.36	0.31	0.60	0.37	0.47	3.7	1.15	0.97	0.58	0.74	0.58	0.39	0.85	2.98	2.58	1.45
ALKALIS[+]	0.65	0.75	0.58	0.60	0.56	0.59	0.90	0.72	0.64	0.40	0.43	0.57	0.75	0.52	0.42	0.44	0.23
L.O.I.	2.22	1.68	1.90	4.60	2.51	2.15	1.2	5.9	4.9	2.3	1.9	16.6	12.2		0.11	<0.01	0.11
Carbon	1.62	1.13	1.59	3.72	1.43	2.65	—	5.13	4.37	1.90	1.47	15.3	11.5	—	0.03	0.02	0.02
Sp. Grav.	2.35	2.43	2.31	2.35	2.65	2.65	3.15	2.50	2.50	2.50	2.50	2.22	2.22	3.15	2.65	2.60	2.49
% >44 μm	19.0	12.0	31.4	26.8	12.6	5.75		7.60	3.55	29.3	28.1	34.6	36.9		12.6	20.6	10.5

REGION 3 / REGION 4 / REGION 5

	3BL.3	3UP.3	3BL.4	CEM.3	4BL.1	4UP.1	4UP.2*	4BL.3	4UP.3	CEM.4	5BL.1	5UP.1	5BL.2	5UP.2	5BL.3	5UP.3	CEM.5
SiO_2	50.2	50.8	34.7	20.5	32.8	30.9	59.3	36.1	35.7	20.0	40.2	40.8	49.8	56.3	52.4	56.3	19.9
Al_2O_3	18.7	18.7	17.4	5.97	17.6	17.3	27.0	19.8	19.3	5.93	17.8	17.5	15.8	19.5	20.5	19.5	6.08
Fe_2O_3	15.8	15.7	6.2	2.15	6.1	5.6	2.8	5.97	5.83	2.53	6.0	6.3	12.8	4.9	4.5	4.9	2.62
CaO	4.9	4.9	27.6	63.4	28.6	30.5	5.40	25.7	26.2	62.7	21.4	23.2	10.4	10.2	11.0	10.2	62.3
MgO	1.0	1.1	5.8	2.6	6.0	6.7	1.7	5.0	5.4	0.9	4.7	4.8	1.1	1.1	2.1	2.1	1.0
SO_3	1.15	1.10	3.85	2.9	3.26	3.20	0.44	1.48	1.66	4.4	1.60	1.90	1.90	0.92	0.70	0.84	4.7
ALKALIS[+]	0.70	0.75	1.54	0.78	1.17	1.35	0.27	0.77	1.02	0.94	0.76	0.65	0.92	1.10	1.10	1.20	0.89
L.O.I.	0.62	0.73	0.31	0.89	0.30	0.50	0.59	0.60	0.37	2.51	0.17	0.20	2.50	2.60	1.20	0.64	1.86
Carbon	0.41	0.53	0.11		0.36	0.93	0.66	0.70	0.43		0.11	0.01	2.47	2.39	1.15	0.48	
Sp. Grav.	2.38	2.36	2.69	3.17	2.69	2.67	2.14	2.65	2.66	3.15	2.63	2.67	2.49	2.47	2.39	2.43	3.14
% >44 μm	20.3	16.8	8.45		13.0	13.3	22.1	13.4	10.0		12.0	11.7	21.0	20.5	21.1	23.2	

*No upset condition sample taken for 3BL.2; no base load sample taken for 4UP.2.

[+]As equiv. Na_2O.

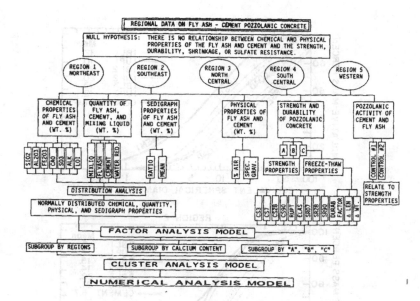

Figure 1. Flowsheet for developing fly ash concrete factor analysis and cluster analysis models.

after 28 days (d) or in some cases 90 d of curing.
The objectives of this study were:

- to reduce the set of independent fly ash/cement chemical and physical properties to a set of normally-distributed properties;
- to standardize the normally distributed input data matrix by conversion to a symmetrical correlation matrix;
- to determine the rank, k, of the components matrix in the factor analysis model based on the eigen values or characteristic roots of the correlation matrix;
- to use the correlation coefficients of the reduced set of k factors as input to the cluster analysis model to determine the "best" partition of "n" fly ash compositions into "j" clusters;
- to evaluate the strength and durability of concrete mixtures containing a wide variety of fly ashes.

EXPERIMENTAL PROCEDURES

Ash Sampling and Testing

Representative fly ash samples were obtained from five regions throughout the U.S. during base load and upset conditions. The power plants were selected with the following constraints: (1) boiler type pulverized coal input; (2) electrostatic precipitator or baghouse fly ash removal system; (3) dry conveyance collection system; (4) different coal source from each plant. Both base load (when generating unit was operating during steady-state conditions) and upset condition (when generating unit was operating under rapidly changing or other unsteady-state conditions) sets of fly ash samples were taken at each

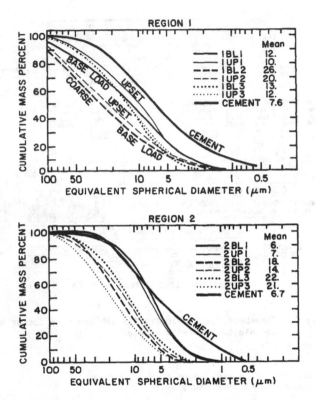

Figure 2. Sedigraph plots for Regions 1 and 2 cement and fly ashes.

plant. Ash sampling conformed to ASTM C 311 procedures. The sampling location was preferably along the conveyance line or directly below the hopper
manifold; however, some samples were taken at the storage silo or from plant
disposal trucks. Fifty-five gallon (0.21 m^3) drums were filled with 23 kg (50
lbs) of discrete samples collected continuously over a 1.5 to 2.5 hour (h)
period. When the base load was compared with the upset conditions, more re-
liance was placed on those ash samples collected along the conveyance line or
from the hopper manifold, rather than on those from the storage silo or from
the disposal trucks. The chemical analyses of the fly ash samples are given
in Table I. Particle size distributions of each of the fly ash samples (shown
in Figures 2 and 3) were measured by a sedimentation method on a Micromeritics
SediGraph utilizing X-ray scattering.
    The following concrete mixtures were prepared from each region [1,5]:

•Category A -- without chemical admixture: (4 reference mixtures without
 fly ash; 2 mixtures with 15% base load fly ash from two of the three
 plants; 2 mixtures with 15% upset load fly ash; 2 mixtures with 30% base
 load fly ash; 2 mixtures with 30% upset fly ash)
•Category B with 2.60 mL water reducer/kg of cement (4 oz./100 lbs):   (2
 reference mixtures without fly ash; 2 mixtures with 15% base load fly
 ash at two of the three plants; 1 mixture with 15% upset fly ash; 2
 mixtures with 30% base load fly ash; 1 at 30% upset fly ash)

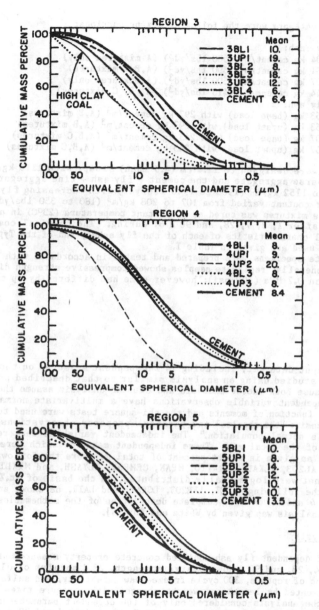

Figure 3. Sedigraph plots for Regions 3, 4 and 5 cement and fly ashes.

·Category C for freeze–thaw experiments with 2.60 mL water reducer/kg cement (4 oz./100 lbs) and 1.95 mL air entraining agent/kg cement (3 oz./100 lbs) (2 reference mixtures; 2 mixtures with 30% base load; 1 mixture with 30% upset fly ash).

Within each category were the following mix proportions:

- no fly ash:
    404 kg cement/m³ (680 lbs/yd³) (A mixtures only)
    351 kg cement/m³ (590 lbs/yd³) (A,B,C mixtures)
    297 kg cement/m³ (500 lbs/yd³) (A mixtures only)
    244 kg cement/m³ (410 lbs/yd³) (A, B, C mixtures)
- with fly ash:
    53 kg (base load) with 297 kg cement/m³ (A,B mixtures)
    53 kg (upset load) with 297 kg cement/m³ (A,B mixtures)
    107 kg (base load) with 243 kg cement/m³ (A,B,C mixtures)
    107 kg (upset load) with 243 kg cement/m³ (A,B,C mixtures)

Mixes were designed to 11.4 cm (4.5 inch) slump, with 1113 kg/m³ (1875 lbs/yd³) coarse aggregate and the cement + fly ash + fine aggregate varied from 1179 to 1152 kg/m³ (1985 to 1940 lbs/yd³) with increasing fly ash content. Water content varied from 107 to 208 kg/m³ (180 to 350 lbs/yd³). Each of the above mixtures was cured at a constant temperature (23°C) in a fog room and tested after 1, 3, 7, 28, and 90 d of curing. The chemical composition and physical characteristics of each of the fly ashes, and of the Type I Portland cement used are given in Table I.

Concrete specimens were prepared and tested in accordance with ASTM procedures. Generally, replicate samples showed compressive strength differences of less than 0.7 MPa (100 psi); however some had differed by up to 3.5 MPa (500 psi).

RESULTS

Normality

The input data set, including 14 independent measures on the 177 mixtures, was studied using an analysis similar to that described previously [3,4]. Because both factor analysis and cluster analysis assume the independent and dependent variable observations have a multivariate normal distribution, the function of moments and the chi-square tests were used to test the null hypothesis: "the chemical and physical property observations were not drawn from a normal population." The independent variables were represented as percent of the total mix. Those independent measures which were normally distributed as given in weight percent of total mix were the following: CAO, SIO2, SO3, AL2O3, ALKALIS, RATIO, MEAN, CEMENT, FLYASH, and MIXLIQ; those measures that were log normally distributed, on the basis of wt.% of total mix, were the following: LLOI, LFE2O3, LGRAV, and LAIR, as shown and defined in Figures 4 and 5. A more complete description of the mathematics for this normality analysis was given by White and Roy [4].

Factor Analysis

The 12 dependent fly ash-modified concrete property measurements included: 3, 7, 14, 28, and 90 d compressive strength, 28 d modulus of elasticity, 28 d modulus of rupture, 300 cycle freeze-thaw durability, and sulfate resistance (represented by length and weight changes with exposure times up to 360 d). When the analysis considered only of the dependent parameters: Factor I was represented by compressive strength at 7, 28, and 90 d and 28 d modulus of rupture; Factor II by the sulfate resistance after 7 and 28 d of curing; Factor III by the 90 d cured sulfate resistance and the modulus of elasticity; and Factor IV by the drying shrinkage after 28 and 90 d and by the early compressive strength (3 d). The 14 independent chemical and physical measurements, including weight percent of the chemical components (cement + fly ash) and values of the physical characteristics relative to the proportion of fly ash or cement in each mixture, were reduced to a set of factors. Factor I

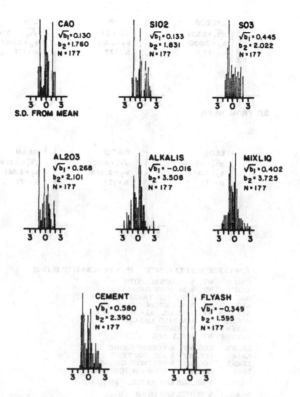

Figure 4. Distribution analyses for chemical properties of fly ash and cement mixtures.

represented the quantity of cement and/or CaO in each mixture, which was negatively related to the quantity of fly ash and/or SiO_2 in each mixture. Factor II was not as well defined. Factor III consistently represented the quantity of air entraining agent added to each mixture, which was negatively related to the quantity of mixing liquid added to each mixture. Others of the independent measures loaded onto these and other Factors I through VII with no consistency. Examples of these results are shown in Figure 6. When the dependent and independent measures were considered together, the strength properties loaded onto Factor II separate from the independent measures as shown in Figure 7, thus showing no clear relationships between the independent fly ash/cement properties and the dependent concrete properties.

Cluster Analysis

Based on the quantity of CaO in the 177 mixtures, the data were subdivided into subgroups based on quantity of CaO in the fly ash alone: (1) 1 to 5%; (2) 10-15%; and (3) >20%. Within the low calcium fly ash mixtures obtained from eastern U.S. power plants in this study, two distinct clusters are defined, based on the combination of 28 d compressive strength and CaO in each mixture. Three groups were defined when the 28 d compressive strength was analyzed with respect to the alkali content. The cluster analysis models were

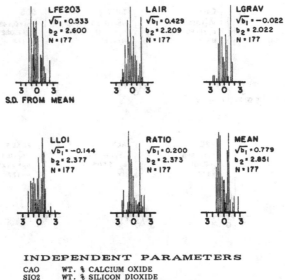

INDEPENDENT PARAMETERS

CAO	WT. % CALCIUM OXIDE
SIO2	WT. % SILICON DIOXIDE
SO3	WT. % SULFER TRIOXIDE
AL2O3	WT. % ALUMINUM OXIDE
ALKALIS	WT. % AVAIL. ALKALIS (Na$_2$O)
MIXLIQ	WT. % MIXING LIQUID
CEMENT	WT. % CEMENT
FLYASH	WT. % FLY ASH
LFE2O3	LOG WT. % FERRIC OXIDE
LAIR	LOG WT. % AIR CONTENT
LGRAV	LOG WT. % SPECIFIC GRAVITY
LLOI	LOG WT. % LOSS ON IGNITION
RATIO	% SEDIGRAPH RATIO: $\frac{1}{2}(D_{95} - D_5)$
MEAN	% SEDIGRAPH MEAN: D_{50}

where D_{95}, D_{50}, and D_5 are the respective diameter of the fly ash/cement at the 95, 50, and 5 cumulative mass percent levels.

Figure 5. Distribution analyses for chemical and physical properties of fly ash and cement mixtures.

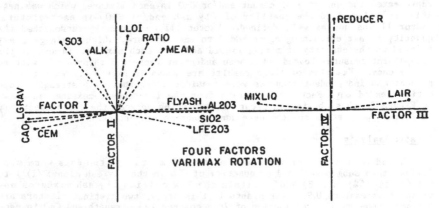

FOUR FACTORS
VARIMAX ROTATION

Figure 6. Factor loadings for chemical and physical properties of uncured fly ash and cement mixtures.

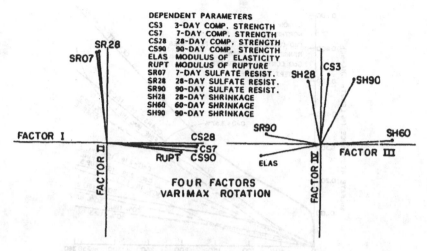

Figure 7. Factor loadings for strength, sulfate resistance, and shrinkage properties of cured fly ash concretes.

limited in their applicability in that only one dependent and one independent parameter could be modeled at the same time.

When studying more closely the properties of the individual cement/fly ash mixtures, effects of CaO content, Fe_2O_3 content, alkali content, and fineness of the fly ash/cement mixture controlled the relative position of the samples within each cluster. An example of these results reported by White, et al. [5] showed that a higher calcium, alkali, and fineness and a lower iron content produced higher strength fly ash concretes.

Numerical Analysis

Numerical analysis of the changes in sulfate resistance or compressive strength with curing time up to 360 d are shown in Figures 8–10. The low calcium fly ashes have greater sulfate resistance (smaller length change) than the pure cement mixtures as shown in Figure 8. The sulfate resistance was less when the CaO content increased from 1.7 to 3.6% or when the fly ash was more coarse. An increase in the quantity of fly ash within each mixture increased the sulfate resistance.

With high calcium fly ashes it was generally necessary to use 30% fly ash to improve the resistance compared with the cement control specimens. A decrease in CaO content of fly ash from greater than 25% to 5% increased the sulfate resistance of the concrete mixtures (Fig. 9). In a complementary study [6], it was found that the decrease in CaO content also increased the resistance to chloride ion diffusion in fly ash/cement pastes.

There was more variability in the sulfate resistance of the high calcium fly ashes. For most fly ash sources an increase from 15 to 30% fly ash increased the sulfate resistance, except for one base load fly ash (4 BL.1). Its complimentary upset fly ash, (4 UP.1), with a very similar chemistry followed the pattern of the others. Excessive clay content in a coal seam is shown to produce a less sulfate resistant concrete (Fig. 10). During collection the fly ash sample failed the autoclave test, probably the result of high lime ash and the minimal quantity of coal in the seam.

In Figures 11–12, the 90 d compressive strengths of the Category A fly ash concretes (without chemical admixtures) from all regions are plotted

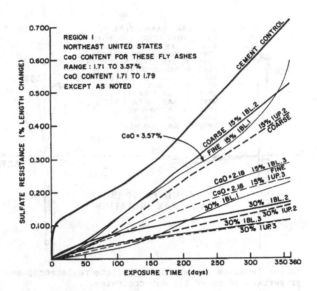

Figure 8.   Sulfate resistance of low calcium fly ash mixtures at 0, 15, and
30% fly ash content.   Note: higher calcium content fly ash (1BL.1)
and coarse fly ash (1BL.2 and 1UP.2) produced a less sulfate
resistant concrete.

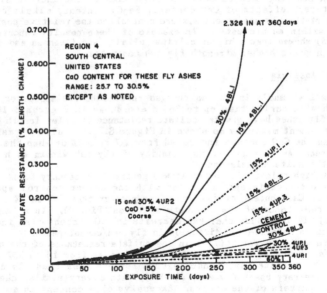

Figure 9.   Sulfate resistance of high calcium fly ash mixtures at 0, 15, 30,
and 60% fly ash content.   Note: the 4UP.2 fly ash with a calcium
content of 5% and 2.8% Fe_2O_3 was a more sulfate resistant
concrete.

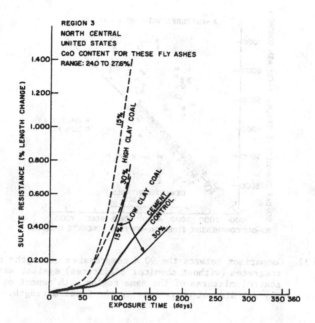

Figure 10. Sulfate resistance based on clay content in the original coal seam.

against strengths of control mixtures of the same region with cement only. The assumption that an equal mass of fly ash would replace an equal mass of cement is based on the assumption that the hydration product (e.g. C-S-H) occupies similar volumes irrespective of whether the source is cement or fly ash. The mix design with 244 and 297 kg/m^3 (410 and 500 lb/yd^3) controls had extra sand when the cement content was low. Figure 11 (case I) contains two groupings, the 15% and 30% fly ash concretes plotted against the 243 and 297 kg cement/m^3 (410 and 500 lb/yd^3) reference mixtures respectively. Data fitting on the diagonal line would indicate the fly ash did not contribute at all to the strength. The vast majority of the data fall in the left/ "above" line portion, indicating that fly ash contributes significantly to the strength. In Figure 12 (case II) the strengths of both 15 and 30% mixtures are plotted against the strength of the appropriate regional reference mixture with 351 kg cement/cubic meter (590 lb/yd^3). In this plot, points on the diagonal would mean that the fly ash behaved as the cement. It is not surprising that the majority of the data lie below/"right" of the diagonal, indicating less contribution to strength than the cement itself, but several of the mixtures lie above/left of the diagonal showing greater contribution to strength than cement in the total mixture.

The same general types of comparisons apply in the mixtures containing a water-reducing admixture, but are much more striking as shown in Figure 13. In the lower figure, case I, all the fly ash data, when plotted against the 243 kg/m^3 (410 lb/yd^3) control mixtures, fall to the left of the diagonal (are stronger), whereas even in case II, the upper figure, the majority of the data plotted against 351 kg/m^3 (590 lb/yd^3) mixtures fall to left of the diagonal.

262

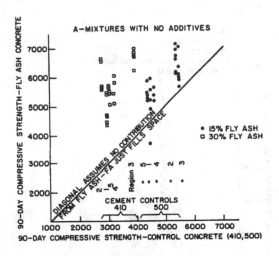

Figure 11. Comparison between the 90 d compressive strengths of fly ash
concretes (without chemical admixtures) against strengths of
control mixtures of the same region with cement only, assuming
fly ash did not contribute to the final strength.

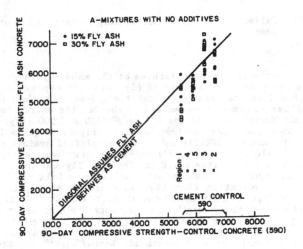

Figure 12. Comparison between the 90 d compressive strengths of fly ash
concretes (without chemical admixtures) against strengths of
control mixtures of the same region with cement only, assuming
fly ash behaved as cement in the final strength of the concrete.

This shows that there is much enhancement in 90 d strength (relative to cement) when the combination fly ash-water reducer is used. The importance of
appropriate control specimens is borne out in each of these examples.

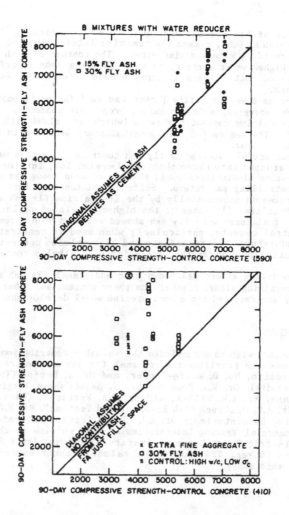

Figure 13. Comparison between the 90 d compressive strengths of fly ash
concretes (with water reducer) against strengths of control
mixtures of the same region with cement only, assuming fly ash
behaved as cement in the upper figure and did not contribute to
the the concrete strength in the lower figure.

CONCLUSIONS

A statistical model was developed to relate the fly ash and cement
characteristics and mix proportions to strength and sulfate resistance characteristics of fly ash concretes. Chemical factors such as high calcium content
(cement plus fly ash) were found to create a more reactive mix which accelerated strength development. Interactions between the reactive constituents
of the cement and fly ash controlled the final product strength, with the
cement controlling the earlier strength and the fly ash the later strength.
Total calcium, as well as the free lime in the cement, relative to the reac-

tive portion of the fly ash was found to be significant. Some of the w/c ratios, as high as 0.86, masked the reactive interactions of the fly ash and cement by providing excess mixing water. The coarser fly ash mixtures tended to require higher water content, as well as being less reactive. Therefore, compositions of such fly ash mixtures should be used with downward-adjusted sand contents.

The freeze-thaw Durability Factor was satisfactory (about 85 to 95) for all fly ash concretes, which contained proper air contents, except for three intermediate calcium content fly ashes (which were mixed with coarser cement) in Region 5. The reason for this unsatisfactory behavior in the three mixes is not totally clear.

The most useful modeling of fly ash concrete properties was possible when ashes were grouped into three chemical levels: low, intermediate, and high calcium. Better distinctions could thereby be made among the effects of other property-controlling parameters. Sulfate resistance of Portland cement mortars was increased substantially by the low calcium fly ash substitution and by the high calcium fly ashes at the higher levels of replacement. A large proportion of mixtures with fly ash showed enhanced 90 d strengths over those of the control cements, particularly when used in combination with water reducing admixtures. Intermediate and high calcium content mixes showed a slight negative 28 d strength correlation with alkalis content and SO3 content.

This modeling effort has shown that additional fly ash characterization such as determining glass, crystalline phase content and other measures of ash reactivity, are required for a more precise model development.

ACKNOWLEDGMENTS

The authors wish to acknowledge the valuable contributions of the following researchers in providing the regional fly ash and concrete test results: Dr. P.H. Licastro, Ms. D. Wolfe-Confer, and Mr. S. Devine (The Pennsylvania State University), Dr. W.J. Head and Mr. J. Sajadi (West Virginia University), Mr. O.E. Manz, Mr. K.W. Wilken, and Mr. B.A. Westacott (University of North Dakota); Mr. A.C. Masbruch (Ash Research and Testing), Ms. P.L. Printy (Western Testing Laboratories); Dr. R.L. Smith (Enreco Laboratories), Mr. O.R. Werner (Commercial Testing Laboratories). Also we wish to thank Mr. Dean M. Golden (Electric Power Research Institute) and Ms. Joyce S. Perri and Mr. Garth Gumtz (Baker, TSA) for their valuable contributions and comments throughout this study.

REFERENCES

1. G. Gumtz, J.S. Perri, D.M. Roy, E.L. White and E. Dunstan, Fly Ash Classification Project Final Report. Fly Ash Cement/Concrete Use Projects. EPRI Project RP 2422-10 (Electric Power Research Institute, Palo Alto, CA, 1986) 230 pp.
2. J.R. Belet Jr., in Fly Ash Utilization Proceedings: Edison Electric Institute - National Coal Association - Bureau of Mines, Pittsburgh, PA. Bureau of Mines Circular 8483 (1967).
3. E.L. White and D.M. Roy, in Fly Ash and Coal Conversion By-Products: Characterization, Utilization and Disposal II, edited by G.J. McCarthy, F.P. Glasser and D.M. Roy, Mat. Res. Soc. Symp. Proc. Vol. 65, (Materials Research Society, Pittsburgh, 1986) pp. 243-254.
4. E.L. White, Water Resources Bull. 11, 676-687 (1975).
5. E.L. White, P.D. Cady, D.M. Roy and S. Devine, Effects of fuel source and plant operations on the properties of fly ash and the resultant concrete. (submitted to National Council for Cement and Building Materials Proceedings, New Dehli, India)
6. R.I.A. Malek, D.M. Roy, P.H. Licastro, this volume.

MODELING OF TEMPERATURES IN CEMENTITIOUS MONOLITHS

S. KAUSHAL, D.M. ROY* and P.H. LICASTRO
Materials Research Laboratory, The Pennsylvania State University, University Park, PA 16802
*Also affiliated with the Department of Materials Science and Engineering

Received 2 February, 1987; Communicated by G.J. McCarthy

ABSTRACT

Temperatures in large cementitious monoliths (works) can rise very high due to unfavorable thermal properties such as low conductivity and high diffusivity of the monolith and the surrounding media. Heat moderation becomes necessary in such situations to avoid excessive thermal stresses. Moderation due to the addition of inert additives such as sand in mortars is compared to that obtained by the addition of reactive but low heat evolution substituents such as Class C and Class F fly ashes. Substitution of cement by slag has also been considered. The hydration temperatures for the extreme conditions (adiabatic) have been experimentally measured and compared to those predicted under real conditions. Such a simulation has been made by measuring the thermal properties and analyzing the temperature distribution due to exothermic reactions as predicted by a finite differences computer model. In general, lower temperatures can be maintained by increasing the thermal conductivity and heat capacity of the hydrating material. This material can be tailored for both heat evolution and setting times by incorporating inert additives such a sand (quartz) and/or reactive additives such as slag and fly ash.

INTRODUCTION

The hydration of cementitious materials is the result of a series of complex exothermic reactions. As a result of these hydraulic reactions, the temperature of a cementitious monolith increases significantly within three days [1]. Similarly high temperature increases have been predicted as well as reported in complex blended cements [2,3].

The increase in temperatures in large works is known to cause earlier strength development. This has led to fears about thermal cracking due to the large temperature differentials between the interior and exterior of concrete bodies [4].

At the same time, the appearance of fly ash as a cheap and abundant mineral source has made the equivalent of 10-20 million tons of cement available in the U.S. This has resulted in a lowering of the energy requirement in the production of concrete and directed more research toward incorporating fly ash in cement. However, very little effort has been directed toward recognizing blends of portland cement and fly ash as integral reacting systems and to study such a system at the more realistic elevated temperatures attained [5].

While it is accepted that the heat and the alkali hydroxides released during cement hydration are the causes for fly ash activation, it is not known whether fly ash causes retardation or acceleration of the kinetics of cement hydration [5,6].

The study of temperatures in a large cementitious work can be broken up into a prediction of the actual temperatures forecast and the simulation of such temperatures in a laboratory. The salient features of such temperature studies are given in Table I. In the present paper we attempt to model the temperatures under adiabatic conditions and the effect that they have on the cement, slag and fly ash in terms of the resultant hydration products and microstructures.

Table I

Study of Temperature in a Large Cementitious Work

Actual Temperatures	Simulated Temperature (Adiabatic Conditions)

– heat of hydration
– thermal conductivity
– thermal diffusivity
– dimensions
– boundary conditions
	– study of hydrated
	  material; possible
	  (XRD, SEM)
	– gives cumulative heat
	– independent of size

Table II

Chemical Analysis (wt.%) of Starting Solids

	Type I Cement	Slag	Fly Ash Class C	Fly Ash Class F
CaO	65.18	39.37	26.14	5.60
SiO_2	20.10	34.70	36.90	50.95
Al_2O_3	5.79	10.70	17.50	26.00
MgO	1.32	11.90	5.05	1.89
Fe_2O_3	3.41	0.41	5.90	7.75
Na_2O	0.14	0.25	1.91	0.51
K_2O	0.42	0.55	0.45	3.33
LOI (950°C)	0.86	----	0.39	1.83
Totals	100.3%	100.36%	99.6%	100.1%

EXPERIMENTAL

Starting Materials Characterization

Table II shows the bulk chemical (oxide) composition in terms of the important elements of the fly ash, slag and the Type I cement used in the present investigation. Both low-calcium (ASTM Class F) and high-calcium (ASTM Class C) fly ash were studied. Additional activation of the fly ash and slag was achieved by the addition of a high alkali (NaOH, $NaNO_3$, $NaNO_2$, $NaAl(OH)_4$, Na_2CO_3, etc.) waste solution [2] containing 31% by weight salts.

Adiabatic Calorimetry

Measurements of adiabatic temperature rise were made using a computer-controlled adiabatic calorimeter developed at the Materials Research Laboratory (MRL) of The Pennsylvania State University. The samples were prevented from losing any heat during hydration by placing them in a container and surrounding it with a glass vessel with silver plating on its inner and outer surfaces. Conditions very close to zero heat exchange with the environment were achieved by placing the assemblage in a water bath which was maintained at the same temperature as the hydrating sample at all times. An oil bath was used for any sample whose temperature was anticipated to rise above 100°C during adiabatic curing. The temperature differential within the sample, as

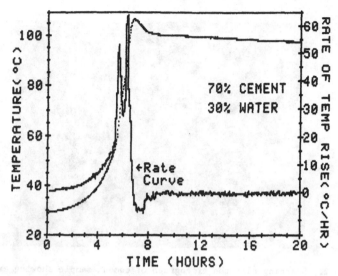

Figure 1. Adiabatic calorimeter results for a cement sample.

well as between the sample and surrounding bath, was within 0.1°C at all times.

Other Related Measurements

All samples were mixed in accordance with ASTM procedure C 305 for the mixing of cement pastes and mortars. At the time of mixing samples were also cast for specific heat and thermal conductivity measurements. For the specific heat measurements, cubes 5.1 cm on an edge were cast in molds and hydrated at 38°C and 95% relative humidity (RH) for seven days (d). Measurements were made by the method of mixtures using kerosene as the liquid (procedure CRD-C 242-54). Cylindrical samples, 5.1 cm in diameter and 1.3 cm high were used for thermal conductivity measurements, which were made with a thermal comparator developed at MRL [7]. The specific heat and thermal conductivity values are needed in order to calculate the heat evolution and to predict the temperatures in an actual cement structure.

The adiabatically cured samples were also analyzed by x-ray diffraction (XRD) and scanning electron microscopy (SEM).

RESULTS AND DISCUSSION

Cement Hydration

The adiabatic temperature rise vs. time for a mixture containing 70% cement and 30% deionized water is plotted in Figure 1. It can be seen that the temperature of the sample rises to above 100°C within 7 hours (h) of mixing. This represents an increase of about 80°C. A rate curve has been generated by a 5-point numerical differentiation of the temperature curve. While the first hydration peak was missed during mixing of the sample, very sharp second and third hydration peaks show up. These correspond to the hydration of tri-calcium silicate (C_3S) and tricalcium aluminate (C_3A), respectively.

Figure 2.   Scanning electron micrograph of cement sample showing extensive cracking.

The "boiling off" of the free water in the not fully hydrated sample has resulted in a lowering of its temperature once it exceeds 100°C. Such intense hydration peaks have resulted in micro-cracking of the cement sample as can be seen in the SEM in Figure 2.

Temperature Moderation

High temperatures would develop in a cement monolith work since the hydrating conditions for the cement would be very close to adiabatic. Thus the situation would be close to that seen in Figure 1. This would lead to high thermal stresses and "boiling off" of the mixing water. In order to moderate the temperatures, one has the following options:

1. Increase thermal conductivity.
2. Decrease thermal diffusivity or increase specific heat.
3. Change the environment around hydrating material, e.g., surrounding material, heat dissipation mechanism, size, etc.
4. Moderate heat of hydration and hydration rate by addition of:
    a. inert additives, e.g., sand, aggregate (as in normal concrete),
    b. low heat evolving reactive additives,
    c. ground granulated blast-furnace slag,
       (1) activated by alkali addition,
       (2) unactivated,
    d. Class C and Class F fly ash,
       (1) activated by alkali addition,
       (2) activated by cement or slag.

Out of the four alternatives above, the first three are usually difficult to achieve or are defined by the nature of the cementitious work. Hence it becomes important to reduce the intensity of the second and third peaks as well as spread the hydration over a longer time. This is achieved by the addition of slag, fly ash and/or unreactive sand and aggregate.

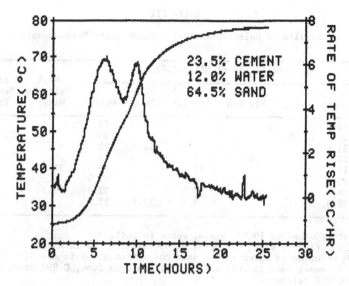

Figure 3.   Temperature moderation by use of an inert additive.

The addition of sand to form a 1:2.75 cement-sand mortar results in a lowering of the adiabatic temperature rise as well as of the C_3S and C_3A hydration peaks, as seen in Figure 3.   It is also seen that the hydration peaks are broadened by the presence of inert silica.   This presents a barrier to the diffusing species and results in a lowering of the second and third hydration peaks by approximately 10 times, a factor greater than that explained by the 73% replacement of cement by an inert additive.

The use of a less reactive additive such as slag or fly ash causes a similar reduction in temperatures.   However, it has been seen that the use of slag instead of cement results in cracking and disintegration of the samples unless it is diluted by inert phases [2].   This has been noticed for both unactivated as well as alkali activated slags.   The heat moderation by slag is not sufficient unless a slag cement mortar is used.   The temperature of the 1:1 cement-slag mix (mixture 2), though not as high as cement alone, rises to above 100°C as seen in Table III.

Hydration of Fly Ash

Applications such as radioactive waste encapsulation require that a sufficient quantity of solids be of a reactive nature so as to potentially fix a maximum amount of waste in the matrix.   Thus, the use of fly ash offers an attractive proposition for a waste encapsulation constituent.

The effect of fly ash is to both retard the hydration of cement and lower the heat of hydration, as is seen in Figure 4.   This is done by a combination effect of the fly ash and the fly ash-activating alkali salts.   The extent of temperature moderation by the fly ashes in mixtures 4 and 5 is comparable to that due to the inert sand, as seen in Table III.   A lesser quantity of cement is used in these than in mixture 3 due to the reactive nature of the fly ash.

The retarding nature of the alkalies on the hydration of fly ash is seen clearly in Figure 5.   The dissolution of lime and tricalcium aluminate in a Class C fly ash in water produces an early reaction within 2 h and an adiabatic temperature increase of 16.7°C.   The alkali activated Class C fly ash

Table III

Results of Selected Adiabatic Calorimetry Measurements

	Composition of Mix (wt.%)						Wt.%	Adiabatic
Mix #	Type I Cement	Slag	Class C Fly Ash	Class F Fly Ash	C109 Sand	Water	Salts in Water	Temperature Rise (°C)
1	70					30		85+
2	35	35				30		80+
3	23.5				64.5	12		55
4	15		40			45	31	51.7
5	15			45		40	31	50.4
6	16.8			50.3		32.9	6.25[1]	31.1
7	16.8			50.3		32.9	19.8[2]	41.6
8[3]	33.8			8.2	33.8	15.9		68

+Temperature exceeded 100°C causing water to boil.
1. NaOH solution in water in the same concentration as in mix 5.
2. NaNO$_3$ solution in water in the same concentration as in mix 5.
3. Type V cement used in this mix with 7.3% silica fume, 0.99% superplasticizer, and 0.005% defoamer.

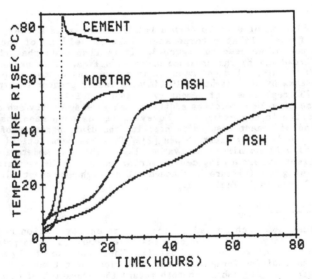

Figure 4. Adiabatic temperature rise for mixes 1, 3, 4, and 5.

did not produce any early peak. Instead it produced a peak at 42 h and an adiabatic temperature increase of 55.5°C. This shows a major reaction of the glassy phase with the alkalies and formation of C-S-H I gel and sodium aluminosilicate reaction products.

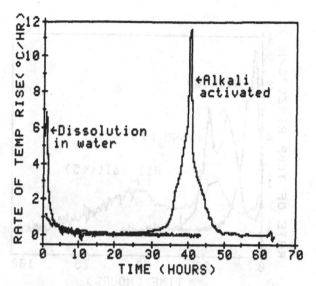

Figure 5. Adiabatic reaction of Class C fly ash in water and by alkali activation.

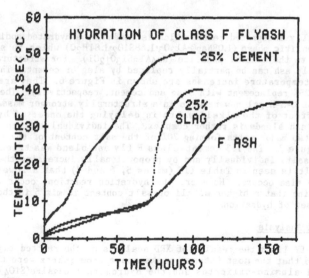

Figure 6. Adiabatic temperature rise in Class F fly ash.

Class F fly ash does not show any appreciable reaction unless it is activated by the alkali solution. The activation is slow and less than that in Class C fly ash. A major hydration peak occurs at 94 h and results in a total adiabatic temperature increase of 36.7°C. The end product is nevertheless

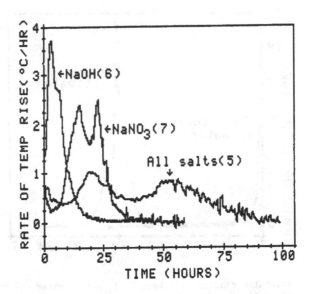

Figure 7. Adiabatic reaction peaks for mixes 5, 6 and 7.

a structurally integrated cement which is bonded by a hydrated sodium alumino-silicate zeolitic phase ($1.08Na_2O \cdot Al_2O_3 \cdot 1.68SiO_2 \cdot 1.81H_2O$) which is structurally similar to the mineral sodalite ($Na_4Al_3Si_3O_{12}Cl$). For structural applications the fly ash can be partially replaced by slag or cement. The resulting adiabatic temperature increases are shown in Figure 6. They are $40.2°C$ and $50.4°C$ for 25% replacement with slag and cement, respectively. The additional formation of C-S-H gel now results in a structurally stronger mass.

The effect of the alkali salts in delaying the onset of hydration in cement-fly ash blends is rather complex. The individual effects of NaOH and NaNO_3, the two major salts forming 70% of the salt content of the solution, is seen in Figure 7. The 1:3 cement-Class F fly ash blend was tried with these two sodium salts individually and by proportionally increasing the other components. It is seen in Table III (mixes 5, 6 and 7) that a lower adiabatic temperature rise occurs. However, the hydration reactions occur sooner (Fig. 7), suggesting that a higher alkali or salt content in mix 5 is the reason for delayed onset of hydration.

XRD and EDX Analysis

Table IV lists the results of XRD analysis on the reacted fly ashes. It was noticed that the most intense peaks apart from quartz were those due to the alkali alumino-silicates and the hydrogarnet $Ca_3AlFe(SiO_4)(OH)_8$. This indicates that major crystalline binding phases resulting from fly ash reaction were present in addition to C-S-H gel.

The electron micrograph in Figure 8 shows the alkali attack on Class F fly ash. The EDX analysis in the light area surrounding the fly ash particle showed very little calcium but much sodium.

Table IV

Phases Identified by XRD in Activated Fly Ash

	Class C	Class F		Class C	Class F
Unreacted Starting Material			**Sodium Alumino-Silicates**		
Quartz	*	*	$N_{1.08}AS_{1.68}H_{1.81}$		*
Mullite	*	*	$NAS_{1.9-2.2}H_{1.9-2.2}$	*	
Merwinite	*		$NA_3S \cdot H$	*	
Melilite	*				
Glassy Phase	*	*	**Other Crystalline Phases**		
Hematite	*	*	Tetra-calcium Aluminate-13		
Ferrite Spinel	*		hydrate (with iron sub)	*	*
			Calcite	*	
Crystalline Matrix Phases			Friedel's Salt	*	
C-S-H gel		*	Hydrotalcite		*
C-S-H I gel	*		Garnet-Hydrogarnet s.s.	*	*
			$C_3(AF)SH_4$	*	

* = crystalline phase detected in sample.

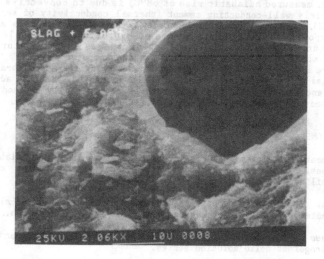

Figure 8. SEM of alkali activated Class F fly ash with 25% slag replacement.
Oxide analysis at point X is Na_2O = 29.6%, Al_2O_3 = 22.7%, SiO_2 =
37.3%, CaO = 4.6%, FeO = 3.3%, and K_2O = 2.6%.

Heat Evolved

The heat evolution quantities per unit mass of the cement, slag, Class C
and Class F fly ash are listed in Table V. It is seen that both Class C and
Class F fly ashes can be used to moderate heat evolution in cementitious monoliths and thereby keep temperatures lower. The longer activation times can
also be useful in dissipating the heat.

Table V

Effect of Alkali Activation on Fly Ash

| | Heat Evolved (cal/g) | | Activation Time (h) |
	Unactivated	Activated	
Cement	70–90	60–80	5–10
Slag	15	40–60	35
Class C	7–8	30–35	42
Class F	<3	15–20	94

Maximum Temperature in an Actual Cement Work

A finite differences computer program [8] was developed at MRL to analyze the actual temperatures in cylindrical cementitious structures. Figure 9 shows a comparison of the adiabatic temperature rise vs. the computer-predicted temperature rise at the center of a 0.8 m borehole in a granite plugged with a blended cement (mix 8). The lower simulation value of temperature rise (42°C vs. measured adiabatic rise of 68°C) is due to convective boundaries, and a fairly well-conducting cement (thermal conductivity of set mix was 1.2 Watts/m–K). A lower thermal conductivity value of 0.8 W/m–K was predicted for this composition during early stages of hydration. This would result in a temperature rise of about 70°C. Such increases in temperature have been measured while doing confirmatory work on the computer model.

More research needs to be done on establishing the actual heat conduction mechanisms for a hydrating cement. Still, it can be said that adiabatic conditions and the resulting high temperatures can be easily reached and, therefore, need moderation by one or more of the additives discussed.

CONCLUSIONS

•Cement paste temperatures may rise to above 100°C in adiabatic conditions. This makes heat moderation necessary to prevent cracking and "boiling off" of the water.

•Heat evolution in cements is moderated by the addition of slag and to a greater extent by inert aggregate such as sand, and fly ash.

•Class C fly ash reacted extensively to form C–S–H I gel and the garnet-hydrogarnet solid solution series.

•Class F fly ash on activation by sodium salts formed sodium aluminum silicate hydrate--a zeolitic structure. This resulted in a low temperature bonded cement containing no calcium.

•Even mixes containing a high percentage of fly ash maintain their structural integrity.

•Increasing the setting time (by addition of fly ash and a combination of sodium salts) and the thermal conductivity (by addition of sand) can help in dissipating the heat and lowering the temperature.

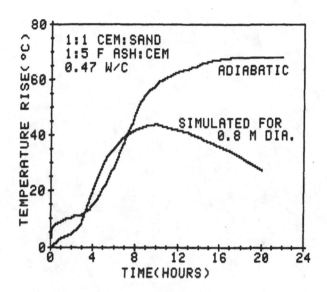

Figure 9. Comparison of a computer simulated temperature rise
with adiabatic temperature rise.

REFERENCES

1. D.M. Roy, K. Luke and S. Diamond, in Fly Ash and Coal Conversion By-Products: Characterization, Utilization and Disposal I, edited by G.J. McCarthy and R.J. Lauf, Mat. Res. Soc. Symp. Proc. Vol. 43 (Materials Research Society, Pittsburgh, 1985), pp. 3–20.
2. S. Kaushal, D.M. Roy, P.H. Licastro and C.A. Langton, in Fly Ash and Coal Conversion By-Products: Characterization, Utilization and Disposal II, edited by G.J. McCarthy, F.P. Glasser, and D.M. Roy, Mat. Res. Soc. Symp. Proc. Vol. 65 (Materials Research Society, Pittsburgh, 1986), pp. 311–320.
3. C. Gotsis, D.M. Roy, P.H. Licastro and S. Kaushal, in Concrete at Early Ages (American Concrete Institute, Detroit, 1986), pp. 49–70.
4. G.M. Idorn and K.R. Henrikson, Cem. Concr. Res. 14, pp. 463–470 (1984).
5. G.M. Idorn, in Fly Ash and Coal Conversion By-Products: Characterization, Utilization and Disposal II, edited by G.J. McCarthy, F.P. Glasser, and D.M. Roy, Mat. Res. Soc. Symp. Proc. Vol. 65 (Materials Research Society, Pittsburgh, 1986), pp. 3–10.
6. K. Luke and F.P. Glasser, in Fly Ash and Coal Conversion By-Products: Characterization, Utilization and Disposal II, edited by G.J. McCarthy, F.P. Glasser, and D.M. Roy, Mat. Res. Soc. Symp. Proc. Vol. 65 (Materials Research Society, Pittsburgh, 1986), pp. 173–180.
7. A.A. Rousan and D.M. Roy, I&EC Product Research and Development 22, 349–351 (1983).
8. C. Gotsis, ATHENAN: Axisymmetric Thermal Nonlinear Analyses--A Computer Program for Cements and Other Chemically Reactive Cylindrical Domains and Associated Computer Graphic Algorithms, Ph.D. Dissertation, The Pennsylvania State University, University Park, PA (1984).

THE EFFECT OF SIMULATED LARGE POUR CURING CONDITIONS ON THE
TEMPERATURE RISE AND STRENGTH GROWTH OF PFA CONTAINING CONCRETE

M.J. COOLE and A.M. HARRISSON
Blue Circle Industries PLC, Group Research, 305 London Road, Greenhithe,
Kent, United Kingdom

Received 20 October, 1986; refereed

ABSTRACT

When concrete is poured in large volumes, it is necessary to be able to
predict the temperature rise which may occur inside the mass because of the
effect this may have on the ultimate properties of the hardened concrete. It
is known that the elevated temperatures generated may have a detrimental
effect on final strengths and that if the difference in temperature between
the centre and the surroundings exceeds 20-25°C, cracking may occur. In order
to study these effects, a calorimetric controlled apparatus has been designed
that is able to simulate the temperature rise profile occurring within any
size of concrete pour. The apparatus is also used to control a curing bath
thus enabling the compressive strength of match cured concrete to be determined. Results have been obtained for both temperature rise and strength growth
at the centre of simulated 0.8, 1.5 and 3 m deep pours, using plain Portland
and Portland PFA cement concrete. These show that in the larger sized pours
the strength of concrete from a Portland PFA cement blend grows, after 2-3
days, at a greater rate than that of pure Portland cement, while giving lower
temperature rises. Comparative strengths at 28 days are 48 Nmm^{-2} for the PFA
cement concrete and 38 Nmm^{-2} for the pure Portland cement concrete. The
influence of temperature on the reactivity of the PFA under these conditions
has been studied using a dilute acid dissolution method. The hydrates formed
and the progress of the pozzolanic reaction within the actual concretes has
been monitored using scanning electron microscopy.

INTRODUCTION

The hydration of cement is an exothermic reaction which, in the case of
large pours, can cause a considerable temperature rise in the concrete placed.
When concrete is poured in a large quantity, it is necessary to be able to
predict the temperature rise which may occur inside the mass because of the
effect this may have on the ultimate properties of the hardened concrete. It
is known that elevated temperatures may have a detrimental effect on the
strength growth pattern of the concrete and that if the difference in temperature between the centre and surroundings of a pour exceeds 20 to 25°C
cracking may occur [1,2]. Whether cracking occurs depends both on the
magnitude of the temperature induced stress and the capacity of the concrete
to accommodate the strain. The need to be able to predict these temperature
rises is greater with composite cements, containing PFA or blast furnace slag,
where the use of these materials introduces an additional uncertainty as to
the effect of increased temperature on the development of strength with time.
In the past, various methods have been employed to try to ascertain these
effects, such as the adiabatic calorimeter and the vacuum flask calorimeter
[3] but none of these methods is able to simulate the typical temperature-time
profile of a large pour of concrete. In situ temperature measurement, using
thermocouples placed at specific depths in the concrete, can be adopted and
may be used to control an auxiliary cube curing bath thereby enabling internal
strength values to be obtained [4,5]. This, of course, is a retrospective
method which can only give an indication of temperature profiles, and their
effect on strength, in a structure when it is too late to undertake any

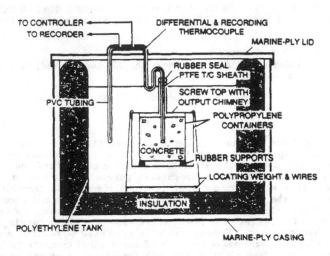

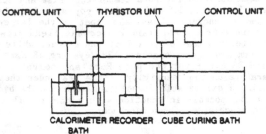

Figure 1.   The calorimeter and the complete system for
matched curing.

corrective action. However, results obtained using this technique agree very
well with results obtained in practice [6,7].

The calorimeter described in this paper is designed to simulate any size
of pour of concrete, and to supply information on the associated temperature
profile to a cube curing bath which follows the temperature of the main
concrete sample by thermocouple control to within 0.05ºC. Cubes from the
curing bath can then be used to determine the development of strength in the
simulated pour. This system (shown in Figure 1) only requires about 4 kg of
cement to supply enough concrete for the control calorimeter and 12 (100 mm)
cubes. The equipment can therefore be used to explore the behaviour of a
range of concrete mixes and composites, with regard to temperature rise and
strength development, before undertaking a large pour on site.

The apparatus is based on a thermally insulated calorimeter. A concrete
sample is contained in a polypropylene cell within a larger polypropylene
container submerged in a well stirred, heated, water bath (Fig. 1). The
principle of the apparatus is to use a thermocouple to sense the temperature
of the hydrating sample held in the calorimeter, and apply the thermal emf to
control the temperature of a water bath surrounding the specimen container.
When the enclosure temperature is maintained equal to the instantaneous
temperature of the hydrating mass in the calorimeter, the heat losses are zero

TABLE I

Chemical and Physical Characteristics of OPC, Phoenix Cement and PFA

	O.P.C.	Phoenix Cement	P.F.A.
SiO_2	20.1	28.4	51.3
I.R.	0.57		
Al_2O_3	4.7	10.4	27.4
Fe_2O_3	2.5	4.2	9.2
Mn_2O_3	0.18	0.13	0.07
P_2O_5	0.12	0.15	0.23
TiO_5	0.22	0.41	0.97
CaO	64.4	48.6	1.2
MgO	2.1	2.0	1.6
SO_3	2.6	2.4	0.92
L.O.I.	1.4	1.3	2.0
K_2O	0.72	1.4	4.1
Na_2O	0.20	0.36	1.2
F	0.10	0.07	–
Free Lime	2.5	1.8	–
Surface Area(m^2/kg)	405	397	–
45 μm residue	9.3	5.3	8.4

and the mass behaves as if it were part of an infinitely large concrete pour. Enclosure temperatures less than that of the mass will permit heat losses from the mass, at rates determined by the magnitude of the temperature differential, and so permit simulation of the thermal history of concrete within pours of different sizes. Thus, it can be shown, by calibration using Schmidt's method [8,9] that a temperature differential ($\Delta\theta$) of 0.08°C corresponds to the behaviour of a pour of 3 m minimum dimension while $\Delta\theta$ = 0.5°C corresponds to a pour size of about 0.8 m.

Because the temperatures differentials, $\Delta\theta$, needed to achieve the low heat losses experienced by the centre of a large concrete pour are small, it is necessary to apply close control of the enclosure temperature so that it maintains a fixed relation to the hydration temperature profile of the sample. This control has been achieved by using five sensing thermocouples in series giving an output of 40 μV/°C each. With the electronic control equipment used being able to set to ±1 μV this gives a theoretical control of ±0.005°C for the enclosure temperature but tests have shown that a more realistic value would be ±0.01°C. A more comprehensive description of the apparatus together with calibration details can be found in a previous paper [10].

EXPERIMENTAL

A pure Portland cement (OPC) and a Portland PFA cement (marketed in the UK as Phoenix) manufactured by intergrinding the same clinker with 25% PFA, were used to produce two 11 cube batches of concrete according to BS 4550 (cement:sand:aggregate:water is 1:2.5:3.5:0.6 by weight). See Table I for chemical and physical characteristics. This provided enough concrete for the control calorimeter (approximately 1 kg) and for two 100 mm cubes to be tested at ages of 1, 3, 7, 28 and 91 days. The tests were carried out on both concretes with the control calorimeter set to give simulated pour sizes of

0.8, 1.5 and 3 m depth. Reference cubes were also cured at 20°C using the BS 4550 test procedure.

Small amounts of the Phoenix cement pastes were also stored with the cubes, in sealed plastic vials, so that the hydration could be stopped at the various curing ages in order to determine the amount of unreacted PFA, by a method of solution in diluted HCl (1:49). This method relies on the fact that PFA derived from bituminous coals is approximately 98% insoluble in a dilute HCl whereas the products of pozzolanic reaction are readily soluble. Hence the extent of reaction can be determined by the change in the insoluble residue. It was also thought useful to simulate the internal pressure experienced in a 3 m pour, arising from the mass of concrete above by mechanically loading the cubes. This was done by removing about 5 mm of concrete from each cube mould and placing a steel plate (13 mm thick), that was just able to slide within the 100 mm mould, on top of the freshly mixed concrete. This plate was then loaded to 70 kPa uniaxially by placing a lead weight on top.

After the strength tests were completed at the various stages of curing, the inside of each cube was sampled and hydration was stopped by dropping the sample into acetone. it was then prepared for examination by scanning electron microscopy. The sample was first dried under vacuum, then fractured, and the fracture surface was mounted on a stub and given a thin gold coating.

RESULTS

Temperature Rise and Compressive Strength

Figure 2 shows the temperature rises obtained at the centre of simulated pours of 0.8, 1.5 and 3 m deep for both the OPC and the Phoenix mixes. As the pour size increased, although the Phoenix mix displayed a lower peak temperature than the OPC mix at all sizes, the time taken for the Phoenix mix to reach this maximum gradually increased, with respect to the OPC, until at 3 m the time taken was nearly twice as long. It appears that once the higher temperatures of a 3 m pour were reached, although the OPC hydration rate was accelerated as expected, the dilution effect of the PFA combined with increased reaction between the PFA and hydration products of the OPC, produced a lower early temperature rise but one which continued as the PFA reactivity increased to produce a much later peak temperature. This was also reflected in the strength growth of the concrete cured under the 3 m simulated pour conditions (Fig. 3) as compared with concrete cured at standard 20°C conditions (Fig. 4).

With the OPC the rapid increase in temperature rise, and hence hydration rate, caused hydrates to grow in a rapid and perhaps disorganized manner. giving higher early strengths but eventually reduced later strengths. In contrast, the Phoenix concrete under these conditions produced a slower, more prolonged reaction which, together with reaction products between the PFA and OPC combined to give strength growth superior to that of the OPC concrete after 2.5 d. Figure 3 also shows the strength growth in intermediate sized pours.

The way in which the temperature has affected the strength growth of the Phoenix concrete, compared with the OPC concrete, is demonstrated in Figure 5, where strength has been plotted against pour size at 1, 3, 7, 28 and 91 d. This shows quite clearly the fall off in strength with increasing pour size for the OPC concrete after 7 d, whereas the Phoenix concrete was only slightly affected after 91 d.

The reaction rate of the PFA was determined at various ages, for each of the curing regimes mentioned, by arresting the hydration with acetone and measuring the amount of unreacted PFA by dissolving the ground sample in 1:49 HCl (about 0.25 N).

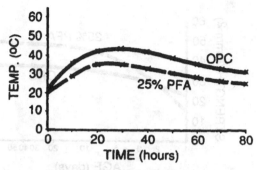

Temperature rise centre of 0.8m pour

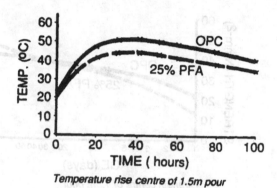

Temperature rise centre of 1.5m pour

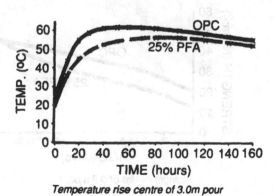

Temperature rise centre of 3.0m pour

Figure 2. Temperature rise at the centre of 0.8, 1.5 and 3.0 m pours.

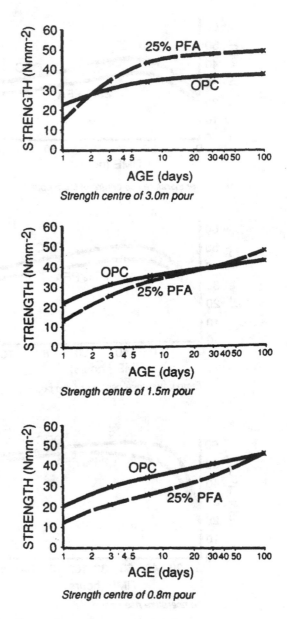

Strength centre of 3.0m pour

Strength centre of 1.5m pour

Strength centre of 0.8m pour

Figure 3.   Strength at the centre of 3.0, 1.5 and 0.8 m pours.

283

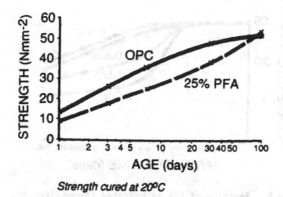

Strength cured at 20°C

Figure 4. Strength after curing at 20°C.

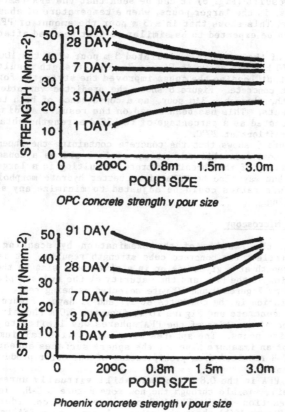

OPC concrete strength v pour size

Phoenix concrete strength v pour size

Figure 5. OPC and Phoenix concrete strength vs. pour size.

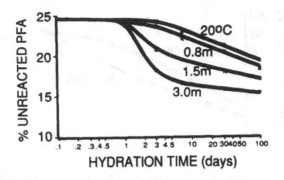

Figure 6.   Reaction of PFA in hydrated Phoenix cement.

From these results (Fig. 6) it can be seen that the PFA reaction rate rapidly increases, in the larger pours, when a temperature of about 40°C or more is attained. This shows that in a 3 m pour the amount of PFA reacted after 2-3 days can be expected to be similar to that reacted after 100 days curing at 20°C.

The results of the tests in a simulated 3 m pour with the 100 mm cubes being loaded to 70 kPa are shown in Figure 7. The strength results at 1, 7 and 28 days show that loading the cubes improved the strengths for both OPC and Phoenix cement concrete. Figure 8 shows the predicted variation in 28 d strength with depth, through a 3 m pour, as a consequence of both temperature and pressure effects. This has been based on the results as displayed in Figure 7 and presented as a percentage of the 28 d strengths obtained from BS 4550 curing conditions at 20°C.

Study of Figure 7 shows that the concrete containing the Phoenix cement seemed to benefit most from this compaction. This could be a consequence of the better workability of the Phoenix concrete, resulting in a larger loss of expressed water, but may also be related to better hydrate morphology under load. Concrete w/c ratios could be adjusted to minimize any associated increase in bleeding.

Scanning Electron Microscopy

In general, the results of the examination by scanning electron microscopy (SEM) reflect the concrete cube strength results. It is possible to relate the strength at a given age or in a given pour size to the observed degree of hydration of the PFA, and the maturity of the hydrate phases. The photographs shown as Figures 9 and 10 were selected as being representative of the degree of hydration in the concretes at 1,7 and 91 days. Figure 9 is from the 0.8 m Phoenix concrete and Figure 10 is from the 3.0 m Phoenix concrete. At 1 d, the degree of reaction of the PFA spheres was limited to a similar extent in both size pours. The spheres were coated with calcium silicate hydrate (C-S-H) of an immature type but the sphere surfaces appeared uncorroded and the C-S-H resulted from precipitation of the outer product from the cement clinker phases.

By 7 d, the PFA in the 0.8 m pour was still virtually unreacted, the smooth surface being visible through the now more mature C-S-H. In the 3.0 m concrete, the reaction of the PFA had proceeded to a considerably more advanced stage. Figure 10(b) shows the spherical shape of a PFA relief which has been almost all replaced by C-S-H, having a radial appearance characteristic of inner product from PFA spheres. The difference in compressive

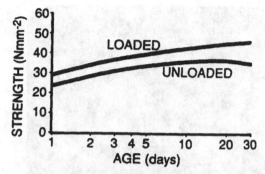

Figure 7(a). OPC concrete unloaded and loaded.

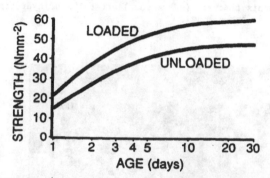

Figure 7(b). Phoenix concrete unloaded and loaded.

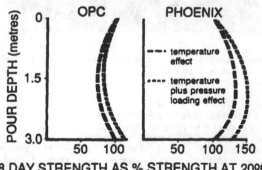

Figure 8. Predicted effect of temperature and pressure loading on 28 day strength.

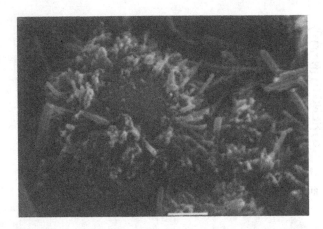

Figure 9(a). Phoenix concrete (0.8 m pour) after 1 d of hydration.
(scale bar = 1 μm).

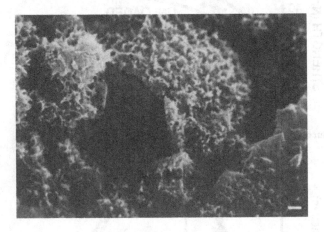

Figure 9(b). Phoenix concrete (0.8 m pour) after 7 d of hydration.
(scale bar = 1 μm).

strengths of the two concretes directly reflects the differences seen in the photographs, the 0.8m concrete having a strength of 26.0 Nmm^{-2} and the 3.0m concrete, 42.1 Nmm^{-2}.

The two 91 d concretes show marked similarities and the two photographs given as Figures 9(c) and 10(c) show that spheres have reacted to a considerable extent in both concretes. The 0.8 m pour has now caught up both in microscopic appearance and strength, the two values being 45.9 Nmm^{-2} for the 0.8 m concrete, and 45.2 Nmm^{-2} for the 3.0 m pour.

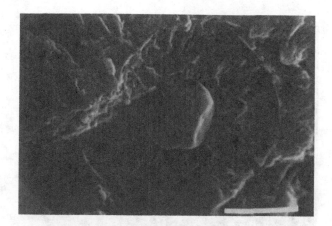

Figure 9(c). Phoenix concrete (0.8 m pour) after 91 d of hydration.
(scale bar = 1 μm).

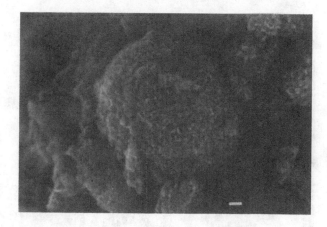

Figure 10(a). Phoenix concrete (3.0 m pour) after 1 d of hydration.
(scale bar = 1 μm).

CONCLUSIONS

1. The results show that Portland PFA cement can produce increased core
   strength in mass concrete pours where the temperature rise would normally
   reduce the strength of an OPC concrete to about 70% of its expected
   value.
2. From the temperature profiles of the simulated pours it can be seen that
   the interground PFA cement produced lower temperature rises, at later

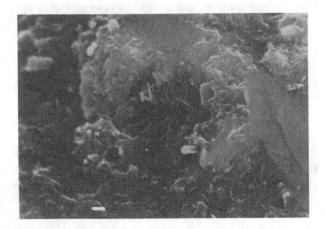

Figure 10(b). Phoenix concrete (3.0 m pour) after 7 d of hydration.
(scale bar = 1 μm).

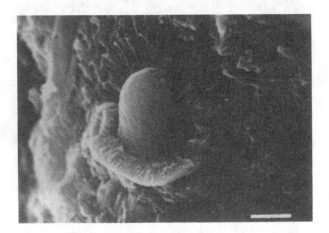

Figure 10(c). Phoenix concrete (3.0 m pour) after 91 d of hydration.
(scale bar = 1 μm).

ages, than the OPC, so reducing the thermally induced stresses within a
large pour.
3. Although the PFA was slow to react with OPC at standard curing conditions, once the curing temperature exceeded 40°C the reaction rate
increased markedly. The extent of this acceleration is such that at the

centre of a 3 m pour, the amount of PFA reacted after 2–3 d can be expected to be similar to that reacted following 100 d curing at 20°C.
4. Simple mechanical loading of the sample cubes showed that, at 28 d, the pressure generated by the mass of concrete above can be expected to reduce the strength loss at the centre of the large OPC pours and further enhance the strength of Portland–PFA pours. Consequently, a full simulation of large pour conditions should include this loading factor.

ACKNOWLEDGEMENTS

The authors would like to thank both Dr. G.K. Moir and J.J. Kollek for their assistance and Blue Circle Industries PLC for permission to publish this work.

REFERENCES

1. M.R.H. Dunstan and P.B. Mitchell, Proc. Inst. Civ. Engrs, 60, Part 1, 27–52 (1976).
2. M.E. Fitzgibbon, Concrete, 10, No. 12,, Dec. (1976).
3. R. Alegre, Revue des Materiaux, Nos. 544–549 (1961).
4. H. Blakey, Concrete, 10, No. 5, May (1976).
5. R.P. Cannon, Silicates Industriels, XLVII, No. 7/8 (1982).
6. P.B. Bamforth, Proc. Instn. Civ. Engrs. 69, Part 2, 777–800 (1980).
7. P.L. Owens, Concrete, August 1986; Concrete, September, 1986.
8. C. Rawhouser, J. Am. Con. Inst., 16, No. 4, (1945).
9. Manual of Concrete Practice, Part 1 SR-34 (American Concrete Institute, Detroit, 1981) p. 207.
10. M. J. Coole, British Ceramic Proceedings 35, 325–401 (1984).

PART IV

Utilization

STATUS OF ASTM AND OTHER NATIONAL STANDARDS FOR THE USE
OF FLY ASH POZZOLANS IN CONCRETE

RICHARD M. MAJKO
American Fly Ash Company, 606 Potter Road, Des Plaines, IL 60016

Received 9 November, 1986; refereed

ABSTRACT

The Subcommittee of ASTM that is responsible for pozzolan specifications,
(C09.03.10) is currently revising C 618, the standard specification, and
C 311, the standard test methods. It is no small task. It is generally
considered that dividing fly ashes into two classifications (Class F and Class
C) is no longer acceptable. The Subcommittee has reached the tentative
conclusion that one class of fly ash pozzolan is more appropriate and less
ambiguous, provided it is accompanied by an optional table outlining the
appropriate limits for sulfate resistance, heat of hydration, hydraulic
properties, etc. That is, the engineer need specify only a fly ash pozzolan.
If the engineer needs more than a "general purpose" fly ash he calls for
special characteristics from the optional requirement table. Any major
revision in the C 618 specification challenges the Subcommittee to look for
better and more useful test methods in C 311. The committee is looking at
C 1012 for a sulfate resistance procedure, conduction calorimetry as a heat of
hydration method, and better methods to evaluate pozzolanic activity. The
committee is aware of the limited usefulness of hydrated lime or sodium
hydroxide in evaluating pozzolanic activity. Perhaps a more suitable term,
i.e. the activity index with cement, is more appropriate. Fly ash properties
that need to be evaluated include water reduction, activation by lime and
alkalies, hydraulic or self-cementing properties and pozzolanic activity. Any
test that tends to compensate for a fly ash's lower reactivity (i.e., accelerated thermal curing at 35°C or 65°C) probably leads to misleading results. A
critical look at the national standards of other countries might allow the
committee to take a visionary yet practical approach toward a new fly ash
specification.

INTRODUCTION

The classification of fly ash into two classes of mineral admixtures as
is done by ASTM C 618 may no longer be useful or appropriate. The French use
the term silico-aluminous (much like Class F), Silico-calcic (much like an
intermediate grade Class F/C), and sulfo-calcic (like Class C), to classify
fly ashes [1]. Some fly ashes appear to be strictly pozzolanic; that is, they
react slowly over time with lime and alkalis. Other ashes are very hydraulic
and rather quickly form calcium aluminate hydrate and ettringite. Another
(intermediate) class, is weakly hydraulic forming ettringite, calcium aluminate, and hydrated gehlenite [1]. It does not appear useful to classify fly
ashes by their hydraulic properties. For example, in the 1960's (before there
was a Class C fly ash designation), the American Fly Ash Company sold a
hydraulic Class F fly ash produced from Southern Illinois bituminous coal. It
was not particularly pozzolanic (high iron content and a coarse size
distribution) yet it set up rapidly.

With every passing year, new coal seams are mined, and the chemical
diversity of the resulting fly ashes increases. As more sophisticated instruments are applied to fly ash, our knowledge of the complexity of fly ash
increases. However, ASTM still needs reasonably simple procedures to determine the appropriateness of a material to be used in concrete. The classifi-
fication of fly ash is hindered by the fact that there always seems to be

Table I

History of Fly Ash Specifications

Requirement	Limit	ASTM SPECIFICATION (Extension indicates year of adoption)											TESTING FREQUENCY (in tons)			
		C350-65T[a]	C618-68T	C618-71	C618-72	C618-73	C618-77	C618-78	C618-80	C618-83	C618-84	C618-85	C311-64T	C311-68	C311-77	C311-85
STANDARD CHEMICAL REQUIREMENTS																
$SiO_2+Al_2O_3+Fe_2O_3$	%Min	70.0	70.0	70.0	70.0	70.0	70[b]/50C[b]	70F/50C	70F/50C	70F/50C	70F/50C	70F/50C	1000[c]	1000[c]	2000	2000
SO_3	%Max	5.0	5.0	5.0	5.0	5.0	5.0	5.0	5.0	5.0	5.0	5.0	100	100	2000	2000
Moisture	%Max	3.0	3.0	3.0	3.0	3.0	3.0	3.0	3.0	3.0	3.0	3.0	100	100	400	400
LOI	%Max	12.0	12.0	12.0	12.0	12.0	12F/6C	12F/6C	12F/6C	12F/6C	6.0[d]	6.0[d]	100	100	400	400
OPTIONAL CHEMICAL REQUIREMENTS																
MgO	%Max	NR[e]	NR	NR	NR	NR	5.0[f]	5.0[f]	5.0[f]	NR	NR	NR	1000	1000	2000	2000
Avaiable Alkalis (as Na_2O)	%Max	1.5	1.5	1.5	1.5	1.5	1.5	1.5	1.5	1.5	1.5	1.5	1000	1000	2000	2000
STANDARD PHYSICAL REQUIREMENTS																
Blaine Surface Area (cm^2/cm^3)	Min	NR	6500	6500	6500	6500	NR	NR	NR	NR	NR	NR	100	100	NR	NR
Mean Particle Diameter (µm)	Max	9	NR	NR	NR	NR	NR	NR	NR	NR	NR	NR	NR	NR	400	400
No. 325 Sieve Residue	%Max	NR	NR	208	208	348	34	34	34	34	34	34	100	100	100	100
Multiple Factor	%Max	NR	NR	150	150	255	NR	NR	NR	NR	NR	NR	100	100	100	100
Compressive Strength (% of Control)	7 d Min	100	100	100	100	100	NR	NR	NR	NR	NR	NR	100	100	100	100
	28 d Min	100	100	100	100	100	NR	NR	NR	NR	NR	NR	100	100	100	100
Drying Shrinkage (28 d)	%Max	0.03	0.03	0.03	0.03	0.03	NR	NR	NR	NR	NR	NR	100	100	100	100
Soundness/Autoclave Expansion/Contraction	%Max	0.5	0.5	0.5	0.5	0.5	0.8	0.8	0.8	0.8	0.8	0.8	100	100	400	400
POZZOLANIC ACTIVITY INDEX																
28 d (% of Control)	%Min	85	85	85	85	85	75	75	75	75	75	75	100	100	2000	2000
7 d	Min	800	800	800	800	800	800	800	800	800	800	800(F)/NR(C)	1000	1000	400	2000
WATER REQUIREMENT (% of Control)	%Max	105	105	105	105	105	105	105	105	105	105	105	100	100	2000	2000
UNIFORMITY REQUIREMENTS (Variation from Moving Average)																
Specific Surface Area	%Max	15	15	15	15	15	NR	NR	NR	NR	NR	NR	100	100	NR	NR
Specific Gravity	%Max	5	5	5	5	5	5	5	5	5	5	5	100	100	400	400
No. 325 Sieve Residue (% pts.)	Max	NR	NR	NR	NR	NR	5	NR	5	5	5	5	NR	NR	400	400
Air Entraining Agent (amount)	%Max	20	20	20	20	20	NR	NR	NR	NR	NR	NR				
OPTIONAL PHYSICAL REQUIREMENTS																
Multiple Factor	%Max						255F	255F	255F	255F	255F	255F				
Drying Shrinkage (28 d)	%Max						0.03	0.03	0.03	0.03	0.03	0.03				
Reactivity with Cement Alkali (14 d Mortar Expansion)	%Max	0.02	0.02	0.02	0.02	0.02	0.02	0.02	0.02	0.02	0.02	0.02	20	20	20	20
OPTIONAL UNIFORMITY REQUIREMENTS																
Air Entraining Agent (amount)	%Max						20	20	20	20	20	20				

exceptions to any rule. We do not completely understand fly ash and its reactions in concrete. The purposes of this paper are to review current standards throughout the world, to review the history of the development of ASTM C 618 and C 311, to discuss ASTM's current approach to a new fly ash specification, and to recommend changes to both C 618 and C 311. It is hoped that the reader will realize there is no final answer in the attempt to come up with a better more workable standard.

HISTORY OF ASTM FLY ASH SPECIFICATIONS

ASTM fly ash specifications have not evolved greatly over the last 20 years. This is somewhat surprising in view of the complexity of this hetero-geneous by-product of coal combustion, especially resulting from the use of more subbituminous and lignite coals. The history of ASTM fly ash specifi-cations is provided in Table I. The major changes that have occurred over in the last 20 years are discussed below.

ASTM C 618

1. The measurement of fineness changed from the Blaine air-permeability to 45 μm (No. 325) wet-wash sieve method.

2. Compressive strengths of mortar cubes at 7 and 28 days (d) were no longer required after 1971. (Control included 500 g cement; test mix included 500 g cement + 125 g fly ash).

3. The drying shrinkage test became optional.

4. The maximum expansion limit for the autoclave soundness test changed from 0.5% to 0.8%.

5. The uniformity requirements for amount of air-entraining agent required in mortar became optional.

6. "Class C" mineral admixture was first specified in 1977 with a 6% maximum loss on ignition and a 50% minimum $SiO_2+Al_2O_3+Fe_2O_3$ requirement.

7. When Class C fly ash was first specified, the purchaser could request a MgO limit of 5% maximum (1977). However, this proved to be meaningless as the autoclave soundness limit was still the determining factor. This optional requirement was eliminated in 1983.

8. Finally, in 1985, the lime-pozzolan specification limit was removed for Class C mineral admixtures. However, the specific characteristics of a Class C fly ash has not been fully determined by the Committee.

9. The term "mineral admixture" was first used in C 618 in 1977. Prior to that the classes were "Pozzolan Classes."

Footnotes to Table I:
  a. T = Tentative
  b. F = Class F (Low-Calcium); C = Class C (High-Calcium)
  c. Frequency testing required for SiO_2 only
  d. LOI can exceed 6% for Class F but must be <12% if sufficient data are available.
  e. NR = Not Required
  f. MgO greater than 5% allowed if autoclave soundness passes.
  g. The change from 20% to 34% retained reflected the change from the Tyler mechanical sieve to an electroformed sieve.

I. **Specification Limits**

A. Fly ash must meet current ASTM C618 physical and chemical requirements (optional requirements applicable only when requested by purchaser).

1. Specific Gravity Uniformity — As according to ASTM C618, an individual value shall not vary by more than 5% from a moving average established by the 10 preceding results.

2. No. 325 Sieve Residue Uniformity — As according to ASTM C618, an individual value shall not vary by more than 5% sieve residue from a moving average established by the 10 preceding results.

B. LOI (Carbon) Uniformity — Applicable to air-entrained concrete only. An individual value shall not vary by more than 1.5% LOI (or C) from a moving average established by the 10 preceding results (i.e., moving average = 3% LOI, then LOI can't exceed 4.5% or be less than 1.5% for next individual value). Note: If variations greater than 1.5% from the moving average do not cause a change in the required air-entraining agent dosage, such variations need not be cause for rejection.

II. **Pre-Qualification**

Each fly ash from a particular power plant and made available for sale, must be pre-qualified before use in Portland Cement concrete. At least six months of test results shall be included in the quality history of a new source.

A. An ASTM C618 certification at least once per month shall be included in the quality history.

B. A quality history also shall include at least 40 of the most recent individual test results, no greater than 2 years old, for each of the following: LOI (or carbon), No. 325 sieve residue, and specific gravity.

C. If the fly ash (individual values) meets the C618 specification limits and the uniformity requirements of (I) above, the fly ash is pre-qualified. This pre-qualification continues, provided the test data required in III(A) continues to meet or exceed specification and uniformity limits.

III. **Post-Qualification**

A. Once a fly ash is pre-qualified, a quality control program is set up to test the most significant fly ash parameters during the course of the project. At a minimum, each fly ash shall be tested daily for LOI (or carbon) and No. 325 sieve residue. If all LOI (C) test values are below 1%, test LOI according to C311 frequency. Specific gravity shall be tested according to C311 frequency or weekly, whichever is more frequent.

B. A complete C618 is performed according to C311 frequency or monthly, whichever is more frequent (optional requirements applicable as requested by buyer).

C. The quality control data records shall be made accessible to the purchaser.

IV. **Qualification of the Lab Performing the Tests**

A. Each laboratory testing for the required complete C618 shall comply with the applicable provisions of ASTM E329 which require laboratory inspection by a qualified national authority.

Figure 1.   Guidelines for a fly ash quality assurance program from the American Coal Ash Association. [Revised February, 1986]

ASTM C 311

1. Frequency of testing for many of the critical parameters changed from 100 to every 2,000 tons (for SO_3) or from every 100 to every 400 tons (for loss on ignition, fineness).

2. Prior to 1968, only SiO_2 was required to be tested frequently (not Al_2O_3 or Fe_2O_3).

DISCUSSION OF ASTM AND OTHER NATIONAL STANDARDS

Table II is a summary of international fly ash standards. It is based on earlier compilations of Berry and Malhotra [2], Manz [3] and Rossouw and Kruger [4]. Specific aspects of these standards are discussed below.

Loss on Ignition (LOI)

There is universal agreement that this parameter must be controlled. The loss on ignition specification in Canada and the U.S. has dropped to 6% maximum LOI for all fly ashes. If the user accepts performance or laboratory data on higher loss on ignition fly ashes, then the limit is 12% maximum. A new concept recently appeared in the U.S. with the publication of a uniformity

Table II

Fly Ash Chemical and Physical Requirements for Use in Concrete
[after references 2–4]

	A.S. 1129 Australia	ONORM B3320 Austria	CAN3–A23.5–M82 Canada Class F	CAN3–A23.5–M82 Canada Class C	Britian B.S. 3892	Britian Draft	I.S. 1344 -1968 India	JIS A 6201 Japan	K.S. L 5405 Korea	Turkey	ASTM C 618 USA Class F	ASTM C 618 USA Class C	Gost. 6269-63 USSR
CHEMICAL REQUIREMENTS													
Free water (% max)	1.5	1.0	3.0	3.0	1.5	0.5		1.0	3.0	3.0	3.0	3.0	
LOI (% max)	8.0	5.0	6.0	6.0	7.0	7.0	12.0	5.0	12.0	10.0	6.0	6.0	10.0
MgO (% max)		5.0			4.0	4.0			5.0	5.0			
Sulphate (as %SO_3 max)	2.5		5.0	5.0	2.5	2.5	3.0		5.0	5.0	5.0	5.0	3.0
Total sulphur (as %SO_3 max)		3.5											
SiO_2 (S) (% min)		42-60											
Al_2O_3 (A) (% min)		16-32											
Fe_2O_3 (F) (% min)		3-12											
Minimum S+A+F							70.0	45.0	70.0	70.0	70.0	50.0	
CaO (% max)		5-20											
Free CaO (% max)		2.0								6.0			
Total Alkalies (as %Na_2O max)									1.5				
Available Alkalies (as %Na_2O max)							1.5				1.5	1.5	
Carbon (% max)		3.0											
Chloride (% max)		0.1											
PHYSICAL REQUIREMENTS													
Specific Surface (Blaine, min, cm²/g)		4–5000					2800	2400					
Avg. Particle Diam. (um)									9.0				
Sieve (max %;um)[a]	10;150	34;45	34;45		12.4;45[b]					0.3;200 8;87	34;45	34;45	

Table II (continued)

Fly Ash Chemical and Physical Requirements for Use in Concrete
[after references 2-4]

	A.S. 1129 Australia	ONORM B3320 Austria	CAN3-A23.5-M82 Canada Class F	Canada Class C	Britian B.S. 3892 Draft	I.S. 1344 -1968 India	JIS A 6201 Japan	K.S. L 5405 Korea	Turkey	ASTM C 618 USA Class F	USA Class C	Gost. 6269-63 USSR
Specific Gravity (min)						2.0						
Pozzolanic Index (cement) (min % of control, 28 d)		80.0	75				60/70	85.0	70.0	75.0	75.0	
(min % of control, 7 d)				68c								
Sand Replacement									100.0			
Water requirement (max % of control)					95.0			102.0	105.0	105.0	105.0	
Pozzolanic Index (with lime) (min MPa) [MPa X 145.45 = psi]			0.8			3.9		5.5		5.5		
Soundness Autoclave expansion (max %) Other tests		PAT	0.8			0.8		0.5	10d	0.8	0.8	
Drying shrinkage (max % at 28 days)						0.1						
Drying shrinkage (increase over control %)			0.0					0.0		.03	.03	
Alkali reactivity (max % expansion at 14 d)								0.0		0.0	0.0	
Alkali reactivity (max % redn. of expansion)			60.0									

a. Sieve requirements are expressed as the maximum permitted quantity of material (as a percentage) retained on a screen of specified aperture (expressed in µm units); e.g., a maximum permitted quantity of 34% retained at 45 µm is expressed as 34;45.
b. 12.4% retained is equivalent to 21.25% retained on an 'electroformed' sieve (which the U.S. uses).
c. Accelerated curing at 65°C; no Class C fly ash used in development of CSA STD.
d. Le Chatelier expansion test.

requirement for LOI, introduced to protect users of fly ash in air-entrained concrete. This requirement was part of a series of guidelines for a fly ash quality assurance program devised by the American Coal Ash Association. This set of guidelines is given as Figure 1.

MgO

There is some concern that high MgO levels might indicate the presence of expansive periclase which would present soundness problems. However, the periclase in fly ash is quite unreactive. Most countries have de-emphasized a MgO limit, particularly when autoclave expansion or a Le Chatelier expansion test is required.

SO_3

Britain, Australia, India and the Soviet Union have more rigid requirements for maximum SO_3. Concern has been expressed that soluble SO_3 may be of more importance [5]. Butler [6] has argued strongly for a water-solubles test. The American Fly Ash Company provided values for soluble SO_3 because of concrete efflorescence concerns. A prominent concrete admixture company argues that soluble SO_3 (from the fly ash) affects chemical admixture performance [19] with portland cement.

SiO_2

Austria and Japan have minimum requirements for SiO_2.

$SiO_2+Al_2O_3+Fe_2O_3$

India, Korea, Turkey and the U.S. specify a minimum total for the sum of these three oxides. However, there is no firm technical basis for this; a fly ash is reactive as a pozzolan because its glass contains high amounts of SiO_2 and Al_2O_3 [8-10]. It is incorrect to assume that a higher total of these three oxides represents more glass or a better quality glass. Numerous workers have shown that that there is little or no correlation between this sum and strength of the resulting concrete and that it is the reactive, more glassy outer portion of fly ash that is important in fly ash performance. (See for example Figure 2.) The only utility of this requirement may be: (1) that one can quickly determine the parameter from routine chemical analyses; (2) if the total exceeds 70%, the fly ash has a higher probability to provide the pozzolanic reaction, and to prevent alkali-aggregate reaction or maintain or increase the sulfate resistance of concrete. However, with regard to this last factor, Mehta [11] cautions that even a very low Dunstan "R factor" [12] fly ash (very high $SiO_2+Al_2O_3+Fe_2O_3$ total) may still hinder the sulfate resistance of cement-fly ash pastes.

CaO (maximum)

Austria and Turkey have this requirement. In the writer's opinion, a specification of a maximum CaO content has little technical basis. It probably came about for those people concerned about setting of concrete and sulfate resistance. Performance tests are far superior in detecting problems like the above.

Available Alkalies

Available alkalies are those alkalies in the fly ash solubilized by reaction with hydrated lime for a specified curing period and temperature. This is specified only by ASTM and India's standard. The test specified in ASTM C 311 is non-precise and it leads to ambiguous results. It is highly dependent on the type of lime used, and the age and temperature of curing.

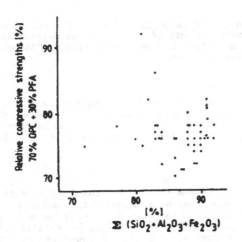

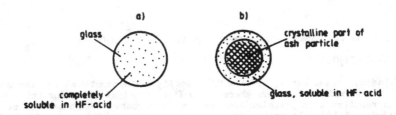

Figure 2. Scatter plot showing the lack of correlation between 28 d
relative compressive strengths vs. $SiO_2+Al_2O_3+Fe_2O_3$ (upper
diagram). The relevant factor is the glass content and
reactivity near the surface of the fly ash grain. (After
Tutti and Fagerlund [10].)

The paper by Lee, Schlorholtz, and Demirel [7] clearly showed that the time of
curing must be lengthened for some Class C fly ashes to maximize alkali solubility. They also showed that accelerated curing at 55°C at short ages does
not correlate well to the long term curing at 38°C. The writer does not
understand what alkali solubility, measured after reaction with hydrated lime,
has to do with portland cement concrete.

Carbon

Austria is the only country specifying maximum carbon content rather than
a maximum loss on ignition. Specifying carbon is preferable, but the loss on
ignition test is simpler and less expensive. Eventually, all specifications
should refer to carbon, as this would more directly analyze the critical component in fly ash capable of adsorption of air-entraining agent.

Fineness —— Specific Surface/Mean Particle Diameter

The indirect measurement of surface area by means of the Blaine air-
permeability apparatus has fallen into considerable disfavor in many coun-

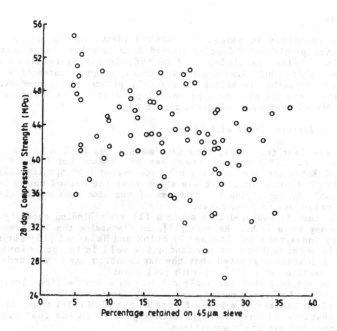

Figure 3. Scatter diagram showing little correlation between retention
on a 45 μm sieve and 28 d cube compressive strength. (After
Cabrera et al. [13].)

tries. Carbon significantly affects the Blaine value. In 1965, the mean
particle diameter was specified by ASTM C 350 to be 9 μm. However, this value
was obtained by running the Blaine air-permeability test method. This
requirement was changed in 1968 to a specific surface of 6500 cm^2/cm^3.
Finally, in 1973, this requirement was eliminated by ASTM.

However, there is considerable interest in developing a measurement and
specification for either a mean particle diameter or the percentage below
10 um, using instruments that determine particle size distribution.

Fineness -- 45 μm (No. 325) Sieve

As noted in Table 2, there is a considerable difference of opinion as to
this requirement. The U.S. and Canada specify that no more than 34% be retained on a 45 μm (No. 325) electroformed sieve. The U.S. specified 20% in
1971 and 1972, which was changed to 34% with the introduction of the electroformed sieve. Britain specifies 12.4% retained on a 45 μm sieve (equivalent
to 21.25% on a U.S. sieve). Australia is more lenient, permitting 10% retained on a 150 μm sieve.

The problem, here, is that the determination of fineness by looking at
what passes one sieve size is inadequate. Figure 3, for example, clearly
shows poor correlation between percentage retained on a 45 μm sieve and 28 d
compressive cube strength. West Germany requires an additional fineness test
at 20 μm. It is hoped that determining fineness with two sieves may be
required in revised ASTM specifications.

Soundness

The autoclave soundness test method identifies the presence of course, unreactive, particles of free unhydrated lime or other phases that expand on hydration. (Finer particles tend to hydrate during testing and curing). Periclase, which could also cause unsoundness, is quite unreactive in fly ash. The autoclave method is probably an overly severe test procedure. However, the test alerts the user to a potential problem. Some European countries specify the alternative Le Chatelier test [8].

Pozzolanic Activity Index with Lime

In the last few years, this test method has fallen into considerable disfavor. In ASTM the requirement was eliminated for Class C fly ashes. India and Korea are now the only other countries that specify this test. In the 1985 ASTM round robin, it was shown that the method could be improved considerably by specifying the fineness of the lime used and by changing from 2x4 in. cylinders to 2 in. cubes.

This test is supposed to measure a fly ash's binding capacity with lime, but this may be in doubt. Helmuth [14], in discussing the low lime-pozzolanic activity index results obtained by Brink and Halstead [15] for several of their fly ashes (ashes that had done quite well in the pozzolanic activity test with cement) suggested that the explanation may be an inadequate lime content specified in C 311. Helmuth [14] says:

"It is apparent that for a fly ash having a high silica-alumina reactive glass, a low lime content, and a high fineness, the pozzolanic reaction will be limited by a lack of lime in the ASTM C 311 test with lime, but probably will have a slight excess of lime in the test with portland cement for typical compositions."

Dunstan [12] has actually found an inverse relationship between 7 d lime-fly ash mortar strength and 28 d concrete compressive strengths.

Pozzolanic Activity Index with Cement

This test is the more widely specified throughout the world than the lime-pozzolanic activity test. Canada was the first nation with a 7 d accelerated cement-pozzolan test. ASTM has tried for years to come up with an accelerated test, but has not been able to reach consensus on the strength index limit. Several problems exist both with the Canadian method (accelerated curing for 7 d at 65°C) and with the current ASTM method (curing for 28 d at 35°C):

1. It is difficult to assure full hydration of the mortar cube specimens. (Cubes are stored in sealed mason jars.)

2. The higher temperature tends to reverse the order of reactive fly ashes, i.e. it compensates for less reactive fly ashes (and cements).

3. In 1985, the ASTM round robin (using the same cement) demonstrated poor precision and considerable variation. This variation is enhanced at temperatures exceeding 23°C.

4. The method is extremely dependent on the cement used. This was realized even in 1962 with the report by Subcommittee III-h on the Cooperative Test Program. The Subcommittee report [16] stated:

"The tests indicate the the effect of fly ash on the strength of mortar or concrete is influenced appreciably by the particular cement with which it is used. In most instances, the differences between results are greater when comparing cements than when comparing fly ashes".

5. It is thought that adjusting the water content to obtain constant flow as measured by the flow table adds variation. Unfortunately, CCRL data confirms that the coefficient of variation between labs is not decreased by constant water in C 109 mortars. Several researchers, including Mehta [8] and Helmuth [14], have suggested eliminating the flow table in tests for pozzolanic activity or making it part of a separate test for water requirement. Then one pozzolan can truly be compared to another in terms of its pozzolanic activity.

6. The name of the test is misleading in that more than pozzolanic activity is taking place. The real purpose of the test is to evaluate a pozzolan's reactivity and interaction with a cement. That involves physical and chemical effects which would be difficult to separate. One of the chief benefits of a fly ash pozzolan is its ability to reduce the water demand. The name of the test should simply be "Activity Index with Cement."

The ASTM Subcommittee feels strongly that water reduction and chemical (pozzolanic) reactivity must be part of the same test method and that is the intent behind the present procedure. Helmuth [17] has argued that for fly ash replacements beyond 20%, little additional water reduction is gained. He described this as, "similar to the saturation effect with a mono-molecular layer of an adsorbed organic admixture." Helmuth's hypothesis is that water reduction has little to do with the spherical nature of fly ash particles:

"...Water reduction is a result of adsorption of very fine fly ash particles on portions of the cement particle surfaces, with resulting dispersion of the cement particles, similar to the action of organic water-reducing admixtures."

ROUND ROBIN PROGRAM CONCERNING C 311

A proposed round robin test program on the activity index with cement will be structured by ASTM in 1987. Factors to be considered in the test are described below.

1. Doubling the batch size specified in C 311 to provide more thorough mixing.

2. Removing 20% of the cement in the control and substituting an equivalent mass of fly pozzolan. Currently ASTM C 311 calls for 35% volume substitution.

3. All labs participating in this round robin will use the identical cement and the same source of Ottawa silica sand.

4. A constant amount of water will be added to the control; the fly ash mixes will be proportioned with water such that they are ±5 of the flow of that control.

5. Three cubes will be cured 7 d at 73°F; three cubes will be cured for 28 d at 73°F. All cubes will be cured in lime-saturated water.

PROPOSED REVISION TO THE ASTM C 618 SPECIFICATION

The following summary of this proposed revision to C 618 is based on the discussion recently presented by Philleo [18]. Instead of three classes of mineral admixtures (Class N, F, C), the C 618 Task Force is proposing two classes of pozzolans: a natural pozzolan (Class 1) and a fly ash pozzolan (Class 2). The user would specify a fly ash pozzolan (Class 2) when he desired a general purpose fly ash pozzolan. If the user desired special properties, he would invoke specific requirements from an optional requirements table. Some of the special properties specified in this table would include: alkali-aggregate reactivity, sulfate resistance, heat of hydration, and hydraulic index. The task force is proposing that either the activity index with cement or the pozzolanic activity index with lime would be used to indicate specification compliance. Both test results would be available in seven days. If the user required high levels of pozzolanic activity, he would specify high totals of $SiO_2+Al_2O_3+Fe_2O_3$ or perhaps a high pozzolanic activity index with lime.

RECOMMENDATIONS

This writer has the following recommendations for consideration by the those developing specifications and for those planning research projects.

1. The first is in support of the following recommendation made by the Rilem Technical Committee 73-SBC meeting held 5/11/84:
   "Drop all chemical requirements except SO_3 and loss on ignition. A limit of maximum 5% SO_3 (ASTM C 618) seems reasonable to prevent contamination with the scrubber-treated ashes, and possible time of set and soundness problems. A limit of maximum 6% loss on ignition seems reasonable to prevent excessive requirement of air-entraining and other admixtures in concrete."

2. The total of $SiO_2+Al_2O_3+Fe_2O_3$ would have benefit only in an optional requirement table, namely, as influencing sulfate resistance. That is, either that total value (>70%) or a sulfate resistance mortar expansion limit would indicate specification compliance. Obviously the former test is quicker and would pre-approve some fly ash pozzolans.

3. LOI, moisture, and SO_3 should be tested daily or for every 100 tons, whichever is more frequent.

4. Research must be performed to determine whether soluble components in fly ash pozzolans are significant.

5. Though no limits should be invoked, the individual values for SiO_2, Al_2O_3, Fe_2O_3, CaO, MgO, SO_3, Na_2O and K_2O should be reported to the user.

6. Drop available alkali content as a requirement.

7. Measure fineness at two sieve sizes: 45 μm and 20 μm. Perhaps a particle size method that determines the mass % below 20 μm could be used to provide this value. Fineness should be measured either daily or for every 100 tons whichever is more frequent.

8. Retain the autoclave soundness test and test it at the frequency specified in ASTM C 311. It informs the user of the possibility of a potentially deleterious reaction.

9. Either the pozzolanic activity index with lime or the activity index with cement should qualify a fly ash pozzolan. The results should be available in 7 d. Even though no limit should be proposed, the ratio of the 7 d strength (at 73°F) to the 28 d strength with same cement (at 73°F) should be reported. This ratio may be interesting to the user of the more hydraulic fly ash pozzolans. No limit on the water requirement should be invoked; it should be reported, however.

10. Suitable limits for sulfate resistance must be established using ASTM C 1012.

11. A suitable test for heat evolution should be developed. Perhaps the conduction calorimeter method being developed by ASTM subcommittee C01.26 would be suitable.

12. Though it is questionable how a consumer might use the hydraulic index to optimize fly ash concrete mixes, the reported value would contribute to the understanding of the material.

13. Finally, a uniformity requirement for loss on ignition should be included with the two present uniformity requirements in C 618 (specific gravity and sieve residue). This would be applicable to air-entrained concrete only.

REFERENCES

1. P.C. Aitcin, F. Autefage and A. Carles-Gibergues, and A. Vaquier, in Second International Conference on Fly Ash, Silica Fume, Slag, and Natural Pozzolans in Concrete, SP-91, (American Concrete Institute, Detroit, 1986) pp. 91-114.
2. E.E Berry and V.M Malhotra, Fly Ash in Concrete, SP85-3 (CANMET, Ottawa, Ontario, 1986) pp. 139-149.
3. O.E. Manz, in Sixth International Ash Utilization Symposium Proceedings DOE/METC/82-52 (Department of Energy and National Ash Association, Washington, DC, 1982) pp. 235-245.
4. E. Rossouw and J. Kruger, in Proceedings First International Conference on the Use of Fly Ash, Silica Fume, Slag, and Other Mineral By-Products in Concrete, SP-79 (American Concrete Institute, Detroit, 1983) pp. 201-220.
5. S. Diamond, in Proceedings International Symposium on The Use of PFA in Concrete. Volume 2, edited by J.G. Cabrera and A.R. Cusens, April, 1982, (Dept. of Civil Eng., Leeds University, UK) pp. 9-28.
6. B. Butler, ASTM Cem. Concr. Aggreg., 4, 68-72 (1982).
7. C. Lee, S. Schlorholtz, and T. Demirel, in Fly Ash and Coal Conversion By-Products: Characterization, Utilization and Disposal III, edited by G.J. McCarthy, F.P. Glasser and D.M. Roy, Mat. Res. Soc. Proc. Vol. 65 (Materials Research Society, Pittsburgh), pp. 125-130.
8. P.K. Mehta, in Second International Conference on Fly Ash, Silica Fume, Slag, and Natural Pozzolans in Concrete, SP-91, (American Concrete Institute, Detroit, 1986) pp. 637-659.
9. S. Diamond, in Proc. Symp. N – Effects of Fly Ash Incorporation in Cement and Concrete, edited by S. Diamond (D.M. Roy, Materials Research Laboratory, Univ. Park, PA 16802, 1982) pp. 12-23.
10. K. Tuutti and G. Fagerlund, in Technology of Concrete When Pozzolans, Slags, and Chemical Admixtures are Used, (Joint Symposium of ACI and RILEM) pp. 105-119.
11. P.K. Mehta, ACI J., November-December, 1986, pp. 994-1000 (1986).
12. E. Dunstan, Fly Ash and Fly Ash Concrete, Bureau of Reclamation Rept. ERC-82-1, May, 1984; ASTM Cem. Concr. Aggreg. 2, 20-30 (1980); in Sulfate Resistance of Concrete, SP-77 (American Concrete Institute, Detroit, 1980).
13. J. Cabrera, C. Hopkins, G. Woolley, R. Lee, J. Shaw, C. Plowman and H. Fox, in Second International Conference on Fly Ash, Silica Fume, Slag, and Natural Pozzolans in Concrete, SP-91, (American Concrete Institute, Detroit, 1986) pp. 115-144.
14. R. Helmuth, ASTM Cem., Concr. Aggreg. 5, 103-110 (1983).
15. R.H. Brink and W.J. Halstead, Proc. ASTM, 56, 1161-1206 (1956).
16. Subcommittee III-h of ASTM Committee C9, "Cooperative Tests of Fly Ash as an Admixture in Portland Cement Concrete", Preprint 19a, ASTM, June 1962.

17. R. Helmuth, in <u>Second International Conference on Fly Ash, Silica Fume, Slag, and Natural Pozzolans in Concrete</u>, <u>SP-91</u>, (American Concrete Institute, Detroit, 1986) pp. 723-740.
18. R.E. Philleo, "Recent Developments in Pozzolan Specifications", Supplementary Papers, Second Int'l Conference on the Use of Fly Ash, Silica Fume, Slag, and Natural Pozzolans in Concrete, Madrid, Spain, 1986, (also available from Craig Cain, Chairman, ASTM Subcommittee CO9.03.10.)
19. G.S. Bobrowski, Master Builders, private communication, August, 1981.

CONSISTENCY OF PERFORMANCE OF CONCRETES WITH AND WITHOUT FLY ASH

WESTON T. HESTER
University of California, Department of Civil Engineering, Berkeley,
California 94720

Received 9 November, 1986; refereed

ABSTRACT

Fly ash and other pozzolans are increasingly used in concretes that must consistently achieve high compressive strength and other measures of performance. In contemporary design and construction practice, however, there is some concern as to whether or not use of commercial fly ash materials contribute to or reduce variations observed in field compressive strengths and properties of the fresh concrete. Regardless of its potential for improving the performance of the concrete, if the daily ongoing use of commercial fly ash will decrease the consistency of the concrete produced, the use of fly ash will be curtailed. In this paper, the experiences of geographically diverse, specific concrete suppliers using their respective sources of fly ash are summarized in detail. The nominal compressive strength of concretes, made with and without fly ash, are analyzed statistically. Special attention is given to moderate- and high-strength concretes. In conclusion, it is shown that concretes made with fly ash may not be more consistent than concretes with only portland cement.

INTRODUCTION

Pozzolans, and particularly fly ash, are increasingly used as a cement replacement and supplement in commercially produced concretes. But many designers and owners continue to express strong reservations about the use of fly ash and its effect on the consistency and performance of the concrete.

Many laboratory test programs have documented that fly ash materials may be used to produce more durable, higher-strength concretes. But, these laboratory results notwithstanding, particular concerns have been raised about the allegedly greater sensitivity of fly ash concretes to curing conditions and to variations in mix materials in commercially produced concretes. Also, some design professionals assert addition of another mix material, e.g., fly ash, introduces more variables to the concrete mix and results in even more inconsistent strength data.

The purpose of this paper is to review selected features of commercially produced fly ash concretes relative to plain concretes, including their relative consistency in strength, uniformity in strength gain, and for anomalies sometimes seen in current practice.

RESULTS AND DISCUSSION

Data Considered

Laboratory tests are ordinarily based on a small number of specimens prepared under specific temperature, slump, and material conditions. In contrast, in commercial concreting practice we encounter a much broader range of conditions, including natural variations in ambient environmental conditions, mix materials, and conditions of placement.

In this paper we refer to data provided by two different ready-mixed concrete producers, one of which produced otherwise comparable plain and fly ash concretes. We also make reference to comparable results observed by others. The principal characteristics of the mix materials and proportions

TABLE 1

Summary of Strength Data

Mix No.	1	3	4	15	5
Specified strength (psi)	5,000	4,000	4,000	5,000	4,000
Type of mix	Plain	Plain	Plain	Fly ash	Fly ash
Sample size	74	195	189	54	517
			7 Day		
Sample mean (psi)	4,540	4,115	4,150	4,120	3,610
Sample mode (psi)	4,460	4,300	3,890	3,890	3,540
Standard deviation (psi)	455	445	380	470	400
Coefficient of variation (%)	10	10.8	9.2	11.4	11.1
Skewness	-0.591	-2.00	0.020	0.222	0.317
Kurtosis	-2.070	+12.4	-0.41	-1.154	0.473
			28 Day		
Sample mean (psi)	5,740	5,310	5,290	5,600	5,110
Sample mode (psi)	5,530	5,700	5,485	5,030	4,780
Standard deviation (psi)	420	445	440	540	540
Coefficient of variation (%)	7.3	8.4	8.3	9.6	10.5
Skewness	-0.200	-0.270	+0.11	+0.328	+0.302
Kurtosis	+0.950	+0.60	0.065	-0.652	-0.427

used are summarized in the following discussion. For all of the data considered here, each producer used the same mix proportions, sources of fine and coarse aggregates, cement and fly ash.

The purpose of presenting field data based on commercially produced concretes is to illustrate the effect that fly ash materials can have on the consistency and performance of production-cast concretes. Such field data are of concern, but are rarely presented because of the difficulty of obtaining and comparing data for concretes cast at the same time under the same conditions, but with and without fly ash.

In the present report several thousand tests on concretes cast on three geographically separate projects are summarized. The unique opportunity to assess a large amount of test data for specific projects makes this data particularly valuable.

Consistency of Field-Cast Concretes

In commercial concreting practice, consistency of the concrete is expressed in terms of mean and standard deviation of the values for measured strength and other properties. Structural designers and producers are particularly concerned that the specified property levels are achieved, and that the achievement of these desired levels is not too sensitive to changes in ambient environmental conditions or field placement practices.

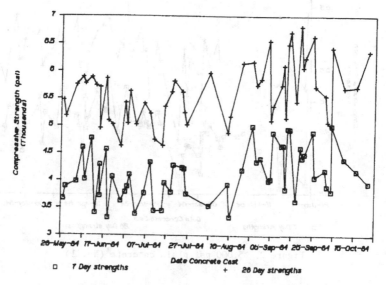

Figure 1.   5,000 psi fly ash concrete (No. 15)

The experience of one producer using a Class C (high lime) fly ash is documented in Table 1. The fly ash is reputedly very consistent, and other mix materials include a siliceous aggregate and Type I cement, but no water-reducing or air-entraining admixture. Where fly ash was used, it was used as a direct cement replacement. These concretes were placed from January of the first year through April of the following year in Minneapolis, MN. The test results represent samples secured approximately uniformly in time over this entire time period.

In Figures 1-3, the 7- and 28-day strengths achieved with selected plain and fly ash concretes are summarized. Note that there are natural seasonal variations in measured strengths (strengths tend to be lower in the warmer, summer months), but that there are no large month-to-month variations which may be attributed to sharp fluctuations in the fly ash or other mix materials.

Referring to Table 1, it is apparent that the plain and fly ash concretes were produced with nominally the same variation in strengths at age 7 days. We may thus conclude, at least tentatively, that use of fly ash as a cement replacement had no significant effect on the variation of measured strengths at early ages. There is, however, measurably more variation in the fly ash concretes than in the plain concretes at 28 days.

A linear regression of 28-day strengths versus 7-day strengths was performed, and two overall trends became clear. First, and not surprisingly, the fly ash concretes gained strength more slowly. However, a second trend also became apparent. The correlation coefficient term, representing the ability to predict 28-day strengths on the basis of 7-day values, is approximately 0.75 for the plain concretes and less than 0.40 for the fly ash concretes. In effect, it is marginally possible to use 7-day strengths to predict 28-day strengths for the plain concretes, but impractical for the fly ash concretes. The relationship between 7- and 28-day strengths for the plain and fly ash concretes is illustrated graphically in Figures 4 and 5.

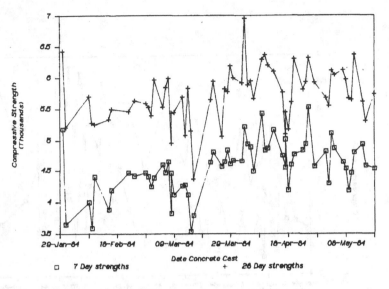

Figure 2. 5,000 psi plain concrete (No. 1)

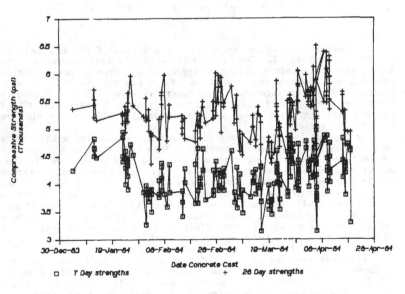

Figure 3. 4,000 psi plain concrete (No. 4)

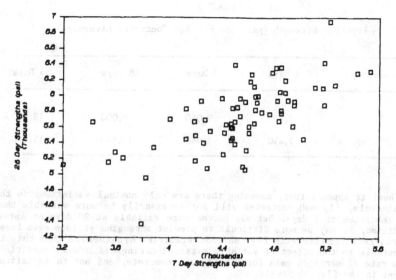

Figure 4. 5,000 psi plain concrete (No. 1)

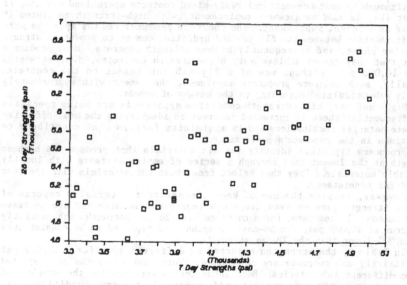

Figure 5. 5,000 psi fly ash concrete (No. 15)

TABLE 2

Reported Strength (psi) for Fly Ash Concretes (Averages)

Laboratory	3 Days	7 Days	28 Days	56 Days
1	7,430	9,040	9,000	13,300
2	7,490	8,940	10,870	11,540

Thus, it appears that, assuming there are only nominal variations in the mix materials, fly ash concretes will not necessarily be more variable than plain concretes at 7 days, but may become more variable at 28 days or later. In addition, it may be more difficult to predict strengths at late ages based on early test results with fly ash concretes. We hypothesize that this is attributable to the effect of variable external curing and ambient conditions on the rate of strength gain for the fly ash concrete, and not to variations inherent in the fly ash itself.

Fly Ash Materials in High-Strength Concretes

Although in most conventional ready-mixed concrete operations fly ash, if used at all, is used as a cement replacement, with high-strength concretes it is used as a cement supplement. The concrete producer relies upon the pozzolanic reaction between the fly ash and hydrating cement to produce a stronger mortar phase, and consequently higher-strength concrete. Most producers report that they cannot achieve very high-strength concretes, i.e., strengths above 10,000 psi, without use of a fly ash supplement to the concrete. Typically, such concrete producers supplement the cement with approximately 20% fly ash calculated according to the weight of cement.

High- and very high-strength concretes apparently are being specified more frequently; there is increased interest in identifying the most effective concrete material combinations and in minimizing factors that contribute to variations in the concrete's performance.

Producers typically identify the mix materials that produce the highest strength at the lowest cost through a series of empirical tests with locally available materials. They then select from these the materials with the best record for consistency.

However, despite the use of known, consistent materials, reports of erratic strength gain or even long-term regression in strength are increasingly common. In one case, for a concrete with 28-day strengths substantially in excess of 10,000 psi, the 56-day strengths were reported to be consistently 20% to 30% lower than the 28-day strengths.

In Table 2, the strengths reported at different ages for a 10,000-psi specified fly ash concrete are summarized. Note that strengths are reported by two different laboratories. Both laboratories were testing the same mix of concrete cast on the same project under nominally the same conditions, although not necessarily out of the same truck of concrete. Each laboratory tested 200 to 350 specimens at each age, depending upon the test age and laboratory.

At least two generalizations can be drawn from the data presented in Table 2. There is relatively little difference between the results reported by the two laboratories for the 3- and 7-day results, but there are large differences at age 28 days and later. There are also large differences in the rates of strength gain for the concretes reported by the two laboratories.

Although there are important differences in testing practices between the two laboratories, it is believed that this alone does not explain the erratic patterns of strength gain noted in Table 2. Such patterns are frequently observed on other high-strength projects as well.

Based upon close observation of the testing practices and material handling procedures used on the above and other projects, it is hypothesized that the manner in which the test specimens are cured has an extraordinarily large effect on the rate of strength gain, and on the ultimate strength achieved by the concrete specimens. As reported by others for low-, moderate- and high-strength concretes without fly ash, [1,2], it appears to be essential to cure the specimens in a lime-water solution rather than a moist room or fog room if continued rapid gains in strength are to be achieved. In the case above, for example, the laboratory specimens cured in a moist room, not in a saturated condition, developed relatively little strength past age 28 days. It appears important, then, to assure long-term strength gain by curing the concrete in a saturated and not simply a fog-cured or comparable condition.

CONCLUSIONS

Based upon the data presented here, it appears that fly ash concretes can achieve consistent strengths largely to the same degree as plain concretes, but they may be more sensitive to improper curing conditions.

It is also concluded that curing conditions clearly influence the strengths developed in high-strength fly ash concretes. This is an important area of concern and warrants additional investigation.

REFERENCES

1. J. M. Plowman, in "Magazine of Concrete Research", 15, 44 (1963).
2. C. D. Pomeroy, in The Effect of Curing Conditions and Cube Size on the Crushing Strength of Concrete (Cement and Concrete Association, Great Britain, 1972), pp. 1-18.

SOME PROPERTIES OF CONTRASTING END-MEMBER HIGH CALCIUM FLY ASHES

SIDNEY DIAMOND and JAN OLEK
School of Civil Engineering, Purdue University, West Lafayette, IN
47907, U.S.A.

Received 1 December, 1986; refereed

ABSTRACT

High-calcium fly ashes with CaO contents >20% do not necessarily contain the glass type found to be characteristic of such ashes in previous results. Instead of this glass (asymmetrical x-ray maximum at 32 deg. CuK-alpha radiation), the fly ash may contain a lower calcium content glass (symmetrical x-ray maximum near 27 deg.) and a substantial content of crystalline CaO. The properties of a representative fly ash of each type are briefly illustrated, and the characteristics of fly ash pastes and of cement-fly ash pastes produced from the two fly ashes are examined.

INTRODUCTION

The properties of very high calcium fly ashes (roughly, those of analytical CaO contents >20%) are still imperfectly established, though they have been actively investigated for some years.

It is well established that analytically-determined CaO contents of such fly ashes represents the sum of calcium present in the glass phase and calcium present in one or more of several well crystallized substances such as crystalline CaO, anhydrite, tricalcium aluminate, and occasionally even tricalcium silicate or dicalcium silicate.

The presence of crystalline calcium-bearing compounds in a given fly ash is readily observed by the usual x-ray diffraction methods, although quantitative analyses may be difficult.

As indicated several years ago by Diamond [1] the x-ray diffraction maximum associated with the glass in fly ash shifts in position as a function of the analytical CaO content of the fly ash. The position of the maximum shifts from that characteristic of vitreous silica (about 23 deg. with copper radiation) to about 27 deg. as the analytical CaO content increases to about 20%. Fly ashes with analytical CaO contents more than about 20% were found to maintain an asymmetrical glass maximum with a fixed position at about 32 deg. The original relationship is presented as Figure 1.

Recently Hemmings and Berry [2] studied the compositions of size and density fractions of a fly ash of an intermediate CaO content, approximately 10% CaO overall. They indicated that two types of glass coexist in the same ash: a "Glass I" low-calcium glass associated with fly ash particles of low specific gravity, which yields a 23 deg. diffraction maximum, and a second type ("Glass II") richer in CaO and associated with high specific gravity particles. This glass yields a diffraction maximum close to 28 deg.

Hemmings and Berry determined glass compositions indirectly, by subtracting the contributions of the amounts of crystalline materials from the overall analysis, and represented them on a triangular CaO – Al_2O_3 – SiO_2 phase diagram. The compositions of the Glass I material fell mostly in the border between the primary phase fields of tridymite (SiO_2) and mullite ($Al_6Si_2O_{13}$), with CaO contents less than about 10%. The Glass II compositions were centered in the anorthite ($CaAl_2Si_2O_8$) primary phase field, with CaO contents around 20%.

The implication of the results of Hemmings and Berry is that the increase in position of the glass maximum with analytical CaO content shown in Figure 1 for fly ashes of up to about 20% overall CaO contents, is due to the presence

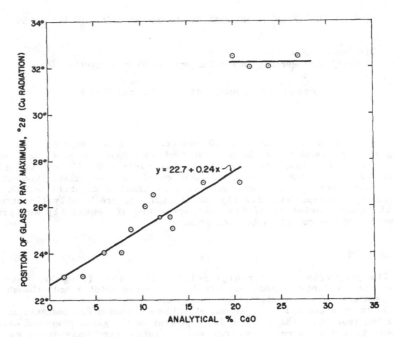

Figure 1. Relation between position of glass x-ray maximum and analytical
CaO content of the fly ash (from Ref. 1).

of increasing proportions of the second type of glass and decreasing proportions of the first type of glass. That is, the shift observed is purely a
statistical phenomenon.

However, as indicated in [1] a third type of glass, with a still higher
CaO content, may occur in very high lime fly ashes and yields the 32 deg.
glass maximum found for such fly ashes. This third type of glass is presumed
to be largely a calcium aluminate glass, as suggested in the original
reference.

It should be specifically recalled that the relationship reproduced in
Figure 1 is with respect to overall analytical CaO content of the fly ash, and
not the CaO content of the glass. Many fly ashes in the higher end of the
linear relationship range did have some indication of the presence of crystalline Ca-bearing compounds, as do fly ashes in the very high CaO content
range in which the third type of glass makes its appearance.

In investigations of a small suite of high calcium fly ashes available in
Indiana, we have found two fly ashes that provide an interesting basis for
comparisons. The two have similar analytical CaO contents (in the 26 - 29%
CaO range). One has nearly all of its CaO in the glass phase, and the glass
is clearly of the third type. The other shows x-ray results indicating that
its glass is of the second type, despite its high analytical CaO content. A
significant part of the analytical CaO is present as crystalline CaO, and
there is also an appreciable content of crystalline anhydrite ($CaSO_4$).

The purpose of this brief paper is (a) to call attention to the existence
of such possible differences in the type of glass and in the content of CaO-
bearing crystalline substances among high calcium fly ashes of similar CaO
contents and (b) to report very preliminary observations of differences in
behavior between the two fly ashes.

TABLE I

Chemical Analyses and Other Characteristics of Fly Ashes Used*

	Rockport	Mitchell
CaO	25.8	28.7
SiO_2	41.3	30.6
Al_2O_3	16.9	14.1
Fe_2O_3	6.8	15.7
SO_3	2.38	3.91
MgO	6.05	7.04
Na_2O	2.36	1.46
K_2O	0.43	0.54
Ignition loss	0.14	0.54
No. 325 sieve residue	3.9	13.4

*Data provided by American Fly Ash Co.

FLY ASH CHARACTERIZATION

The two fly ashes to be discussed in this paper were respectively produced by (a) the Rockport Station of the Indiana and Michigan Electric Co., Spencer County, Indiana, and (b) the Mitchell Station of the Northern Indiana Public Service Co., Lake County, Indiana. Both were obtained through the courtesy of the American Fly Ash Co., Des Plaines, IL, in the summer of 1985.

Table I indicates chemical analyses obtained in routine quality control determinations supplied by the American Fly Ash Co., which markets both fly ashes. As mentioned previously, the CaO contents are similar (26% for Rockport vs. 29% for Mitchell). Otherwise, the Rockport ash has more silica (41% compared with 31%) and less iron oxide (7% vs. 16%); the sulfate contents of both are moderate (2.4% and 3.9%), and the Rockport ash has a bit more Na_2O and a bit less MgO. The loss-on-ignition is extremely low for both fly ashes.

X-ray diffraction patterns of the two ashes are provided in Figure 2. The differences are evident. The Mitchell fly ash has a large content of crystalline CaO (estimated at slightly over 10%), a substantial content of anhydrite (peaks marked "A"), possibly some C_3A, and the usual quartz (marked "Q") and hematite-like iron oxide (marked "H"). The crystalline suite in the Rockport fly ash is qualitatively similar, but there is only a trace of free lime and a much smaller anhydrite content, although the C_3A content may be slightly greater. Nevertheless, the calcium-bearing crystalline components in the Rockport fly likely total to no more than a couple of percent, and nearly all of the high content of CaO must be resident in the glass.

Both fly ashes have significant MgO contents, and appreciable x-ray peaks for periclase, beyond the range shown in the charts.

Figure 3 shows x-ray diffraction patterns taken at a full scale of only 200 cps on the original charts, so as to permit clear indications of the glass response. The Mitchell ash shows the second-type glass maximum, which is characteristically symmetrical and here is centered near 27 deg. In contrast, the Rockport ash shows the characteristic maximum for third type of glass which is characteristically asymmetrical and rises to a maximum near 32 deg.

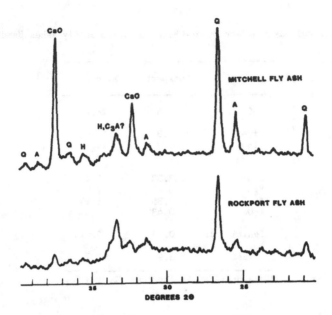

Figure 2. X-Ray diffraction patterns of Mitchell and Rockport
fly ashes.

Scanning electron microscope examination of both ashes was carried out
using the procedure of Lovell and Diamond [3]. Both fly ashes appear to be
almost entirely composed of solid spherical particles without surface
deposits. The Mitchell ash shows an unusually high content of the finest
spheres, as indicated in Figure 4a. However, at the other end of the size
range, there are appreciable numbers of oversize and non-spherical particles
as shown in Figure 4b. As suggested by the micrograph of Figure 5, the
Rockport ash has a more nearly well-graded size distribution, with neither
significant contents of oversize particles, nor a large proportion of very
fine spheres. Neither fly ash has an appreciable content of noticeable carbon
particles, as expected from the very low ignition loss values of Table 1.

PROPERTIES OF PASTES OF FLY ASH AND WATER ALONE

Both fly ashes were found to harden and develop strength when mixed with
water in the absence of cement. However, the behavior of the two fly ashes
differed significantly.

In informal trials carried out by mixing 300 g of fly ash with water at a
water:solids ratio of 0.28 in an open container, the Mitchell ash was found to
display an almost immediate temperature increase of about 6°C. The
temperature increase ceased about 3 minutes after mixing, but the elevated
temperature was maintained for a considerable period, indicating continued
slow exothermic reaction. Despite the heat evolution, there was no immediate
setting response, and set was delayed for some hours.

An x-ray diffraction pattern of the hardened fly ash paste taken the
following day is shown in Figure 6. It is evident that after 1 day only part
of the free CaO had reacted, significant diffraction peaks for unreacted CaO

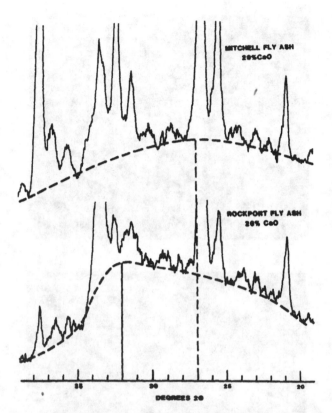

Figure 3. X-Ray diffraction patterns showing locations and shapes
of glass x-ray diffraction maxima for Mitchell and
Rockport fly ashes.

remaining in the pattern. Weak peaks for $Ca(OH)_2$, the expected hydration
product of CaO, are observed in the pattern of Figure 6.

In similar mixing trials the Rockport ash was found not to produce any
immediate temperature rise at all. However, the mix started to set in about
25 minutes, and set about five minutes thereafter. The setting process was
accompanied by a sudden temperature rise of about 8°C. However, the
exothermic process involved soon came to a halt and the temperature decayed to
room temperature within about 30 min. after set.

Strength measurements were carried out on 3 in. x 6 in. cylinder compression specimens cast from the Rockport fly ash at a water:solids ratio of 0.28.
A strength level of about 2300 psi was attained by 7 days, and the quite
respectable compressive strength of 3500 psi was reached at 28 days. Corresponding values for the specific Mitchell ash discussed here were not
available.

320

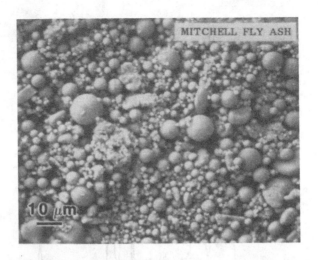

Figure 4. Scanning electron micrographs showing typical fields
for Mitchell fly ash.

CEMENT PASTE CHARACTERISTICS

    Cement pastes containing fly ash (in the weight proportion of 24% of the
total solids) were prepared at a water:solids ratio of 0.49. The cement used
was a Type I portland cement with chemical and physical characteristics near
the middle of the usual range for modern Type I cements.
    Temperature-time curves were monitored for approximately 250 g specimens
of these pastes placed in a nearly adiabatic environment immediately after
mixing. Mixing was carried out in a standard Hobart mixer following the paste
procedure outline in ASTM C 305.

Figure 5. Scanning electron micrograph showing a typical field for Rockport fly ash.

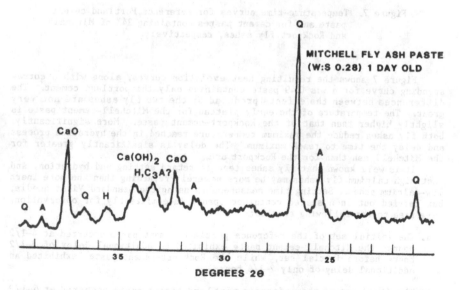

Figure 6. X-Ray diffraction pattern for fly ash:water paste (Mitchell fly ash) after 1 day of auto-hydration.

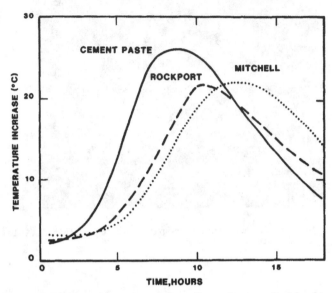

Figure 7. Temperature–time curves for reference Portland cement
paste and for cement pastes containing 24% of Mitchell
and Rockport fly ashes, respectively.

Figure 7 shows the resulting heat evolution curves, along with a corresponding curve for a w:s 0.49 paste containing only the portland cement. The differences between the effects produced by the two fly ashes are not very great. The temperature of the early plateau for the Mitchell–cement paste is slightly higher than that of the Rockport–cement paste. More significantly, both fly ashes reduce the maximum temperature reached in the hydration process and delay the time to reach maximum. The delay is significantly greater for the Mitchell ash than for the Rockport ash.

It is well known that fly ashes tend to retard setting and hydration, and that high calcium fly ashes may be more severely retarding than the more inert low–calcium ashes. Setting time measurements using the standard Vicat needle, but carried out in a sealed container opened intermittantly for observation, gave rise to the following results:

a. The initial set of the reference portland cement paste occurred at 4-1/2 hours; the Mitchell–cement paste exhibited an additional delay of 3-1/2 hours before initial set, while the Rockport–cement paste exhibited an additional delay of only 2-1/4 hours.

b. The final set of the reference portland cement paste occurred at 6-1/2 hours. Again the Mitchell–cement paste exhibited an exceedingly long delay, final set requiring 4-1/4 hours more time than the reference paste; the Rockport–cement paste exhibited a more modest retardation of only 1-3/4 hours.

It is not known whether the extra retardation associated with the Mitchell ash reflects the effect of its glass type, its free CaO content, or some other factor entirely. It would be helpful to monitor the Ca^{++} and OH^- ion contents in the mix water during the period preceding set, but this was not done in the present experiments.

DISCUSSION

Despite fairly widespread current use of high calcium fly ashes in concrete, the details of their reactions have not been extensively studied. It is hoped that this brief contribution may help stimulate such studies.

In order to understand the responses occurring, it is necessary to appreciate the make-up of the starting material.

It appears that very high calcium fly ashes do not necessarily contain the characteristic high-calcium glass with its diffraction maximum near 32 deg., as implied by the Figure 1 relationship. Rather, a combination of the lower CaO content glass giving the symmetrical 27 deg. maximum plus a large component of CaO-bearing crystalline material may occur, as in the present Mitchell fly ash.

For such fly ashes, crystalline CaO is not necessarily easily reacted, as suggested by the results of Figure 6. This raises the question of possible unsoundness, should the delayed reaction result in expansion. While this has not been reported as a practical problem, it may have occurred and have been overlooked.

A systematic study of possible differences in behavior in concrete between high calcium fly ashes of the several different possible characteristics is surely desirable.

CONCLUSIONS

1. Fly ashes of high analytical CaO content (>25%) may have glass of a type showing an asymmetrical x-ray diffraction maximum peaking near 32 deg. (Cu K-alpha radiation) or the glass may be of the lower CaO type that shows a symmetrical maximum near 27 deg. In the latter type a significant portion of the CaO present is necessarily found in crystalline Ca-bearing components rather than in the glass.
2. The Mitchell fly ash examined here is one illustration of the second type.
3. If this fly ash is representative of the class, some immediate heat evolution is expected to occur on addition of water, but much of the CaO may be shielded from early reaction.
4. If the Rockport fly ash examined here is characteristic of fly ashes with the high-calcium (32 deg. maximum) glass type, one can expect significant reaction with water leading to early set, and also very high strength development even in the absence of cement.
5. Differences between the two types of fly ash may not influence the early behavior of cement paste mixtures appreciably, although variations in the retarding effect may be found.

ACKNOWLEDGEMENTS

We thank the American Fly Ash Co. for supplying the fly ashes studied here and for allowing us to quote the analytical data of Table 1. Thanks are also due to Mr. Sheng Qizhong, who carried out the determinations of setting and heat evolution of fly ash-cement pastes.

324

REFERENCES

1. S. Diamond, Cem. Concr. Res. <u>13</u>, 459-464 (1984).
2. R.T. Hemmings and E.E. Berry, in <u>Fly Ash and Coal Conversion By-Products:</u>
   <u>Characterization, Utilization and Disposal II,</u> Edited by G.J. McCarthy,
   D.M. Roy and F.P. Glasser, Mat. Res. Soc. Symp. Proc. Vol. 65 (Materials
   Research Society, Pittsburgh, PA, 1986), pp. 91-104.
3. J. Lovell and S. Diamond, ibid., in <u>Fly Ash and Coal Conversion By-</u>
   <u>Products: Characterization, Utilization and Disposal II,</u> Edited by G.J.
   McCarthy, D.M. Roy and F.P. Glasser, Mat. Res. Soc. Symp. Proc. Vol. 65
   (Materials Research Society, Pittsburgh, PA, 1986), pp. 131-136.

VARIABILITY AND TRENDS IN IOWA FLY ASHES

SCOTT SCHLORHOLTZ, KEN BERGESON and TURGUT DEMIREL
Department of Civil Engineering, Iowa State University, Ames, IA, 50011

Received 1 November, 1986; refereed

ABSTRACT

An investigation has been made of the variability of physical and chemical properties of high-calcium (Class C) fly ashes from four Iowa power plants. The investigation summarizes results obtained from three years (1983 through 1985) of monitoring of the various power plants. All four of the power plants burn low-sulfur, sub-bituminous coal from Wyoming. Fly ash samples were obtained from the power plants in accordance to the procedures described in ASTM C 311. Laboratory testing methods were similar to those specified by ASTM C 311. During the three year period, 102 samples were subjected to chemical and physical analysis while an additional 349 samples were subjected to physical analysis only. In general, the four power plants produce fly ashes of similar mineralogy and chemical composition. The observed time variation of the chemical composition of fly ash from a single power plant was quite small. The sulfur content consistently showed the largest coefficient of variation of the 10 elements studied. Physical characteristics of the fly ashes (as measured by ASTM tests) were also fairly uniform over long periods of time, when considered on an individual power plant basis. Fineness, when measured by wet washing using a 325 mesh sieve, consistently exhibited the largest coefficient of variation of any of the physical properties studied.

INTRODUCTION

The characterization of high-calcium fly ashes has steadily improved in the last few years. Before the late 1970's little was known about the chemical and physical properties of such fly ashes [1,2]. Today the literature abounds with papers concerning high-calcium fly ashes (see for example MRS Symposia 1981, 1984, 1985, and International Ash Symposia 1976, 1979, 1982, 1985). General agreement has been reached on the major crystalline compounds normally encountered in fly ashes of different origins [2-7]. Glass composition and content are currently being investigated. Chemical analysis of fly ash has suffered from two basic problems: (1) lack of adequate standard reference materials for calibration of analytical techniques; (2) misinterpretation of the results obtained from such chemical analyses. It is hoped that the first problem will be solved by late 1986 when the National Bureau of Standards releases three new fly ash standards [8]. The second problem is simply one of confusion about the type of data obtained from the chemical analyses specified in ASTM C 311 [9]; it is hoped that future work will help to eliminate such misunderstandings.

Simple correlations between the chemical composition of fly ash and its influence on the strength and durability characteristics of portland cement pastes, mortars and concretes have been pursued for many years [10-17], but to date these have not been adequately defined.

The purpose of this paper is to present the results obtained from three years of routine monitoring of fly ashes from four Iowa power plants. The sampling scheme used is that described in ASTM C 311 [9]. Statistical analysis methods have been employed, when applicable, to check possible correlations between the measured chemical composition of fly ashes and their observed physical characteristics.

Table I

Power Plant Details

Power Plant	Boiler Type	Generating Capacity (NET MW)	Annual Ash Production (Tons/yr)	Precipitator Type	Year on Line	Ash Silo Storage Capacity (Tons)	Coal Source
Council Bluffs #3	Babcock-Wilcox	700	90,000	H-ESP (Na$_2$CO$_3$ added to aid precipitator)	1978	2,500	Wyoming (Eagle Butte & Bell Ayr mines)
Lansing #4	Rilley-Stoker	260	28,000	H-ESP	1977	200	Wyoming (Eagle Butte & Bell Ayr mines)
Ottumwa	Combustion Engineering	675	80,000	H-ESP (Na$_2$CO$_3$ added to aid precipitator)	1981	3,500	Wyoming (Sunedco-Cordero mines)
Port Neal #4	Foster-Wheeler	600	80,000	H-ESP (Na$_2$CO$_3$ added to aid precipitator)	1979	5,000	Wyoming (Rawhide mine)

H-ESP = Hot side electrostatic precipitator.

GENERAL INFORMATION

Fly Ash Sources

Fly ash is a by product of burning coal. It is logical to assume that both coal composition and power station boiler and precipitator configuration should influence the chemical and physical properties of fly ash. Table I defines these pertinent details for the fly ashes discussed in this paper. All 4 power plants burn low-sulfur, sub-bituminous coal from the Powder River Basin near Gillette, Wyoming. Several of the power plants routinely add sodium containing compounds (typically sodium carbonate) to the raw coal feed to enhance the performance of their electrostatic precipitators. We have found that this can influence both the mineralogy and bulk chemical composition of the fly ash.

Laboratory Testing Program

Fly Ash samples were tested for chemical and physical properties by using methods similar to those specified in ASTM C 311 [9]. The major deviations from the methods described in C 311 were: (1) we did not measure the lime pozzolanic activity of the fly ashes; (2) we did not use gravimetric and/or volumetric methods of chemical analysis for measuring the chemical composition of the fly ashes.

We have found that the lime pozzolan test described in ASTM C 311, tends to be biased against Class C fly ashes that contain large concentrations of calcium [18]. Others have also mentioned such an anomaly [19]. Commercial grade calcitic lime (i.e., technical grade calcium hydroxide) can be substituted for the reagent grade calcium hydroxide with good results, but it seems unproductive to change sources of lime until one finds a source which allows the majority of fly ashes to meet or surpass the 800 psi specification. Instead, we have adopted an interim specification (M 295-84I) drafted by the American Association of State Highway and Transportation Officials

(AASHTO) [20]. This specification requires a 7 day (d) pozzolanic activity test using a mortar containing fly ash and portland cement. The specification recommends that the laboratory testing program should use the same type and source of cement as will be used for the actual construction project. We feel that this is a rational step toward a workable specification that may help to enhance fly ash utilization.

Quantitative X-ray fluorescence spectrometry (QXRF) was used to measure the major and minor elements present in fly ash [21, 22]. We have adopted QXRF because the analysis is quick and the method allows for the independent determination of all elements from fluorine to uranium. One of the major problems encountered with the techniques listed in ASTM C 311 is that they were initially designed to analyze portland cement. Class F fly ashes can be analyzed with these methods with reasonable (?) success because Class F fly ashes typically contain only small to negligible amounts of Ti, P, Sr and Ba. Class C fly ashes often contain significant (>0.5 wt.%) amounts of each. The Ti and P tend to accumulate in the R_2O_3 group and thus, lead to a systematic error in the gravimetric Al_2O_3 determination [23,24]. This error can be eliminated by determining the amounts of P_2O_5 and TiO_2 in the R_2O_3 group and applying a correction (i.e., $\%Al_2O_3 = \%R_2O_3 -$ (residual $\%SiO_2 + \%TiO_2 + \%Fe_2O_3 + \%P_2O_5$). Likewise, the Sr and Ba tend to accumulate in the portion of sample used for the determination of calcium, and one actually measures the sum of these three elements (commonly reported as oxides) when using gravimetric techniques. Cost quickly rules out the use of wet chemical techniques as the number of elements needing to be determined increases. Instrumental techniques (i.e., QXRF or atomic absorption spectroscopy (AAS)) are not subject to errors of this kind but suffer from their own inherent limitations. Both QXRF and AAS are calibration type analyses and rely heavily on good standards. Few good fly ash standards are presently commercially available. We have avoided this difficulty by generating several "in-house" standards which cover a sufficient range for reliable calibration (see refs. 21 and 22 for technical details). Currently, we estimate the relative error in our fly ash analyses at about 3 to 5% of the amount reported for major elements, and about 5 to 10% for minor elements. We are presently finishing work on a much more sophisticated QXRF analytical scheme which should yield nearly a total analysis of our fly ashes (i.e., determination of 13 elements, final oxide sum between 99.0 to 100.5% by weight). When the new method has been tested and its accuracy verified, all of the analyses reported in this paper will be repeated.

Fly Ash Mineralogy

A detailed description of the mineralogy and and its variation in the various fly ashes is well beyond the scope of this paper. It is important, however, to briefly address the question of mineralogy because it helps to explain some of the perturbations that are observed in the physical properties of the various fly ashes, and also it helps to justify one of the assumptions that was made in utilizing QXRF for chemical analysis.

Figure 1 illustrates typical X-ray diffractograms of fly ashes obtained from the four different power plants. The crystalline compounds present in all of the fly ashes were: quartz, anhydrite, lime (CaO), periclase (MgO) and a phase similar to tricalcium aluminate. The fly ashes may also contain small amounts of a compound similar to tetracalcium trialuminate sulfate. The major type of glass found in all of the fly ashes had an amorphous scattering hump above 30 degrees 2-theta (Cu K-alpha radiation). It is obvious from Figure 1, that the fly ashes exhibit very similar mineralogies, the differences being mostly due to relative amounts rather than mineral structure. One of the major assumptions that had been made when we developed the calibration curves for our fly ash samples was that we were dealing with a consistent mineralogy (i.e., we ignored the mineralogical effect). Figure 1 indicates that we made a valid assumption.

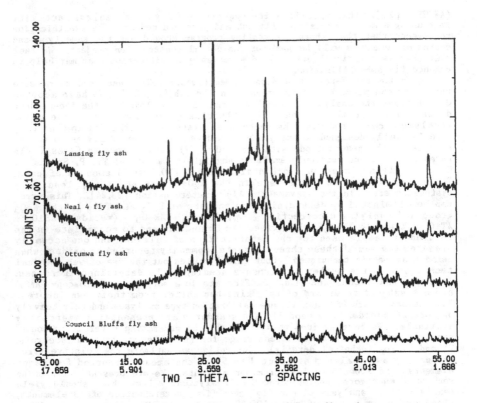

Figure 1. X-ray diffractograms of Iowa fly ashes.

RESULTS

Table II summarizes the data obtained from the fly ash samples subjected to chemical and physical analysis. Table III summarizes the data obtained from the physical testing conducted on each 400 ton lot of fly ash. The data listed in Tables II and III were obtained from fly ash samples collected from the various power plants at unequal intervals from 1983 to 1985. The majority of the samples were obtained between April and October (i.e., the portland cement concrete construction season). In Tables II and III, $\bar{X}$ refers to the arithmetic mean, S refers to the standard deviation, R refers to the arithmetic range and n refers to the number of samples. In general, the analytical results listed in the two tables have been reported using a format suggested by the ASTM [25]. This was done so that additional statistical quantities (i.e., coefficient of variation, etc.) can be calculated from the tables without experiencing severe rounding errors. Table IV lists the important details about the various Type I portland cements used for evaluating the pozzolanic activity of the fly ash samples. Note, that during 1984 and 1985 two different lots of cement were used each year for evaluating pozzolanic activity. The cements were from the same manufacturer but they were obtained at different times. The compressive strength values listed in Table IV are the averages of the control samples for that year.

Table II

Summary of Results of ASTM C 618 Testing of Fly Ashes from 1983 to 1986

Council Bluffs Power Plant

Year →	1983 n = 4			1984 n = 7			1985 n = 9		
Test	$\bar{X}$	S	R	$\bar{X}$	S	R	$\bar{X}$	S	R
Moisture content	0.07	0.02	0.03	0.06	0.03	0.07	0.09	0.06	0.20
Loss on Ignition	0.32	0.13	0.29	0.45	0.08	0.22	0.47	0.14	0.42
Fineness	10.82	3.20	6.86	11.73	2.05	6.77	13.30	3.13	9.90
7 Day Pozzolan	Not Determined			89.7	6.4	19.0	86.7	4.6	12.0
Autoclave Exp.	0.14	0.01	0.03	0.06	0.01	0.04	0.11	0.03	0.09
Specific Gravity	2.71	0.03	0.06	2.65	0.05	0.11	2.71	0.03	0.10
28-Day Pozzolan	98.8	4.0	9.0	100.9	8.3	24.0	87.9	3.7	10.0
H_2O Required	91.5	5.3	10.0	90.3	0.76	2.0	88.8	1.4	5.0
SiO_2	31.48	0.65	1.51	33.64	1.67	5.24	30.81	1.64	4.50
Al_2O_3	16.90	0.26	0.55	17.15	0.57	1.67	15.82	0.62	1.80
Fe_2O_3	5.15	0.16	0.33	5.06	0.24	0.77	5.40	0.45	1.37
SO_3	3.06	0.30	0.64	2.77	0.30	0.81	3.78	0.50	1.29
CaO	27.90	0.56	1.36	26.83	0.86	2.19	28.12	0.42	1.10
MgO	6.65	0.16	0.35	5.67	0.21	0.60	5.80	0.49	1.45
P_2O_5	0.87	0.14	0.78[*]	1.24	0.18	0.54	1.00	0.18	0.65
K_2O	0.33	0.04	0.08	0.34	0.04	0.13	0.28	0.04	0.11
Na_2O	1.78	0.08	0.19	1.98	0.16	0.44	1.91	0.16	0.49
TiO_2	1.36	0.04	0.08[*]	1.33	0.07	0.22	1.24	0.14	0.41
Avail. Alk.	1.31	0.09	0.22	1.28	0.16	0.52	1.34	0.15	0.42

Lansing Power Plant

Year →	1983 n = 4			1984 n = 4			1985 n = 7		
Test	$\bar{X}$	S	R	$\bar{X}$	S	R	$\bar{X}$	S	R
Moisture Content	0.04	0.03	0.06	0.04	0.04	0.07	0.05	0.05	0.14
Loss on Ignition	0.44	0.27	0.56	0.29	0.05	0.11	0.47	0.18	0.56
Fineness	12.95	2.35	5.23	11.17	2.82	6.18	12.77	1.92	6.80
7-Day Pozzolan	Not Required			90.3	2.2	5.0	90.0	5.1	13.0
Autoclave Exp.	0.11	0.02	0.04	0.07	0.01	0.02	0.10	0.03	0.08
Specific Gravity	2.77	0.04	0.09	2.78	0.02	0.05	2.79	0.02	0.07
28-Day Pozzolan	85.8	7.7	18.0	91.2	8.0	19.0	86.9	5.1	16.0
H_2O Required	95.5	5.4	12.0	90.0	0.0	0.0	89.4	1.0	3.0
SiO_2	35.72	3.68	7.70	34.32	3.19	7.37	31.50	1.41	4.00
Al_2O_3	16.72	0.66	1.60	15.58	0.17	0.38	15.53	0.46	1.40
Fe_2O_3	5.54	0.22	0.52	5.68	0.42	0.97	5.94	0.38	1.20
SO_3	3.66	0.68	1.63	4.29	0.70	1.52	4.35	0.36	1.03
CaO	26.72	0.68	1.44	26.82	1.07	2.39	27.66	0.64	1.60
MgO	6.63	0.51	1.16	6.06	0.37	0.87	5.77	0.30	0.89
P_2O_5	1.00	0.30	0.57[*]	0.84	0.08	0.19	0.86	0.19	0.64
K_2O	0.38	0.02	0.04	0.40	0.12	0.27	0.29	0.04	0.10
Na_2O	2.05	0.20	0.45	1.88	0.05	0.10	2.06	0.24	0.71
TiO_2	1.29	0.03	0.06[*]	1.20	0.04	0.07	1.20	0.10	0.33
Avail. Alk.	1.42	0.11	0.25	1.33	0.11	0.24	1.44	0.22	0.57

[*]Denotes n = 3

Table II (continued)

Ottumwa Power Plant

Test	1983 n = 8 $\bar{X}$	S	R	1984 n = 17 $\bar{X}$	S	R	1985 n = 17 $\bar{X}$	S	R
Moisture Content	0.04	0.02	0.05	0.03	0.01	0.05	0.03	0.02	0.06
Loss on Ignition	0.24	0.05	0.14	0.26	0.05	0.21	0.25	0.06	0.21
Fineness	10.22	0.32	0.94	10.41	0.75	2.74	9.99	0.69	2.50
7-Day Pozzolan	Not Required			90.2	6.3	23.0	91.9	4.0	16.0
Autoclave Exp.	0.05	0.02	0.06	0.03	0.01	0.04	0.06	0.02	0.05
Specific Gravity	2.61	0.02	0.06	2.60	0.03	0.12	2.65	0.03	0.11
28-Day Pozzolan	103.1	10.6	30.0	97.6	5.7	20.0	94.2	6.3	23.0
H_2O Required	89.1	3.4	10.0	90.2	0.8	3.0	86.8	1.9	9.0
SiO_2	34.48	1.49	4.86	35.33	1.42	5.05	32.23	1.64	6.48
Al_2O_3	19.98	0.41	1.20	18.36	0.35	1.60	18.33	0.34	1.29
Fe_2O_3	5.23	0.14	0.44	5.19	0.16	0.63	5.44	0.40	1.28
SO_3	1.67	0.22	0.65	2.16	0.37	1.32	2.56	0.44	1.76
CaO	24.72	0.68	1.89	23.77	0.70	2.17	25.11	0.54	2.03
MgO	4.94	0.19	0.57	4.63	0.16	0.57	4.92	0.13	0.54
P_2O_5	1.41	0.30	0.92*	1.80	0.23	0.84	1.59	0.38	1.20
K_2O	0.40	0.03	0.09	0.40	0.04	0.12	0.38	0.03	0.09
Na_2O	1.96	0.24	0.67	2.58	0.16	0.55	2.13	0.52	1.95
TiO_2	1.47	0.05	0.16*	1.37	0.05	0.16	1.42	0.04	0.13
Avail. Alk.	1.41	0.18	0.59	1.54	0.30	0.84	1.54	0.33	1.32

*Denotes n = 7

Neal #4 Power Plant

Test	1983 n = 4 $\bar{X}$	S	R	1984 n = 6 $\bar{X}$	S	R	1985[a] n = 15 $\bar{X}$	S	R
Moisture Content	0.02	0.01	0.03	0.03	0.03	0.07	0.03	0.02	0.06
Loss on Ignition	0.17	0.01	0.03	0.31	0.06	0.14	0.31	0.04	0.14
Fineness	7.49	2.56	5.79	11.57	0.69	2.06	11.42	2.20	7.10
7-Day Pozzolan	Not Required			88.4	6.1	16.0	92.7	5.1	20.0
Autoclave Exp.	0.08	0.01	0.02	0.06	0.02	0.05	0.07	0.02	0.07
Specific Gravity	2.69	0.02	0.04	2.66	0.04	0.11	2.59	0.08	0.28
28-Day Pozzolan	104.2	8.8	19.0	90.3	5.2	14.0	95.3	6.8	25.0
H_2O Required	88.2	0.5	1.0	91.8	4.0	10.0	88.5	1.0	4.0
SiO_2	35.20	0.97	2.19	33.63	1.00	2.74	35.23	2.52	9.98
Al_2O_3	15.68	0.25	0.58	15.69	0.55	1.61	16.24	0.91	3.11
Fe_2O_3	6.20	0.13	0.24	5.83	0.26	0.68	5.59	0.50	1.67
SO_3	3.33	0.28	0.60	3.82	0.60	1.36	3.25	0.74	2.56
CaO	25.89	0.41	0.90	25.88	0.57	1.71	25.45	1.62	4.89
MgO	6.04	0.22	0.50	5.81	0.22	0.52	5.65	0.41	1.34
P_2O_5	0.76	0.05	0.09*	0.97	0.20	0.51	0.99	0.19	0.74
K_2O	0.29	0.05	0.12	0.30	0.03	0.08	0.32	0.07	0.22
Na_2O	2.08	0.12	0.29	2.54	0.19	0.43	2.20	0.23	0.78
TiO_2	1.02	0.02	0.04	1.04	0.06	0.16	1.06	0.07	0.11
Avail. Alk.	1.46	0.08	0.18*	1.57	0.16	0.39	1.39	0.27	0.89

[a]Denotes that two different coal sources were used in 1985.

*Denotes n = 3

Table III

Summary of Physical Testing of Fly Ash

Council Bluffs Power Plant

Year →	1984			1985		
	n = 27			n = 24		
Test	X̄	S	R	X̄	S	R
Moisture Content	0.05	0.05	0.22	0.13	0.14	0.59
Loss on Ignition	0.46	0.12	0.45	0.48	0.29	1.35
Fineness	12.56	1.46	5.91	12.55	2.37	10.50
7-Day Pozzolan	91.5	6.8	29.0	88.6	5.2	22.0
Autoclave Exp.	0.07	0.02	0.07	0.10	0.02	0.08
Specific Gravity	2.65	0.05	0.19	2.71	0.03	0.14

Lansing Power Plant

Year →	1984			1985		
	n = 13			n = 15		
Test	X̄	S	R	X̄	S	R
Moisture Content	0.06	0.04	0.14	0.05	0.03	0.14
Loss on Ignition	0.27	0.08	0.29	0.48	0.14	0.50
Fineness	9.46	1.18	3.86	12.18	1.66	5.80
7-Day Pozzolan	87.5	5.9	19.0	86.8	4.2	15.0
Autoclave Exp.	0.07	0.02	0.06	0.11	0.02	0.09
Specific Gravity	2.78	0.03	0.09	2.79	0.03	0.10

Neal #4 Power Plant

Year →	1984			1985*		
	n = 14			n = 54		
Test	X̄	S	R	X̄	S	R
Moisture Content	0.03	0.02	0.08	0.04	0.03	0.16
Loss on Ignition	0.30	0.06	0.20	0.31	0.07	0.31
Fineness	11.30	1.56	4.97	11.32	2.14	8.30
7-Day Pozzolan	87.6	5.0	20.0	92.2	6.2	25.0
Autoclave Exp.	0.07	0.01	0.04	0.07	0.02	0.08
Specific Gravity	2.64	0.03	0.11	2.59	0.08	0.34

Ottumwa Power Plant

Year →	1983			1984			1985		
	n = 39			n = 78			n = 85		
Test	X̄	S	R	X̄	S	R	X̄	S	R
Moisture Content	0.06	0.03	0.11	0.02	0.01	0.05	0.03	0.02	0.10
Loss on Ignition	0.23	0.08	0.44	0.24	0.06	0.31	0.24	0.06	0.24
Fineness	10.39	0.95	3.80	10.53	1.07	5.06	9.83	0.81	3.90
7-Day Pozzolan	Not Required			92.1	7.9	55.0	93.8	5.6	32.0
Autoclave Exp.	0.05	0.02	0.08	0.03	0.01	0.05	0.06	0.02	0.07
Specific Gravity	2.61	0.04	0.17	2.59	0.04	0.21	2.65	0.03	0.16

*Two different coal sources were used in 1985.

Table IV

Type I Portland Cements used for pozzolanic activity testing

Year Oxide	1983 wt%	1984 wt%			1985 wt%		
		A	B	AVG.	A	B	AVG.
CaO	63.0	62.8	62.4	62.6	63.9	63.3	63.6
SiO_2	21.3	21.9	22.2	22.0	21.7	22.3	22.0
Al_2O_3	4.29	4.03	4.32	4.18	4.32	4.50	4.41
Fe_2O_3	3.01	2.97	1.62	2.29	1.64	1.70	1.67
SO_3	2.65	2.37	2.71	2.54	2.57	2.69	2.63
MgO	2.32	2.58	2.23	2.40	3.03	2.63	2.83
K_2O	0.57	0.42	0.57	0.50	0.48	0.59	0.54
Na_2O	0.16	0.26	0.36	0.31	0.28	0.26	0.27
TiO_2	0.22	0.24	0.23	0.24	0.23	0.24	0.24

Average Compressive Strength (psi)

curing time							
7 d	--			4700			4800
28 d	5500			6000			6100

DISCUSSION

Chemical Tests

Overall, the variation of the mean chemical composition for each individual power plant as a function of time is not very large. The overall coefficient of variation for the major elements (Si, Al and Ca; expressed as oxides) was typically about 10% during the three years of monitoring. Sulfur, sodium and phosphorus exhibited the largest coefficients of variation of the 10 elements (expressed as oxides) measured in this study.

There appears to be a general trend which indicates an increase in the concentration of sulfur in most of the fly ashes from 1983 to 1985. Also, sulfur exhibits a fairly wide range of concentrations at any given power plant during a single year.

The average concentration of sodium is fairly constant at each power plant over the three year monitoring period. The constancy of the average, which may be considered as somewhat misleading because it masks cyclical changes within each yearly period, can be attributed to the fact that three of the four power plants add sodium carbonate to the raw coal feed to enhance the efficiency of their electrostatic precipitators. The cycle starts with the addition of small amounts (or none) of sodium carbonate to the raw coal feed when the precipitator plates are clean and highly efficient. As the precipitator plates become less efficient the amount of sodium carbonate added to the raw coal feed is increased to enhance the precipitator's efficiency. When the upper limit of sodium carbonate addition is reached (i.e., boiler slagging becomes excessive) the precipitator plates are cleaned and process starts again. It appears that in the majority of the Iowa power plants studied, this results in a bulk annual (average) concentration of about 1.9 to 2.1 percent sodium oxide.

Presently, the significance of the phosphorus content of our high-calcium fly ashes is not clear. However, it is apparent from Table II, that the concentration of P varies somewhat in the fly ashes studied. Phosphorus pentoxide is one of the glass network forming oxides and hence, we speculate that it should contribute to the amorphous portion of a given fly ash.

The available alkali test results indicate that the fly ashes contributed about 65% of their total alkali oxide concentration to the test pore solution during the 28 d of curing. The results from several of the fly ashes show the same cyclical pattern that was observed for the total sodium concentration. We currently disagree with the logic behind the available alkali test. Research at our laboratory indicates that for all four of the fly ashes discussed in this paper, the alkalis in the fly ash are still contributing significantly to the alkalis measured in the pore solution at curing times of well over 100 d [26]. Hence, the required 28 d curing period is quite arbitrary and really does not reflect the total amount of alkalis that can be dissolved from the fly ash into the pore solution. This same observation was made in 1956 by Brink and Halstead [13].

Physical Tests

In the ASTM requirements for fly ash testing, physical tests are performed on a much more frequent basis than are chemical tests. It must be kept in mind that the majority of samples summarized in Table II are simply composites (i.e., linear combinations) of the samples summarized in Table III. Hence, we should expect to see a general agreement between the results of physical tests that are common to both tables.

Moisture content and loss on ignition (LOI) results indicate fairly large variations around the mean values. Although the variations were large, they were also insignificant from a practical point of view (i.e., as compared to Class F fly ashes) simply because the observed values were so small. The moisture content was typically between 0 and 0.1% for all of the four power plants monitored. The LOI values were typically between 0.2 and 0.5%, rarely (if ever) greater than 1%. The mean value for either moisture content or loss on ignition was fairly stable from year to year for each power plant. One important point to note about the data is how poorly the arithmetic ranges (denoted as R in the tables) obtained from the a large number of samples (Table III) agree with those obtained from a smaller number of samples (Table II). This is not uncommon and tends to imply that the range of values observed for the chemical composition may be much lower than the "true" range because of the small number of observations that were made. We are currently performing chemical analysis on many of the samples initially subjected only to physical analysis to quantify this inference.

All four of the power plants studied in this investigation typically produced fly ashes that had fineness values below 15% (when expressed as percent retained on a 325 mesh (45 μm) sieve). Each power plant exhibited a fairly large variability (coefficient of variation roughly 10 to 20%) during any given year. However, for the majority of the power plants, the mean fineness observed for each year was fairly constant. Again, there appears to be good agreement between the mean fineness values reported in Table II (mean determined from a small number of observations) when compared to those reported in Table III (mean determined from a large number of observations). The range statistic, however, again shows poor agreement between the two tables. It appears that the act of combining equal portions of 5 fly ash samples to create a single composite sample tends to reduce the range of values observed during testing.

The results of the autoclave test show that none of the fly ash samples produced from any of the four power plants during 1983 to 1985 have failed the test. The year to year variability indicated in Tables II and III can be attributed to the fact that a different source of portland cement was used each year. As mentioned earlier, in the years of 1984 and 1985, two different lots of cement were used during each year. Hence, the rather large range statistics were also influenced by a change in portland cement rather than simply a variation in the expansive tendencies of the fly ashes.

The measured specific gravity of fly ash samples obtained from a single power plant shows little variation about the mean during a given year. Year to year variations in specific gravity appear to be correlated with variations in the Ca content of the fly ash or inversely correlated with variations in the Si content of the fly ash. Previous research [10,11] had indicated a multiple correlation of specific gravity with Fe concentration, Ca concentration and #325 sieve fineness. We have not been able to find significant correlations between specific gravity and either fineness or Fe content for our fly ashes. We believe that the positive correlation between specific gravity and Fe content observed by other researchers is real, and that the major reason that the trend does not appear in our data was due to the fact that our fly ashes exhibit a very narrow range of Fe concentration.

Pozzolanic activity index testing of fly ash-cement mortars produces three important pieces of information for engineers interested in using fly ash for construction projects: (1) water demand of the mortar mix (i.e., H_2O required); (2) short term compressive strength of the mortar (i.e., 7 d pozzolan test); (3) the 28 d compressive strength of the mortar. One fact that must be kept in mind when comparing the pozzolanic activity results (listed in Tables II and III), is that the mortar mixtures are proportioned on the basis of absolute volume. Each test mixture contains 35% fly ash (by volume) and thus, as the specific gravity of a fly ash increases so does the weight fraction of the fly ash in the mix. The 35% (by volume) replacement of fly ash for cement, corresponds to a fly ash replacement of between 28.2% and 32.4% (by weight), if we assume a fly ash specific gravity of 2.3 to 2.8, respectively. We think that it may be more consistent (and also more convenient) to proportion the mixes on the basis of weight percent. Other researchers have also expressed similar concerns [16].

In general, the fly ashes from all four power plants behave like low-range water reducers when incorporated into the mortar mix. These fly ashes, when replacing 35% (by volume) of cement, typically reduced the water demand of the test mortar by 5% to 12% (by weight). The mean water demand value for each power plant was fairly stable and the coefficient of variation was typically less than about 5% (both within a single year and throughout the 3 year monitoring period). In no instance, did a test mortar (containing fly ash) require more water to meet the specified flow criteria than did the control mortar.

Results of the 7 d pozzolan test indicated that the Class C fly ashes studied in this investigation, consistently reached about 90% of the compressive strength of the portland cement control samples. The coefficients of variation of the test results from the various power plants were typically less than 8% (within a single year of monitoring).

The results of the 28 d pozzolanic activity test were not as encouraging as those obtained from the 7 d tests. The four different power plants showed different trends in their 28 d pozzolanic activities. The Ottumwa, Council Bluffs and Neal 4 power plants typically had pozzolanic activities of about 90% to 100% of the portland cement control mortar. There was little (if any) significant increase in relative strength gain when compared to the 7 d test results. Lansing power plant had a 28 d pozzolanic activity that was slightly lower than the other fly ashes. Again, the relative strength gain during the extra curing period was negligible. We have noticed a slight trend in our data, which indicated that the 28 d pozzolanic activity test may be sensitive to the amount of sulfur (expressed as the trioxide) present in a given fly ash. Typically, fly ashes that had high sulfur contents also had low 28 d pozzolanic activity test values. Additional work is currently being done to investigate this anomaly in more detail. The research has been complicated by the fact that the four fly ash sources described in this paper, have rather high contents of Ba and Sr. Hence, it may be important to correct the observed sulfur trioxide content of the fly ashes for the presence of these two nearly insoluble sulfates. Previous concrete work at our laboratory [17] and at the Portland Cement Association (PCA) [14], indicated that a similar trend (i.e., lower rate of strength gain with increasing time) does not occur in moist cured concrete specimens containing fly ash (Class C or Class F).

Hence, the anomaly observed in the mortar mixes may be a remnant of the test procedure used to measure the 28 d pozzolanic activity. The PCA also reported a poor correlation between the 28 d pozzolanic activity test and compressive strengths observed in concretes containing fly ash [14].

SUMMARY AND CONCLUSIONS

An examination of the variability and trends in fly ashes from four Iowa power plants, monitored from 1983 through 1985, has been made. Sampling and testing procedures were similar to those specified by ASTM C 311. The following comments can be made:

Variation in the mean chemical composition for each individual power plant was not very large during the monitoring period. Sulfur, sodium and phosphorus, among the 10 elements studied, exhibited the largest coefficients of variation.

Moisture content and LOI tests exhibited stable mean values, but rather large variability. The variability is not significant from a practical stand point.

Fineness tests consistently exhibited the largest coefficients of variation for any of the physical tests that produced significant results. The mean fineness values, however, were fairly stable over the monitoring period.

None of the fly ash samples tested during the 3 year monitoring period failed the autoclave test.

The specific gravity of fly ash samples obtained from a single power plant have stable means and small coefficients of variation during any single year. Specific gravity of a fly ash, however, may change significantly from year to year.

Pozzolanic activity tests (both 7 and 28 d) indicate that mortars containing the various fly ashes reach about 90% of the strength of the portland cement control samples.

Mortars containing 35% (by volume) replacement of fly ash for cement showed reduced mixing water requirements when compared to the portland cement control mortars. The reduction in mixing water was typically 5% to 10%.

ACKNOWLEDGMENTS

This research program has been supported by Midwest Fly Ash and Materials, Inc., Sioux City, IA.

REFERENCES

1. W.H. Price, J. of ACI, 72, pp. 225–232 (1975).
2. W.R. Roy, R.G. Thiery, R.M. Schuller and J.J. Suloway, Coal Fly Ash: A Review of the Literature and Proposed Classification System with Emphasis on Environmental Impacts, Illinois State Geol. Survey, ENG 96, Champaign, IL 61820 (1981).
3. S. Diamond, in Effects of Fly Ash Incorporation in Cement and Concrete, edited by S. Diamond, Mat. Res. Soc. Symp. N. (Materials Research Society, University Park, 1981) pp. 12–23.

4. D.M. Roy, K. Luke and S. Diamond, in Fly Ash and Coal Conversion By-Products: Characterization, Utilization and Disposal I, edited by G.J. McCarthy and R.J. Lauf, Mat. Res. Soc. Symp. Proc. Vol. 43, (Materials Research Society, Pittsburg, 1985) pp. 3-20.
5. S. Diamond, Cem. Conc. Res. 14, pp. 455-462 (1984).
6. G.J. McCarthy, Cem. Conc. Res. 14, pp. 471-478 (1984).
7. M.L. Mings, S. Schlorholtz, J.M. Pitt and T. Demirel, Transportation Research Record, 941, pp. 5-11 (1983).
8. H. Kanare, in Fly Ash and Coal Conversion By-Products: Characterization, Utilization and Disposal II, edited by G.J. McCarthy, F.P. Glasser and D.M. Roy, Mat. Res. Soc. Symp. Proc. Vol. 65, (Materials Research Society, Pittsburg, 1986) pp. 159-160.
9. American Society for Testing and Materials, 1985 Book of ASTM Standards, Vol. 4.02 (ASTM, Philadelphia, 1985).
10. (a) R.E. Davis, R.W. Carlson, J.W. Kelley and H.E. Davis, J of ACI, 33, pp. 577-612 (May-June 1937). (b) R.E. Davis, H.E. Davis and J.W. Kelley, J. of ACI 37, pp. 281-293 (Jan. 1941).
11. (a) L.J. Minnick, Proc of ASTM, ASTEA, 54, pp. 1129-1158 (1954). (b) L.J. Minnick, W.C. Webster and E.J. Purdy, Jr., J. of Mat., JMLSA, 6 (1), pp. 163-187 (1971).
12. J.D. Watt and D.J. Thorne, J. of App. Chem. 15 (12), pp. 585-594 (1965); 15 (12), pp. 594-604 (1965); 16(2), pp. 33-39 (1966).
13. R.H. Brink and W.J. Halstead, Proc. of ASTM, ASTEA, 56, pp. 1161-1206 (1956).
14. S.H. Gebler and P. Klieger, Effect of Fly Ash on Some of the Physical Properties of Concrete, Portland Cement Association Research and Development Bulletin RD089.01T, Skokie, IL (1986); Effect of Fly Ash on the Durability of Air-Entrained Concrete, ibid., RD090.01T.
15. R.C. Joshi and V.M. Malhotra, in Fly Ash and Coal Conversion By-Products. Characterization, Utilization and Disposal II, Mat. Res. Soc. Symp. Proc., Vol. 65, edited by G.J. McCarthy, F.P. Glasser and D.M. Roy (Materials Research Society, Pittsburg, 1986) pp. 167-170.
16. P.K. Mehta, Testing and Correlation of Fly Ash Properties with Respect to Pozzolanic Behavior, EPRI CS-3314, Final Report, prepared for Electric Power Research Institute, Palo Alto, CA (1984).
17. J.M. Pitt, S. Schlorholtz, R.J. Allenstein, R.J. Hammerberg and T. Demirel, Characterization of Fly Ash for use in Concrete, Final Report for IOWA DOT Project HR-225, Engineering Research Institute, Iowa State University, Ames, IA (1983).
18. S. Schlorholtz, J.M. Pitt and T.Demirel, Cem. Conc. Res., 14, pp. 499-504 (1984).
19. W.B. Butler, Cem., Conc. and Agg., CCAGDP, 4(2), pp. 68-79 (1982).
20. American Association of State Highway and Transportation Officials, Interim Specifications and Methods of Sampling and Testing Adopted by the AASHTO Subcommittee on Materials, (AASHTO, Washington, D.C., 1984).
21. S. Schlorholtz and M. Boybay, in Advances in X-ray Analysis, 27, edited by Cohen, Russ, Leyden, Barrett and Predecki (Plenum Publishing Corporation, New York, 1984) pp. 497-504.
22. S. Schlorholtz and T. Demirel, submitted for publication in Advances in X-ray Analysis, 30.
23. American Society for Testing and Materials, 1985 Book of ASTM Standards, Vol. 4.01 (ASTM, Philadelphia, 1985).
24. H. Kirschenbaum, The Classical Analysis of Silicate Rocks - The Old and The New, USGS Bulletin 1547, U.S. Government Printing Office, Washington, D.C., 1983.
25. American Society for Testing and Materials, ASTM Manual on Presentation of Data and Control Chart Analysis, STP 15D (ASTM, Philadelphia, 1976).
26. C. Lee, S. Schlorholtz and T. Demirel, in Fly Ash and Coal Conversion By-Products: Characterization, Utilization and Disposal II, Mat. Res. Soc. Symp., Vol. 65, edited by G.J. McCarthy, F.P. Glasser and D.M. Roy (Materials Research Society, Pittsburg, 1986) pp. 125-130.

MONITORING OF FLUCTUATIONS IN THE PHYSICAL PROPERTIES OF A CLASS C FLY ASH

SCOTT SCHLORHOLTZ, KEN BERGESON and TURGUT DEMIREL
Department of Civil Engineering, Iowa State University, Ames, IA, 50011

Received 1 November, 1986; refereed

ABSTRACT

The "quality" of fly ash produced during 1985 at Ottumwa Generation Station, was evaluated by two different experimental programs. The first consisted of the physical tests specified in ASTM C 311; these results are applicable to the use of fly ash as an admixture to portland cement concrete. The second consisted of monitoring the changes in the physical properties of fly ash pastes; these results would be applicable to the use of fly ash as a grout or a soil base stabilization agent. The physical properties monitored during the testing program were compressive strength, volume stability and setting time. In general, the results obtained from the two testing programs were quite different. When using testing procedures defined by ASTM C 311 the fly ash appeared quite uniform, but results obtained from the fly ash pastes were quite erratic. It was found that compressive strengths of the pastes can vary by a factor of five in rather short periods of time.

INTRODUCTION

An investigation has been made of various physical properties of fly ash produced during 1985 at a coal fired power plant located near Ottumwa, Iowa. The fly ash has a high-calcium content (ASTM Class C) and is self-cementitious. Power plant details and bulk ash properties have been presented in another paper at this volume [1]. The purpose of the investigation was to: (1) measure the physical properties (i.e., strength, volume stability, etc.) as a function of sampling time in an effort to define the variability of the fly ash; (2) assess the potential of the fly ash as a construction material when it is used as the primary cementing agent (i.e., grouts or road base stabilization).

EXPERIMENTAL PROGRAM

Fly ash samples were obtained from Ottumwa Generating Station (OGS) using a sampling frequency as prescribed in ASTM C 311 [2]. Grab samples were taken from each ash truck exiting the power plant (i.e., each grab sample represented about 20 tons of ash). The trucks were loaded pneumatically from a single silo (3500 ton capacity). After 20 grab samples had accumulated, they were combined to form a composite sample which represented 400 tons of fly ash. The final sample volume was approximately one gallon.

Each sample received by the Materials Analysis and Research Laboratory (MARL) was subjected to physical testing as described in ASTM C 311 [2]. The physical tests required by C 311 are moisture content, loss on ignition, pozzolanic activity, specific gravity, fineness and autoclave expansion. As described in an earlier paper at this volume [1], we no longer use the lime pozzolanic activity test (described in ASTM C 311) for monitoring the pozzolanic activity of high-calcium (Class C) fly ashes. Instead, AASHTO specification M 295-84I [3] has been adopted and Type I portland cement is being used to assess the pozzolanic activity of the OGS fly ash. A total of 85 samples were tested during 1985.

In addition to the tests required by ASTM C 311, a series of tests was performed on fly ash - water mixtures (i.e., pastes) to monitor the self-cementing capabilities of the OGS fly ash:

1. Compressive strength: measured on 1 x 1 inch cubes, after 4 hours, 1 day (d), 3 d, 7 d, 14 d and 28 d of moist curing.
2. Volume stability: measured on 1 x 1 x 11 inch prisms. Two specimens were made for each sample. One specimen was moist cured, the other was allowed to cure at room humidity (approximately 50 to 60% RH). Length measurements were made in accordance with ASTM C 490 [2]. Specimen length changes were monitored for a minimum time of 28 d.
3. Setting time: measured using a pocket penetrometer (Humboldt Mfg. Co., catalog number H-4200). Initial set was defined as the first discontinuity in the pressure versus time curve. Final set was arbitrarily defined as the maximum bearing pressure allowed by the pocket penetrometer. Setting time specimens were not stored in a humidity cabinet between penetration measurements. Setting times were of such a short duration that this probably had little influence on the results.
4. Temperature rise (ΔT, $^{\circ}C$): measured on a fly ash water slurry (10 g fly ash to 28.5 g water) that was sealed in a Dewar flask. The test is neither strictly adiabatic nor strictly isothermal, but it does produce interesting information about the early hydration reactions occurring in the fly ash - water system.

All of the paste specimens were made at a water/fly ash ratio of 0.27. About 50 samples of the OGS fly ash were subjected to this testing scheme. The samples were selected, when possible, in a manner that ensured that at least one fly ash sample would be tested for every two weeks of sampling time. This testing frequency could not be maintained during the late fall, winter or early spring months because the demand for fly ash was so low. During the peak of the construction season (roughly June through September) it was possible to test one to two OGS fly ash samples per week.

RESULTS AND DISCUSSION

Table I summarizes the raw statistical results of the tests that were used to monitor the self-cementing properties of the OGS fly ash. Table II summarizes the results obtained by using the tests defined in ASTM C 311. In Tables I and II, X refers to the arithmetic mean, S refers to the standard deviation, R refers to the range and n refers to the number of samples. It is obvious from the data listed in Table I, that the OGS fly ash exhibits a high degree of variability in its self-cementitious properties. The variability is so large that gross statistical information (i.e., mean and standard deviation) is of little use in helping to establish design guidelines and/or specifications for the utilization of the OGS fly ash. In contrast, results obtained from the ASTM tests (listed in Table II) do not show large variations in the measured physical properties. An attempt was made to correlate the variables listed in Table I to those listed in Table II (i.e., correlation of the fly ash paste data to the results obtained from the physical testing described in ASTM C 311). No significant correlations were observed. The maximum correlation coefficient, r, obtained from the analysis was 0.416. From this point on, only the data obtained from the fly ash pastes will be discussed in detail.

Figure 1 illustrates the variation of compressive strength with sampling date for the OGS fly ash. The compressive strength values shown in Figure 1 are for paste specimens that had been cured for 7 d. Similar trends were observed for specimens cured for other periods. It is obvious from Figure 1, that the relationship between compressive strength and fly ash sampling date

Table I

Summary of Statistics Obtained for OGS Fly Ash—Water Pastes

TEST	$\bar{X}$	S	R	n
Compressive strength (psi)				
4 hour	354	114	531	56
1 d	718	471	2190	51
3 d	1240	915	3600	56
7 d	1570	1140	4410	57
14 d	1800	1300	4770	57
28 d	2080	1510	5570	56
Volume Stability (% Expansion)				
Air-cured	-0.04	0.03	0.06	52
Humid—cured	0.00	0.03	0.14	52
Setting time (minutes)				
Initial set	21	8	28	55
Final set	34	15	95	55
Temperature Rise				
ΔT (°C)	5.4	1.2	5.5	56
Time to reach peak temperature				
t (min.)	45	15	62	52

Table II

Results Obtained from Physical Testing as Described in ASTM C 311
Ottumwa Generating Station - 1985, n = 85

TEST	$\bar{X}$	S	R
Moisture Content (%)	0.03	0.02	0.10
Loss on Ignition (%)	0.24	0.06	0.24
Fineness (%)	9.83	0.81	3.90
7 d Pozzolan (%)	93.8	5.6	32.0
Autoclave Expansion (%)	0.06	0.02	0.07
Specific Gravity	2.65	0.03	0.16

is quite complex. Hence, the gross statistical information listed in Table 1 does a poor job of providing information that would be useful in defining the compressive strength at any given time. In contrast, time series analysis may be quite useful in helping to model the fluctuations in compressive strength as a function of sampling date. We are currently developing a time series model for the OGS fly ash by utilizing bulk elemental analysis information and power plant operating parameters.

Figure 2 shows a compressive strength versus curing time plot for several of the OGS fly ash samples. Figure 2 indicates that strength development was

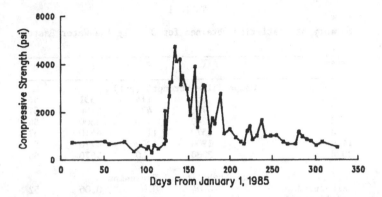

Figure 1. Compressive strength at 7 d vs. sampling date for OGS fly ash.

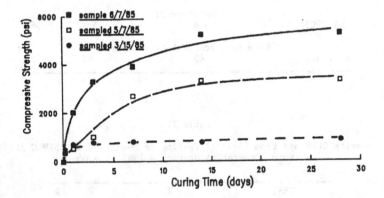

Figure 2. Compressive strength vs. curing time for samples of OGS
sampled on three dates.

initially quite rapid but that after about 14 d of moist curing there was
little gain in compressive strength.

Both Figures illustrate the large variation in compressive strength that
was observed at different sampling times. This is particularly troublesome
from a practical point of view because the strength variations are large and
they occur in relatively short periods of time (i.e., strength can vary by a
factor of about 5 in less than two weeks).

Compressive strengths of fly ash pastes cured for short periods of time
(i.e., 1-7 d) correlated well with compressive strengths observed for the
samples cured for both 14 and 28 d (see Table III). The best correlation
observed for the compressive strength data was between the 7 d and the 28 d
strength of the fly ash pastes, which had a correlation coefficient, r, of
0.947.

Table III

Pearson Correlation Coefficients for the Variables Listed in Table I

TEST VARIABLE	4 hour	1 d	3 d	7 d	14 d	28 d
	Test variable: Compressive Strength					
Expansion (Humid cure)	−0.152	0.637	0.786	0.731	0.751	0.817
Expansion (air cure)	−0.132	−0.389	−0.464	−0.442	−0.465	−0.474
Initial Set	−0.130	0.022	0.090	0.104	0.041	−0.045
Final Set	−0.258	−0.226	−0.227	−0.291	−0.300	−0.326
ΔT	0.640	0.370	0.307	0.229	0.246	0.211
Time to peak	0.196	−0.111	−0.242	−0.229	−0.341	−0.321
Compressive Strength	--	--	--	--	--	--
4 hour	1.0					
1 d	0.321	1.0	symmetric			
3 d	0.143	0.904	1.0			
7 d	0.109	0.800	0.909	1.0		
14 d	0.131	0.767	0.816	0.881	1.0	
28 d	0.076	0.782	0.910	0.947	0.912	1.0

Volume stability of the humid-cured fly ash pastes correlated fairly well to compressive strength. (See table III; maximum r = 0.817). In general, the test specimens with moderate to high compressive strengths exhibited a slight expansion during the 28 d curing period (maximum observed expansion less than 0.10%). Specimens that had low compressive strengths displayed negligible expansion or slight shrinkage (maximum observed shrinkage of 0.04%). The expansion of the air-cured specimens did not show a strong correlation to compressive strength. In fact, the air-cured specimens showed a slight negative correlation to compressive strength. (See Table III; note that the humid-cured specimens indicated a positive correlation to compressive strength). Typically, the specimens that had the largest expansion when subjected to humid-curing also displayed the greatest drying shrinkage when subjected to air-curing. This suggests fly ash pastes could be susceptible to severe volume changes during wetting-drying cycles. The air-cured specimens exhibited a large range of expansive tendencies. The largest expansion observed in the air-cured specimens was nearly equivalent to that observed for the humid-cured specimens, but the majority of the air-cured specimens showed a slight shrinkage during the curing period.

Many of the air-cured specimens exhibited modest to severe efflorescence tendencies during the first week of curing. The white powder was carefully scraped from several of the specimens and subjected to X-ray diffraction analysis, which indicated that the powder was sodium sulfate.

The time of set data was highly variable. The average values (listed in Table I) are very poor estimates of the initial and final set times for a given sample of the OGS fly ash. Several of the samples studied in this experiment had a tendency to flash set. In such mixtures, final set occurred in less than 15 minutes from the time that water was added to the fly ash. Other OGS samples had final set times of nearly 120 minutes, this large sample to sample variation of setting time makes field utilization of the fly ash tricky. The setting times, however, were quite rapid and could easily be monitored in the field, on a lot by lot basis if needed, using inexpensive equipment. Also, initial set showed a modest correlation to final set (r = 0.675) so the pastes may only need to be monitored until initial set occurred. Only poor (if any) correlations existed between setting time and the other variables studied in this research.

The temperature rise (ΔT) data showed a correlation to the 4 hour compressive strength data ($r = 0.640$). We think that this correlation may be real but one must remember that the temperature rise test should also be quite dependent on both the specific surface area and the chemical composition of the fly ash. Hence, we recommend that further work should be done before drawing any firm conclusions. The time required to reach the peak temperature in the fly ash – water system showed a modest correlation ($r = 0.626$) to the final set time observed via the pocket penetrometer test.

SUMMARY AND CONCLUSIONS

The physical properties of a single source of high–calcium fly ash have been studied as a function of sampling date. When the fly ash samples were subjected to physical testing, defined in ASTM C 311, the fly ash appeared to exhibit only minor variations in physical properties. The same fly ash samples were then used to make fly ash – water pastes (water/fly ash ratio = 0.27) to study the self–cementitious properties of the fly ash. The fly ash paste specimens displayed large variations in physical properties as a function of sampling date. The variations in compressive strength, volume stability and setting time, were so large as to suggest the possibility of severe problems when attempting to define field construction specifications for the utilization of high–calcium fly ash when it is used as the sole cementing agent.

Several of the physical properties of the fly ash pastes show correlations to one another. Hence, only two or three quick physical tests may be needed to obtain reasonable estimates of the potential compressive strength, volume stability and setting time for a given high–calcium fly ash.

Thus, it appears that high–calcium fly ash produced by a single power generating station may show significant variability which can limit its direct utilization as a single cementitious material. However, the variability does not appear to affect its potential for utilization as an admixture to portland cement concrete.

We are currently expanding our testing scheme to include high–calcium fly ashes from other Iowa power plants. We have also started to add chemical and mineralogical data to supplement the physical information reported in this paper. This additional information is pertinent to our goal of expanding the current fly ash classification scheme to better describe the observed characteristics of high–calcium fly ashes.

ACKNOWLEDGEMENTS

This research has been supported by Midwest Fly Ash and Materials, Inc., Sioux City, IA, and the following Iowa utility companies: Iowa-Illinois Gas and Electric, Iowa Public Service Company, Interstate Power Company, Muscatine Power and Water, Iowa Power and Light Company and the City of Ames Municipal Electric System.

REFERENCES

1. S. Schlorholtz, K. Bergeson and T. Demirel, "Variability and Trends in Iowa Fly Ashes," this volume.
2. American Society for Testing and Materials, <u>1985 Annual Book of ASTM Standards</u>, Vol. 4.02 (ASTM, Philadelphia, Pa, 1985).
3. American Association of State Highway and Transportation Officials, <u>Interim Specification and Methods of Sampling and Testing Adopted by the AASHTO Subcommittee on Materials, 1984</u>, (AASHTO, Washington, D.C., 1984).

RECENT DEVELOPMENTS IN THE UTILIZATION OF WESTERN U.S. COAL CONVERSION ASH

O.E. MANZ* and DENNIS L. LAUDAL**
*Coal By-Products Utilization Laboratory, Mining and Mineral Resources Research Institute, **Energy Research Center, University of North Dakota, Grand Forks, ND 58202

Received 21 October, 1986; refereed

ABSTRACT

The utilization of ashes from combustion or gasification of western U.S. coals offers many possibilities for useful products. Of the possible uses, the following have been identified in an earlier study [1] as having sufficient potential for laboratory development and testing: mineral wool, sulfur concrete, high flexural-strength ceramics, replacement of cement in concrete, and road stabilization. Three lignite-derived ash products from the Beulah, ND, site were used in the present study: fixed-bed gasification ash; a dry scrubber ash; a combination bottom ash/economizer ash from an electrical power plant, Where possible, ASTM fabrication and testing procedures were used. Mineral wool of similar physical character to commercial wool and at lower potential cost was produced using 100 percent of various western ashes. Sulfur concrete utilizing 80% ash and 20% modified sulfur developed flexural and compressive strengths in excess of 2,250 and 5,000 psi, respectively. An economically competitive vitrified ceramic product with flexural strength above 7,800 psi was produced from a mixture of 50% ash, 45% sand, and 5% clay. By using a total ash mixture of 26% gasifier ash and 74% combustion ash, a very satisfactory, economical and durable roadbed material was developed. The replacement of up to 50% of the cement in concrete with western fly ash produces economical, high strength concrete.

INTRODUCTION

Coal gasification waste products, including those from fixed-bed gasification, have different properties from the combustion ashes, especially with respect to mineralogy [1,2]. To date, comparatively little effort has been directed toward the investigation of bulk utilization of this class of ash. A Gas Research Institute study [3] was directed toward correction of that deficiency by matching properties of lignite-derived fixed-bed ash from the Beulah, North Dakota, gasification plant (originally known as the Great Plains Gasification Associates (GPGA) plant) and combustion ash from the nearby power plant, the Antelope Valley Station (AVS), with existing practical economic possibilities. Manz [4] and Severson et al. [5] have described some of the possible applica-ations for utilization of western U.S. coal ash, including ash from the Beulah site.

This paper summarizes three years of bulk utilization research on the GPGA gasifier ash and the AVS scrubber and bottom ash. The following options were explored: mineral wool, sulfur concrete, high-flexural-strength ceramics, dual concrete replacement, and road stabilization. Where possible, standard ASTM procedures were followed. For most of the options investigated, considerable time was saved by drawing on previous experience at the University of North Dakota (UND) Coal By-Products Utilization Laboratory (CBUL). Efforts were concentrated on utilization of the ashes in products that would be economical on a full-scale basis [5]. Mineral wool was formed by blowing air on molten ash generated with an outdoor pilot plant cupola, as well as an electric arc rocking laboratory furnace. Modified sulfur and ash mixtures were heated, mixed, and formed into suitable specimens for flexure and compression testing. High-flexural-strength ceramic specimens were formed

Table I

Chemical Compositions (wt.%) of Ash By-Products from the Beulah Plants

	Gasification Ash	Bottom Ash	Scrubber Ash
SiO_2	26.3	36.8	30.7
Al_2O_3	12.7	13.7	11.4
Fe_2O_3	9.5	9.2	5.1
CaO	20.5	20.9	21.1
MgO	7.9	7.3	5.6
Na_2O	8.8	7.4	5.3
K_2O	0.6	0.5	1.0
SO_3	1.6	1.6	12.7

with a laboratory extruder, followed by firing in an electric kiln and subsequent testing for flexure and absorption. Road stabilization cylindrical specimens were compacted, cured and tested in compression. Actual laboratory size concrete mixes were made replacing Portland cement with AVS scrubber ash as well as the aggregate with GPGA gasifier ash.

MATERIALS

The research was aimed at finding applications for the by-product ashes from electrical utility plants and coal gasification plants producing substitute natural gas. The ashes used were from the Beulah, North Dakota, site. At this site, substitute natural gas is being produced by ANG Coal Gasification Company and Basin Electric Cooperative operates the Antelope Valley electrical generating station (AVS). Both plants obtain their lignite from the adjoining mine and share coal handling facilities. The following ashes have been studied:

1. Gasification ash (referred to in this paper by its earlier designation as GPGA ash). A granular product from the Lurgi fixed-bed gasifiers.
2. Scrubber ash from the AVS. The dry SO_2 scrubbers at the AVS use the alkaline fly ash with some supplementary $Ca(OH)_2$ as the active material. The product consists of fly ash with substantial surface deposits of Ca sulfates (gypsum and bassanite) and sulfite (the hemihydrate).
3. Bottom ash/economizer ash from the AVS. Ash removed from the economizers of the furnaces at the AVS is mixed with bottom ash clinker prior to disposal. This ash will be referred to as bottom ash.

Chemical analyses of these ashes are given in Table I. Mineralogy and its correlation to bulk composition for these lignite-derived ashes have been reported by McCarthy, et al. [2,6-9] and Stevenson et al. [1,10,11] in earlier papers of this proceedings series.

RESULTS AND DISCUSSION

Mineral Wool

Mineral wool, a fibrous material made by melting rock and/or metallurgical slag, is used in building insulation in the form of batts, board, granulated nodules or loose fibers, and is the bulk material of ceiling tile. It is durable, fireproof, and insect- and vermin-proof. It is currently produced from waste slag obtained from steel mills. In an earlier study [4], bottom ash from cyclone or other "wet bottom" furnaces burning lignite was shown to be a viable substitute raw material for blast furnace slag. The lignite gasifier ash used in the present study is similar, physically and chemically (although not mineralogically), to bottom ash and consequently, is a potentially attractive raw material for mineral wool.

It has been found that for best operation, the "acid-base ratio," i.e., silica plus alumina to calcium oxide plus magnesia ratio [12], in the ash should range from 0.8 to 1.2 to facilitate melting and fiberization [4]. Lignite scrubber ash from the Antelope Valley Station has an "acid-base ratio" of 0.97, and the "acid-base ratio" of GPGA gasifier ash is 1.0. Both are in the middle of the acceptable range.

During the last three years, mineral wool has been fabricated at the CBUL from two of the lignite ashes from the Beulah, North Dakota site. Until December of 1984, most of the mineral wool was made in the pilot-plant sized cupola [4]. After December of 1984, an electric arc rocking furnace (a 10 lb. Detroit Electric Rocking Furnace) was used.

To prepare mineral wool using the cupola, the cementitious scrubber ash, added as a binder, and the gasifier ash were first mixed with water and formed into balls with an approximate diameter of 2 in., and dried in an oven at 200°F. The only other raw material required was coke which was used to fuel the cupola. The coke was ignited and heated to 2200-3000°F. The ash balls were layered into the cupola with additional coke at a 3:1 ratio. When the ash balls had melted, a stream of molten ash was fed from the cupola and bombarded with air to form mineral wool fibers.

Substitution of the the rocking furnace offered additional versatility in testing. Unlike the cupola, this furnace did not require that the ash charge be be pelletized and granular/powdered ash could be used. Without the need to pelletize, addition of AVS scrubber ash as a binder for the balls was no longer required. Thus, mineral wool could be made from 100% GPGA gasifier ash. Once the ash became molten, the fiberization process was the same as that used for the cupola.

Based on a 1265°C eutectic in the $CaO-Al_2O_3-SiO_2$ system, the use of the GPGA Beulah-Zap coal gasification or combustion ashes would be suitable for mineral wool production since they fall within the "satisfactory wool" range of the phase diagram [12].

High quality mineral wool has been produced at the CBUL from a mix of the GPGA gasifier ash and the AVS scrubber ash. The characteristics of this product, as measured by a mineral wool manufacturer, are compared with with the commercial product in Table II. "Shot" means beads of slag that formed instead of fibers. The coal ash mineral wool is of very similar physical character to the commercial product.

Sulfur Concrete

In sulfur concrete melted sulfur is used as a replacement for Portland cement [13]. Unlike Portland cement concrete, sulfur concrete is unaffected by sulfates, salts, and most acids. It is nearly impervious to water, and therefore, resistant to freeze-thaw damage. It can be repeatedly loaded to 95% of its ultimate strength with no evidence of fatigue, whereas cured Portland cement concrete will exhibit fatigue after repetitive loading to 60% of its ultimate strength. It is remeltable and reusable without altering the material. Cast sulfur concrete commonly achieves compressive strengths to 6,000-10,000 psi in two to three years, as compared to the 2,000-2,500 psi compressive strength typically developed by Portland cement in the same time period. Sulfur concrete becomes competitive with Portland cement concrete in severe (especially acid) environments or when superior strengths are required.

Table II

Result of Commercial Test* Comparing Mineral Wool Produced from
Coal Conversion Ash with Commercial Wool

	Commercial Wool Made with Blast Furnace Slag	Experimental Wool Made with Coal Conversion Ash
Percent of shot on 50 mesh sieve	15	16
Percent of shot on 325 mesh sieve	40	36
Rebound from compression test	30 - 35	34
Percent of shot	55	52
Visual quality of wool	very fine, short fibers	very fine, short fibers

*USG Corporation, Red Wing, MN

Table III

Formulations and Flexural Strengths* of Sulfur Concrete Mixes

S-86-	01	02	03	04	05	06	07	08	09	10
FORMULATION (wt.%)										
Gasification Ash	65	70	70	70	65	75	60	60	50	60
Scrubber Ash	15	10	15	5	10	10	15	30	25	20
Modified Sulfur	20	20	15	25	25	15	25	10	25	20
FLEXURAL STRENGTH (psi)*	1011	1459	1267	1408	1596	1571	1455	1102	1150	2252

*Average two tests. Flexural strength determined according to ASTM C 573.

Sulfur concrete was produced at the CBUL by adding sulfur, modified with a plasticizer, to preheated (400°F), -3/8 in., GPGA gasifier ash 3/8 inch). These ingredients were mixed in a preheated mixer until the modified sulfur was completely melted. At this point, the mineral filler (AVS scrubber ash) was added, and the mixing continued until a homogeneous substance was attained. The mix was then molded, vibrated (to remove air voids), cooled, and removed from the mold for testing within 7 d of fabrication. Test specimens with dimensions 3 x 3 x 12 in. tested in flexure and compression. Results of the flexural strength tests are given in Table III. All test specimens had strengths greater than 1000 psi. The mix with the highest flexural strength, S-86-10, was made with 60% of a GPGA gasification ash, 20% AVS scrubber ash and 20% modified sulfur.

High-Flexural-Strength Ceramics

A major focus of this research was to establish the technical feasibility of producing extruded ceramics having high-flexural-strength. High-flexural-strength extrusions consisted of a ceramic material made from AVS scrubber ash, crushed glass or concrete sand, and North Dakota clay. There are several

Table IV

Formulations and Flexural Strengths* of Extruded Ceramic Mixes

	1	2	3	4
FORMULATION (wt.%)				
AVS Scrubber Ash	60	50	60	50
Glass (-100 mesh)	35	45	–	–
Clay	5	5	5	5
Concrete Sand (-100 mesh)	–		35	45
Water (g/1000 g of dry components)	250	180	240	217
FLEXURAL STRENGTH (psi)	7475	7850	7100	7555

*Average of two or more test results. Flexural strength determined according to ASTM C 192.

potential applications for such products: railroad ties, fence posts, and parking lot curbs. Emphasis was placed on the testing of hollow poles and railroad ties.

There is currently world-wide interest in using concrete railroad ties in place of wood ties [14]. Concrete ties have a longer service life and are not as affected by climate as the wooden ties. The concrete ties also give more support, so fewer ties are needed in the same length of track. It is possible that the ceramic ties, with their higher flexural strength, would be even more efficient than concrete. For example, the extruded ceramic has a flexural strength of 5000 psi compared to 800 psi for concrete when tested in three point loading. Power line poles have been made from this material [15].

For the current project, a small laboratory (Dearing) extruder was used to fabricate both rectangular specimens 3/8 in. by 1.5 in. wide with variable lengths, and cylinders 1 in. in diameter. Table IV gives the mix proportions used in the study. As called for in the original work of Williams and Ali [15], ground beverage bottle waste glass was used as a mix component at the beginning of this study. Later it was found that concrete sand could be substituted. All components were ground to -100 mesh and dry mixed. When a homogeneous mixture was obtained, water was added and the material was fed into the extruder and run through once. Before putting the material through a second time, the vacuum on the extruder was turned on. (The vacuum removed air from the extrusion, thus increasing its density and, ultimately, its strength). The specimen was then run through the extruder four times (enough to remove most of the remaining the air). As the specimen went through the die of the extruder for the final time, it was cut at the desired lengths and put on a tray for air drying.

All specimens were air-dried for 24 h and then oven-dried for an 24 h. Each extrusion was fired in an electric kiln to Orton Cone 1, i.e. approximately 2100°F (1150°C). The resulting specimens were tested in flexure to determine the modulus of rupture and results are given in Table IV. They were also tested for water absorption according to ASTM C 373. The specimens were weighed when dry, after 24 h in a cold water bath, and after 5 h in a boiling hot water bath. All specimens showed near zero absorption of water.

The results shown in Table IV indicate that a good quality high-flexural-strength ceramics can be produced completely from waste material (scrubber ash) and the inexpensive local materials (sand and clay). Because there is little difference in the flexural strength values for the four mixes, and because scrubber ash and sand are even less expensive than waste glass (which

Table V

Formulations and Compressive Strengths in Roadbed Stabilization Mixes

	845–80C	845–81C	845–82C	845–83C
FORMULATION (wt.%)				
Mercer County Roadbed				
Aggregate #84–668	95	85	88	83
Hydrated Lime	5	3	2	2
AVS Scrubber Ash	—	12	10	15
Maximum Dry				
density (lbs/ft^3)*	117.8	120.5	120.9	121.8
Optimum Moisture (%)	12.1	10.5	11.1	9.9
COMPRESSIVE STRENGTH*	205	603	395	436

*Determined according to ASTM C 593 (10 lb. hammer, 18 in. drop, 25 blows/layer, 3 layers). Specimens were sealed for 7 d at 100°F before being given a 4 h soak.

requires grinding), the most economical mix is No. 3 with 60 percent scrubber ash.

Roadbed Stabilization

Manz and Manz [16] have described the use of western U.S. fly ash, in combination with hydrated lime, in roadbed stabilization. This utilization option was also explored for the by-products from the Beulah, ND, site.

In one test for a Mercer ND road job, detailed in Table V, a mixture of clay and aggregate was sampled from a gravel road in the county. The sample was mixed with hydrated lime as well as with combinations of lime and AVS scrubber ash. Mix 845–80C used 5% lime as the stabilizing material, whereas the other three mixes used combinations of lime and scrubber ash. The three mixes that included the scrubber ash had compressive strengths that were two to three times as high as the mix with lime alone. Two of these mixes met or exceeded the ASTM C 593 compressive strength requirement (for freeze-thaw durability) of 400 psi (2.76 MPa), and the third was just below it. Considering that the price of lime would be at least five times that of the ash delivered to the road job, there is a marked cost-strength benefit with the use of scrubber ash. The results from this test series assisted Mercer County in selecting the most economical mixture for this particular road job.

In another test, detailed in Table VI, the three ashes from the Beulah, ND, site (gasification ash, scrubber ash and bottom/economizer ash) were used in the weight proportion in which they are generated at the Beulah plants. In two of the tests, compactive effort was varied from the ASTM specifications to see if this variable would affect compressive strength. The reference test result (No. 1 in Table VI) exceeded the 400 psi requirement for the vacuum saturation test. The other two entries in Table VI show that compactive effort does affect compressive strength.

Cement and Aggregate Replacement in Concrete

The ideal situation in utilization is an application that uses all by-products from a particular site. The sulfur concrete and roadbed stabilization applications described above use all of the ash by-products from the

Table VI

Roadbed Stabilization Tests with Ashes from the Beulah, ND, Site*

Test**	Optimum Moisture Content (%)	Dry Density (lb/ft³)	Compressive Strength		
			Cured 7 d at 100°F Vac. Sat. (psi)	7 d	28 d
				Cured at 70°F (psi)	
1	13.5	116.9	926	1094	749
2	15.8	110.5	---	1300	1900
3	16.5	108.7	---	1000	1400

*61 wt.% Scrubber Ash (84-633); 26% GPGA Gasifier Ash (84-229); 13% Bottom Ash (83-654).
**Compactive Effort:
  1. 10 lb Hammer; 18 in. Drop; 25 Blows/Layer; 3 Layers
  2. 5.5 lb Hammer; 12 in. Drop; 25 Blows/Layer; 3 Layers
  3. 5.5 lb Hammer; 12 in. Drop; 12 Blows/Layer; 3 Layers

Beulah, ND, site. Another possible application would be use of the coarser ash materials as aggregate and scrubber ash as a cement replacement in concrete (dual replacement). Such an application was explored during the GRI study. The mixes and corresponding compressive strengths are given in Table VII.

Both 15 and 30 percent replacement of Portland cement with AVS Scrubber ash produced concrete with greater 28 d strength than the control concrete, and these 28 d strengths were greater than the 90 d strength for the control. The mixes with total replacement of the concrete aggregate with GPGA gasifier ash (GA) absorbed excessive water and setup very rapidly. The strengths obtained were low compared to the mixes with those obtained with conventional aggregate. It was suspected that the fine portions of the GA were behaving as an active component of the mix, instead of an inert aggregate. By replacing the finer, -4 mesh, portion of the gasification ash with concrete sand, a more workable mix was obtained; however the strength was even lower than that obtained with the original ash.

These tests indicate that for the lignite ashes of the Beulah site, dual replacement with the gasifier ash in the role of aggregate in the concrete is not promising. The tests did indicate that the use of the scrubber ash alone as a cement replacement in concrete is very promising. Gasification ash differs from the usual inert aggregate in being reactive with water. Hassett et al. [17] have shown that GA produces high-pH and high-Na solutions very quickly on contact with water. It is possible that extractable components (especially the Na) from the GA upset the chemical balance of the concrete pore solutions and degrade the cementitious reactions. In the sulfur concrete and roadbed applications, this GA behavior did not affect the fabrication or properties of the product and use of all three Beulah site ash products in one product was feasible.

ECONOMICS

Severson et al. [5] have presented a preliminary economic analysis of the Beulah site ash utilization products. Additional study of the materials, the fabrication costs and the products has led to the updated estimates discussed below.

Table VII

Concrete Dual Replacement Test Formulations and Physical Properties

	Control Concrete Mix	15% AVS	30% AVS	GA[a] Rinsed	GA Dry -4 Mesh	60% GA
MIX (lbs.)						
Portland Cement (Type 1)	590	500	410	590	590	590
Scrubber Ash	–	90	180	–	–	–
Fine Aggregate[b]	1380	1380	1380	–	–	1302
Coarse Aggregate[c]	1875	1875	1875	–	–	–
GPGA (Rinsed)	–	–	–	3255	–	–
GPGA (Dry)	–	–	–	–	3255	1953
Slump (in)	4.5	4.5	3.5	2.5		
Air Content	5.6	5.75	6.0	5.2		
Unit Weight	145.7	44.35	44.0	38.9		
COMPRESSIVE STRENGTH[d]						
1 d	2180					442
7 d	4280	4461	–	1345	1504	
28 d	4440	5892	5662	1929		
90 d	5626					

a. Dry GPGA was saturated with water and allowed to drain.
b. ASTM C 33
c. ASTM C 33, Size #67
d. in psi; determined according to ASTM C 873

Mineral Wool

A high quality mineral wool that is similar to commercial wool in all required physical properties can be produced from various combinations of the ash by-products from the Beulah site. Cost estimates have been prepared for a plant producing 100,000 tons/year of either loose mineral wool or batts. For loose wool, the total capital investment would be $189.06/ton and the total product cost $213.16/ton, resulting in a net profit of $33.43/ton and giving a rate of return on investment of 13%. For batts, the total capital investment would be $291.85/ton and a total product cost of $299.82/ton. This would produce a net profit of $100.10 per ton and rate of return on investment of 21.4%.

Sulfur Concrete

The economic evaluation of sulfur concrete was based on a product made from 80% gasification ash and 20% sulfur, and on both the ash and the sulfur (from H_2S removal at the gasification plant) being available at no cost from the Beulah site. The evaluation was made in comparison to conventional sulfur concrete, not with a comparable volume of Portland cement concrete. A plant designed to produce 47,800 yd^3 of sulfur concrete would have a total capital investment of $127.99/$yd^3$. If the sulfur were available at no cost, the total product cost would be $147.02/$yd^3$, giving a net profit of $61.49 and a rate of return on the investment of 24.6%. If the sulfur had to be purchased at market value, the total product cost is $192.15/$yd^3$, giving a net profit of $38.93/$yd^3$, with a rate of return on investment of 15.2%.

High-Flexural-Strength Ceramics

Using a mixture of 50% AVS scrubber ash, 45% sand, and 5% clay, a vitrified ceramic material can be made that has a flexural strength above 7,800 psi. It is assumed that the ash is free and the sand and clay are obtained locally at prevailing prices. A conceptual plant designed to produce 300,000 railroad ties and 50,000 hollow poles would have a total capital investment of $46.82 per unit and a total product cost of $36.36 per railroad tie and $57.44 per pole. The net profit for each railroad tie would be $4.32 and $21.28 for each pole. The rate of return on investment for the plant would be 16.8%.

Roadbed Stabilization

By using a total ash mixture of 86% roadbed aggretate, 12% AVS scrubber ash, and 2% lime a very satisfactory, economical, durable, roadbed material was developed. A savings of 55% ($82,500 per mile) is estimated over conventional roadbed materials.

Cement Replacement in Concrete

The replacement of up to 50% of the cement in concrete with AVS scrubber ash produces an economical high strength product. The total product cost is $43.98/yd^3 compared to a current selling price of $48.50/yd^3.

CONCLUSIONS

The objective of this research was to develop and evaluate technologies that would allow for bulk utilization of the solid wastes from an integrated gasification facility (including an associated power plant). The general approach was to match the properties of the raw ash and blends of the gasifier and combustion ashes with existing practical economic possibilities. The following options were explored: mineral wool, sulfur concrete, high-flexural-strength ceramic products, road stabilization, and replacement of Portland cement and aggregate in concrete. Except for the dual replacement of cement and aggregate in concrete, the technical and economical viability of all of these options for the utilization of lignite-derived ash from the Beulah, ND, plants was demonstrated.

ACKNOWLEDGEMENTS

This research was supported by GRI Contract No. 5082-235-0771 through the North Dakota Mining and Mineral Resources Research Institute. USG Corporation is thanked for supplying the cupola and rocking furnace and for testing the experimental mineral wool. D.M. Roy and G.J. McCarthy are thanked for assistance in editing this paper.

REFERENCES

1. R.J. Stevenson and G.J. McCarthy, in Fly Ash and Coal Conversion By-Products: Characterization, Utilization and Disposal II, edited by G.J. McCarthy and F.P. Glasser and D.M. Roy, Mat. Res. Soc. Symp. Proc. Vol. 65 (Materials Research Society, Pittsburgh, 1986), pp. 77-90.
2. G.J. McCarthy, Powder Diffraction 1, 50-56 (1986).
3. O.E. Manz, D.J. Hassett, D.F. Hassett, D.L. Laudal, and R.C. Ellman, Technical and Economic Feasibility of Bulk Utilization and Metal Recovery for Ashes from an Integrated Coal Gasification Facility, GRI 86/0203.1 Final Report, Vol. III (Gas Research Institute, Chicago, 1986) 130 pp.
4. O.E. Manz, Cem. Concr. Res. 14, 513-520 (1984).

352

5. D.E. Severson, O.E. Manz and M.J. Mithcell, in Fly Ash and Coal Conversion By-Products: Characterization, Utilization and Disposal I, edited by G.J. McCarthy and R.J Lauf, Mat. Res. Soc. Symp. Proc. Vol. 43 (Materials Research Society, Pittsburgh, 1985) pp. 195-210.
6. G.J. McCarthy, L.P. Keller, P.J. Schields, M.P. Elless and K.C. Galbreath, Cem. Concr. Res., 14, 479-484 (1984).
7. G.J. McCarthy, L.P. Keller, R.J. Stevenson, K.C. Galbreath and A.M. Steinwand, in Fly Ash and Coal Conversion By-Products: Characterization, Utilization and Disposal I, edited by G.J. McCarthy and R.J. Lauf, Mat. Res. Soc. Proc. Vol. 43 (Materials Research Society, Pittsburgh, 1985) pp. 165-176.
8. G.J. McCarthy, D.J. Hassett, O.E. Manz, G.H. Groenewold, R.J. Stevenson, K.R. Henke, P. Kumarathasan, in Fly Ash and Coal Conversion By-Products: Characterization, Utilization and Disposal II, edited by G.J. McCarthy and F.P. Glasser and D.M. Roy, Mat. Res. Soc. Symp. Proc. Vol. 65 (Materials Research Society, Pittsburgh, 1986), pp. 310-310.
9. G.J. McCarthy, K.D. Swanson, L.P. Keller and W.C. Blatter, Cem. Concr. Res. 14, 471-478 (1984).
10. R.J. Stevenson, Cem. Concr. Res. 14, 485-490 (1984).
11. R.J. Stevenson and R.A. Larsen, in Fly Ash and Coal Conversion By-Products: Characterization, Utilization and Disposal I, edited by G.J. McCarthy and R.J. Lauf, Mat. Res. Soc. Proc. Vol. 43 (Materials Research Society, Pittsburgh, 1985) pp. 177-186.
12. J.E. Lamar, Rock Wool from Illinois Mineral Resources, Bulletin 61, Illinois State Geological Survey, 262 pp. (1934).
13. T.A. Sullivan, in Corrosion-Resistand Sulphur Concrete — Design and Construction Manual, (Sulphur Institute, 1986) pp. 6-20.
14. J.G. White, Modern Railroads, October, 1984, pp. 44-47.
15. T.D. Williams and A.Ali, Development of Power Poles from Fly Ash. Phase II, Electric Power Research Institute, Project 851-1, September, 1979.
16. O.E. Manz and B.A. Manz, in Fly Ash and Coal Conversion By-Products: Characterization, Utilization and Disposal I, edited by G.J. McCarthy and R.J. Lauf, Mat. Res. Soc. Proc. Vol. 43 (Materials Research Society, Pittsburgh, 1985) pp. 129-140.
17. D.J. Hassett, G. J. McCarthy and K.R. Henke, in Fly Ash and Coal Conversion By-Products: Characterization, Utilization and Disposal II, edited by G.J. McCarthy and F.P. Glasser and D.M. Roy, Mat. Res. Soc. Symp. Proc. Vol. 65 (Materials Research Society, Pittsburgh, 1986), pp. 285-300.

EVALUATION OF POTENTIAL USES OF AFBC SOLID WASTES

E.E. BERRY* and E.J. ANTHONY**
*P.O. Box 7261, Oakville, Ontario, Canada
**CANMET, Energy Mines and Resources Canada, Ottawa, Ontario, Canada

Received 19 January, 1987; Communicated by G.J. McCarthy

ABSTRACT

This paper presents a brief review of the literature dealing with utilization of atmospheric-pressure fluidized bed combustion (AFBC) solid wastes. The uses that have been proposed for AFBC residues include the following: agricultural lime, waste neutralization and stabilization, low-strength backfill, soil cementing and asphaltic concrete aggregate. An evaluation of a high-Ca waste from a Canadian AFBC installation is discussed. The waste was found to be unsuitable for applications in Port-land cement concrete because of poor strength development and expansion in mortars. The waste was compatible with Portland cement for soil cementing purposes but the resulting mixes were not resistant to freezing and thawing. Applications in asphaltic concrete were found to be successful in the laboratory and a small field trial is in progress.

INTRODUCTION

Atmospheric-pressure fluidized bed combustion is rapidly gaining acceptance in North America as a means to use "problem" fuels, such as low-grade or high-sulphur coals, for thermal power generation. In AFBC, coarsely crushed fuel is suspended in a combustion chamber with a bed-material, such as sand or limestone. The suspension is sustained by fluidizing the mass with air injected from below the bed. Fluidized bed combustion can be conducted in two modes at atmospheric pressure: in a bubbling or dense-phase bed; or in a circulating bed. In a companion paper [1] the authors have reviewed the principal differences between these two systems and the properties of the products of fluid bed combustion using both limestone and inert bed materials. This paper is directed to the potential utilization of high-Ca waste materials from both bubbling and circulating limestone beds typical of the residues expected to be produced by future commercial utility-scale AFBC in North America.

Although the authors are not aware of any commercial utilization of high-Ca AFBC residues, a considerable body of research and demonstration data is available on their potential industrial and agricultural applications. A broad range of proposed industrial and construction uses for high-Ca AFBC residues have been investigated by Minnick [2,3] and others [4-9]. A critical review of these early studies was prepared by Berry [10] in 1984.

More recently Rose [11] has conducted new investigations of high-Ca AFBC residues in low performance concretes and Berry [12] reported on the examination of high-Ca residues in soil-cement applications. Boesmans et al. [13-15] and Van er Widjven [16] have conducted general studies of applications for high-Ca residues in construction.

For some years, the Tennessee Valley Authority has been involved in studies of utilization of residues from limestone based AFBC. A summary of the scope of this work was recently presented by Harness, Milligan and Cruikshank [17], who note that a broad range of applications are under consideration, including: low-strength concrete, stabilized road base, agricultural lime, Portland cement/H_2SO_4 co-production, pollution control and carbon recovery. With the exception of the work by Rose [11], none of the data from the TVA project has been made publicly available to date [18].

Potential uses for high-Ca residues in agriculture have been extensively investigated by Bennett et al. [19-32] and by a number of other researchers

[33–41].

Reported uses for low-Ca AFBC residues have been in applications as construction materials [42–46] and as sources of aluminum [47–49].

The main applications considered for high-Ca AFBC wastes can be grouped as follows:

- Uses in construction as: aggregate; fill; supplementary cementing material; soil stabilizing agent; soil cement; a filler for asphalt mixes.
- Uses in pollution control: as a source of lime for neutralization of acid waste; as a source of lime for FGD scrubbers; for waste solidification and stabilization.
- Use as an industrial raw material for: cement production (possibly with co-production of sulphuric acid); carbon recovery;
- Use in agriculture as: a soil amending agent; a lime substitute; a source of nutrient values.

In 1984, the authors conducted an extensive investigation of the nature and utilization of high-Ca residues from two 18 t/h steam raising, bubbling AFBC units at Canadian Forces Base (CFB) Summerside, Prince Edward Island (PEI). The installation has been designed to burn high-sulphur (5%) coals, using New Brunswick limestone as a bed material [50]. A summary of the findings from this study is presented in this paper, details of the investigation have been presented in two reports [10,12].

EXPERIMENTAL

The installation at Summerside produces two primary wastes: a bed-drain material, and baghouse fines (baghouse ash). These are discharged to a silo prior to disposal and thus form a mixed third waste, silo ash, the composition of which varies as the relative quantities of the two components change.

All three waste streams were examined in this investigation. Samples were taken at the plant when Minto coal was being burned with Havelock limestone as bed material. The samples were examined in the laboratory for chemical, physical and engineering properties.

A preliminary consideration of the published information on utilization of AFBC residues and local economic factors led to the conclusion that the following applications may be appropriate for the Summerside materials:

- Soil modification and cementing (this was considered especially appropriate in Prince Edward Island, where problems exist in stabilization of sandy soils during highway construction);
- Portland cement concretes;
- Asphaltic concrete.

The following program of laboratory testing was conducted to provide a preliminary overview of the Summerside residues as potential construction materials:

- Chemical analyses;
- Standard soils testing in conjunction with a soil from PEI;
- Examination of the residues in asphaltic concrete mixes;
- Preliminary examination of the residues in Portland cement systems.

RESULTS AND DISCUSSION

Chemical Properties

Chemical analyses of the wastes are given in Table I and show that some separation of constituent elements does occur between the two primary wastes.

Table I

Chemical Analyses (% by mass) Of High-Ca AFBC Residues

| Major | Residue | |
Oxides	Bed Drain	Baghouse
SiO_2	6.6	11.7
Al_2O_3	2.7	5.8
Fe_2O_3	2.4	10.9
CaO	62.7	41.8
MgO	0.1	1.5
K_2O	0.3	0.7
Na_2O	0.2	0.3
SO_3	28.6	17.2
Estimated Phase Composition		
$CaSO_4$	48.7	29.3
CaO	42.6	29.8
Other[a]	8.7	41.0

a. Includes carbon and coal ash components

Thus, the bed drain material is largely derived from the calcined and partially sulphated limestone, whereas elements predominantly derived from coal-ash, such as Si, Al and Fe are more concentrated in the baghouse residue. An estimate was made of the CaO and $CaSO_4$ content of the wastes based on the total Ca and S content. The results are given in Table I and confirm that much of the sulphated limestone remains in the bed material.

Kalmanovich et al. [51] have examined the microstructure of the wastes from this source in some detail. Their work has shown that the particles of residue comprise an inner core of calcined limestone (CaO) surrounded by a layer of anhydrite ($CaSO_4$). In some instances a thin outer layer, rich in Si and Fe, is also present.

The layer structure of the residues has a significant affect upon their reactivity with water and acidic media. Alkalinity (neutralization potential) was determined on the wastes by two methods: a standard technique (ASTM C400), where exposure to acid is limited to 30 min; a method developed by Environment Canada [52], where an extended period of exposure to acid is used. As can be seen from the data in Table II the results from the extended exposure showed the material to have higher neutralization capacity, from which it is concluded that the rate of reaction of the residues was reduced in comparison to that of chemical lime. The data from the extended exposure tests are consistent with the estimates of CaO and $CaSO_4$ content made from the chemical analyses.

Physical and Engineering Properties

Grain-size distributions of the three Summerside wastes and the soil selected for soil-cement evaluation, as determined by dry screening, are shown in Figure 1. The following were noted:

a. The bed-drain material is narrowly graded and coarse, with 90% of the particles between 1.5 and 0.5mm in diameter. The size and grading of this material largely reflects the properties of the limestone feed.
b. The baghouse ash is a fine dust, with 90% of the grains less than 0.1mm in diameter.

Table II

Neutralization Capacity of Lime And High-Ca AFBC Residues

Source of Alkali	Neutralization Capacity (g H_2SO_4/100g)	Relative Activity
CaO (Reagent)	175.0	100.0
Bed Drain[a]	15.8	9.0
Baghouse[a]	16.9	9.7
Bed Drain[b]	71.1	40.6
Baghouse[b]	59.3	33.9

a. Determined by ASTM C400
b. Determined by Extended-exposure Method

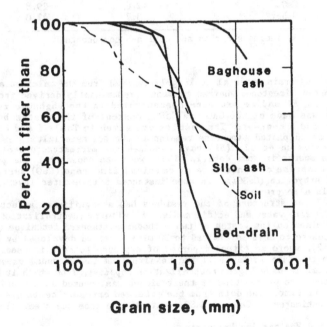

Figure 1. Grain-size distributions of AFBC residues and soil from Summerside P.E.I.

c. As would be expected the silo ash, being a mixture of the two primary residues, showed a bimodal grain size distribution. The size distribution of the particular sample illustrated in Figure 1 indicates that the silo ash comprised approximately equal parts of bed drain and baghouse ash.

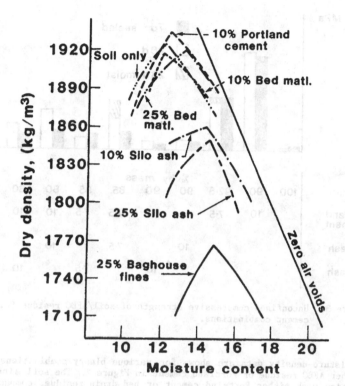

Figure 2. Moisture-density relationships for soil/AFBC residue/Portland cement combinations.

The soil was a poorly-graded sandy-silt, with some coarse gravel and an excess of fines. Atterberg tests showed it to be non-plastic. Such soils, when used in soil-cemented sub-bases, frequently shrink and induce reflection cracking in pavement structures.

From the perspective of soil properties, the bed drain material would be classified as equivalent to a narrowly-graded coarse sand; the baghouse ash is primarily of silt-sized composition.

Soil-cementing Properties

To be of use as soil cementing aids, the AFBC residues must meet at least the following criteria:

a. They must assist (or not reduce) compactability of the soil as determined by maximum dry-density, preferably such compaction should be available without an increase in the water demand of the soil/cement system.
b. The addition of the residue must increase the strength of the soil, as determined by unconfined compression.
c. In a climate such as prevails in PEI, the soil-cement mixture must be stable to cycles of freezing and thawing.
d. The soil cement mixture must also be dimensionally stable.

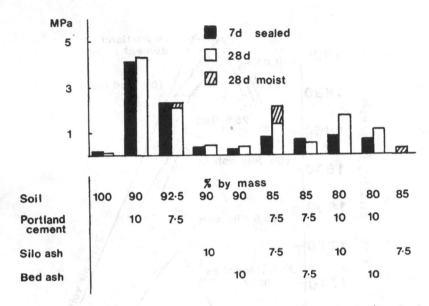

Figure 3. Unconfined compressive strength of soil/AFBC residue/Portland cement combinations.

Moisture–density data are shown for various binary combinations of soil and either AFBC residue or Portland cement in Figure 2. The soil alone, or in combination with either Portalnd cement or bed drain residue, compacted to a maximum density of approximately 1900 kg/m³ at an optimum moisture demand of approximately 12%. Addition of baghouse ash, either alone or as a component of silo-ash, substantially increased the water requirement at optimum density and reduced the compactability of the mass.

In Figure 3, data are presented showing the unconfined compressive strengths of various combinations of soil, Portland cement and the Summerside AFBC residues. Two curing regimes were investigated:

·Type 1 curing – Water was added to the mix in the quantity indicated for optimum compaction, the samples were then sealed in plastic containers (no further moisture available) until they were tested.
·Type 2 curing – A similar quantity of mix water was added, however, the samples were wrapped in cloth and stored at 100% relative humidity (RH) until tested. Further moisture could enter the mass during curing.

The curing regimes produced significantly different results, from which the following conclusions were made:

a. Adequate strength (an acceptable value being approx. 2 MPa at 28 days) is readily attained with 7.5% of Portland cement in combination with the soil examined.
b. Neither of the AFBC residues alone significantly contributed to strength development in soil/residue mixes.
c. In combination with Portland cement, when water was not added during the curing period, both residues interfered with the progress of strength development. This is presumed to have resulted from the removal of free

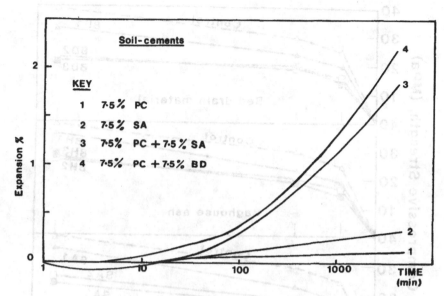

Figure 4.   Dimensional changes of soil/AFBC residue/Portland cement
            combinations. PC = Portland cement, SA = silo ash,
            BD = bed-drain residue.

water from the system (by hydration of CaO and CaSO₄), with the consequent inhibition of Portland cement hydration.
d. When water was provided during curing, the contribution to strength made
   by the Portland cement component of the mixes was unaffected by the
   presence of the residues, but no additional contribution was made by the
   bed drain or silo ash.

Dimensional stability of soil cement compositions was investigated on
four mixes under free-swelling conditions. The mixes were prepared and compacted with the quantity of moisture required for maximum compaction and were
stored at 100% RH with free access to moisture at the upper surface. As can
be seen from Figure 4, mixes containing both Portland cement and AFBC residue
expanded considerably during the test.
    Preliminary evaluation of the effects of freeze-thaw cycles on the soil-
cement mixes used to determine compressive strength indicated that Portland
cement and bed-drain material both improved the soil to an acceptable level.
Mixes containing silo ash were not satisfactory.
    Overall, the AFBC residues from Summerside were not suitable as soil-
cement materials. They made no significant contribution to strength development. The bed drain material, in combination with Portland cement could be
used as a soil modification mixture where expansion was desirable. Careful
control of strength development would be required if freeze-thaw resistance
was to be adequate.

AFBC Residues in Portland Cement Systems

If AFBC residues are to be used in construction materials based on Portland cement, the following criteria must be observed:

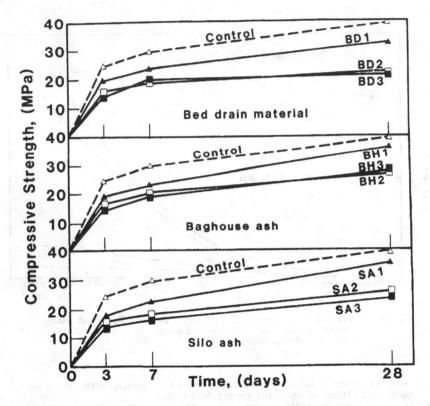

Figure 5. Strength development of Portland cement mortars containing
AFBC residues.

a. Either the AFBC residue must be pozzolanic, or it must be an effective
   aggregate, or it must bring some other performance value to the product.
b. It must achieve the above without adversely influencing the desired
   properties of Portland cement, such as set-time, strength development and
   dimensional stability.

   Standard tests showed that the residues were not pozzolanic. In mortars
they showed some decrease of cement set-time probably as a consequence of
dilution. No severe retardation or false-setting was evident as a consequence
of the use of the wastes.

   Tests were conducted on standard mortar mixes (ASTM C109) in which silo
ash was substituted for both cement and sand, baghouse ash was substituted for
cement and bed residue was substituted for sand. Strength development data
are summarized in Figure 5. The following conclusions were made:

a. The baghouse ash, though not disruptive of mortars when used as a cement
   replacement, did not contribute to strength development and must be
   considered as an inert diluent.
b. The bed-drain material had an adverse effect on strength development when
   used as a sand replacement or when present in silo ash as a replacement

for both sand and cement. This may be attributed either to its being a weaker aggregate than sand or to expansion during hydration leading to the formation of micro-cracks in the hardening mortar.

Mortar expansion was observed with all of the residues. The expansion was minimal with baghouse ash, and was severe with bed-drain material. This is presumed to have resulted from the volume changes associated with the hydration of free-CaO.

It can be concluded that the residues are only suitable for use in Portland cement systems under circumstances where low-strength and expansion are acceptable. Such conditions might be appropriate to their use as mine backfill as has been investigated by Rose et al. [11].

AFBC Residues in Asphaltic Concretes

The objections that limit the use of the residues in Portland cement-based systems do not apply in asphalt paving. Expansion of the residue should be less destructive than in cemented systems where a brittle solid is formed. Leaching to the immediate surroundings is unlikely and sulphate attack on the binder matrix will not be an issue as it may be for cementitious applications. The presence of free-CaO should not be deleterious as lime is frequently used as a filler in asphaltic mixes.

Experimental examinations were made of a limited range of asphaltic concrete materials using Summerside residues. The results indicated no particular technical advantages or disadvantages to their use [10]. Their selection as asphaltic concrete aggregate was expected to depend largely on the economics of transportation. In 1983 a 100 m test section of road at CFB Summerside was paved with asphalt employing 3% of silo ash in the mix. At the time of writing (August 1986) the section was reported to be in good condition and has shown no signs of deterioration [53,54].

CONCLUSIONS

High-Ca residues, from both bubbling and circulating AFBC, are unique among the solid products of coal combustion. Unlike pulverized fuel ash (fly ash), they are not solely composed of the elements introduced in the coal, nor are they pozzolanic glassy particles. They resemble the wastes produced by FGD systems, except that they are dry and contain lime and calcium sulphate in anhydrous forms. The particles are primarily a heterogeneous mixture of CaO and $CaSO_4$, with minor and trace inclusions of the elements normally found in coal ashes and limestone.

As potentially useful materials they present three classes of exploitable properties, being:

·potential sources of lime;
·granular;
·somewhat cementitious, although non-hydraulic.

Their heterogeneous nature, lack of pozzolanic properties and the elevated sulphate content, clearly limit their application (especially in construction) to a few very specific potential uses. The most promising are:

·in agriculture (providing that environmental, health and plant growth requirements are met);
·as a lime substitute for acidic waste neutralization;
·as a lime source in waste stabilization and solidification;
·as a low-strength backfill (with or without Portland cement);
·in soil-cementing and stabilization;
·as an aggregate/filler in asphaltic concrete.

362

ACKNOWLEDGEMENT

We wish to acknowledge the financial support of the Coal Division, Conservation and Alternative Energy Branch of Energy Mines and Resources Canada, for funding of two studies on AFBC Waste Evaluation upon which much of this paper is based.

REFERENCES

1. E.E. Berry, and E.J. Anthony, this volume.
2. J.L. Minnick, Development of Potential Uses for the Residue from Fluidized Bed Combustion Processes, Reports Under Contract No. EF-77-C-01-2549 (US Dept. of Energy, 1977).
3. J.L. Minnick, Development of Potential Uses for the Residue from Fluidized Bed Combustion Processes, Reports Under Contract No. DE-AC21-77-ET10415 (US Dept. of Energy, 1980).
4. J.W. Nebgen, J.G. Edwards and D. Conway; Evaluation of Sulphate-Bearing Material from Fluidized Bed Combustion of Coals for Soil Stabilization, Final Report, September 1977. Contract No. EX-67-A-01-2491. Report No. FHWA-RD-77-136, 69pp. Available from NTIS.
5. R.H. Miller, in Proc. 5th. Int. Conf. Fluidized Bed Combustion, Vol. II (MITRE Corp., McClean, Virginia, 1978) pp. 800-816.
6. R.J. Collins, J. Testing and Evaluation 8, 259-264 (1980).
7. C.C. Sun, C.H. Peterson and D.L. Keairns, Experimental/Engineering Support for EPA's FBC Program: Final Report Vol. III, Solid Residues Study, EPA-600/7-80-015c, January 1980.
8. INTEG Ltd., Investigation of the Utilization and Disposal of Boiler Ash from C.F.B. Summerside P.E.I., CANMET CONTRACT 78-9037-1, Interim Report, Sept. 1979.
9. R. Stone and R.L. Kahle, Environmental Assessment of Solid Residues from Fluidized Bed Combustion Processes, Report PB-282-940, June 1978.
10. E.E. Berry, An Evaluation of Uses for AFBC Solid Wastes, Final Report, DSS Contract OSQ83-00077, 1984.
11. J.G. Rose, Laboratory Testing of Potential Utilization of AFBC Wastes and Other Fossil Fuel Waste in the Production of Low Strength Concretes for Mining Applications, TVA Report TVA/PUB--86/4, March 1985.
12. E.E. Berry, An Evaluation of Uses for AFBC Solid Wastes - Phase II, Examination of Waste from CFB Summerside as a Potential Soil-cement Component, Final Report, DSS Contract OSQ84-00037, Sept. 1984.
13. B. Boesmans and R. Gerritsen, Applications of AFBC-coal Ashes in Cement, Asphalt Filler and Artificial Gravel, Report 1983. TNO-HMT-83-09557, 32pp. (in Dutch).
14. J.L. Walpot, B. Boesmans, B.G. Ten Dam and A.O. Hanstveit, in Proc. 3rd European Coal Utilization Conference, 4, (Ind. Presentations Group, Rotterdam, Neth., 1983) 31-55.
15. B. Boesmans, in Proc. ASHTEC 84, (CEGB, London, 1984) pp. 677-683.
16. A. Van der Wijdeven, Composition and Application of Ash from coal-fired Fluid Bed Boilers, Report 1984, INTRON-84227, 44 pp.
17. J.L. Harness, M.W. Milligan and K.A. Cruikshank, in Proc. 1986 Joint Symp. Dry SO_2 and Simultaneous SO_2/NO_x Control Technologies, (U.S. DOE and EPRI, Raleigh, N.C. June 2-6, 1986).
18. J.L. Harness, Personal Communication, August 1986.
19. O.L. Bennett, W.L. Stout, J.L. Hern and R.C. Sidle, in Energy and the Environment: Proc. 5th National Conf. (Am. Inst. Chem. Eng., Dayton, Ohio, 1977) pp. 84-89.
20. J.L. Hern, W.L. Stout, R.C. Sidle and O.L. Bennett, in Proc. 5th. Int. Conf. Fluidized Bed Combustion, Vol. II (MITRE Corp., McClean, Virginia, 1978) pp. 833-839.
21. R.C. Sidle, W.L. Stout, J.L. Hern and O.L. Bennett, J. Environ. Qual. 8 (2), 236-241, (1979).

22. W.L. Stout, R.C. Sidle, J.L. Hern and O.L. Bennett, J. Agronomy 71 (4), 662–665 (1979).
23. W.L. Stout, R.C. Sidle, J.L. Hern and O.L. Bennett, in Solid Waste Research and Development Needs for Emerging Coal Technologies, ASCE, New York, 1979.
24. O.L. Bennett, W.L. Stout, J.L. Hern and R.L. Reid, in Proc. 6th. Int. Conf. Fluidized Bed Combustion, Vol. III (US Dept. of Energy, CONF-800428, 1980) pp. 885–891.
25. W.L. Stout, E. Fashandi, M.K. Head, R.L. Reid and O.L. Bennett, in Proc. 6th Int. Conf. Fluidized Bed Combustion, Vol. III (US Dept. of Energy, CONF-800428, 1980) pp. 892–898.
26. O.L. Bennett, W.L. Stout J.L. and Hern, in Proc: DOE/WVU Conf. Fluidized-bed Combustion System Design and Operation, (METC, Morganstown, West Virginia, 1980) pp. 402–427.
27. O.L. Bennett, J.L. Hern and W.L. Stout, in Proc. Gov. Conf. Expanding Use Coal N.Y. State: Probl. Issues, edited by M.H. Tress and J.C. Dawson (Res. Found. State Univ. N.Y., 1981) pp. 43–48.
28. T. Whitsel, D.L. Stadmore, O.L. Bennett, W.L. Stout and R.L. Reid, in Proc. Joint Ann. Meeting, Am. Soc. Animal Sci. and Canadian Soc. Animal Sci. (1982).
29. O.L. Bennett, R.L. Reid, D.L. Mays, T. Whitsel, D.M. Mitchell, W.L. Stout and J.L. Hern, in Proc. 7th Int. Conf. Fluidized Bed Combustion Vol. I, (US Dept. of Energy, CONF-821064, 1983) pp. 559–566.
30. W.L. Stout, H.A. Menser, O.L. Bennett and W.M. Winant, Reclamation and Revegetation Research 1, 203–211 (1982).
31. D.M. Mitchell, J.D. May and O.L. Bennett, Poultry Sci. 62, 2378–2382 (1983).
32. E.F. Fashandi, R.L. Reid, W.L. Stout, J.L. Hern and O.L. Bennett, Qual. Plant Plant Foods Hum. Nutr. 35, 359–374 (1985).
33. G.L. Terman, V.J. Kilmer, C.M. Hunt and W. Buchana, J. Environ. Qual. 7, 147–50 (1978).
34. R.F. Korcak, HortScience 14 (2), 163–4 (1979).
35. R.F. Korcak, J. Environ. Qual. 9 (2), 147–151 (1980).
36. R.F. Korcak, Soil Science and Plant Analysis, 11 (6), 571–585 (1980).
37. R.F. Korcak. J. Am. Soc. Hort. Sci. 107 (6), 1138–1142 (1982).
38. J.J. Wrubel, R.F. Korcak and N. Childers, Soil Science and Plant Analysis 13 (12), 1071–1080 (1982).
39. R.F. Korcak, Soil Science and Plant Analysis 15 (8), 879–891 (1984).
40. E.H. Quigley and G.A. Jung, Soil Science and Plant Analysis 15 (3), 213–226 (1984).
41. D.A.A. Arthursson and K. Valdmaa, in Proc. 6th Int. Conf. Fluidized Bed Combustion, Vol. III (US Dept. of Energy, CONF-800428, 1980) pp. 939–940.
42. J. Lotze, and G. Wargalla, Zem.-Kalk-Gips., Ed. B, 38 (5), 239 (1985, in German). English Translation: Zem.-Kalk-Gips, Ed. B, 38 (7), 168 (1985). Zem.-Kalk-Gips, Ed. B, 38 (7), 374 (1985, in German). English Translation: Zem.-Kalk-Gips, Ed. B, 38 (9), 240 (1985).
43. K.W. Belting, World Coal (London), Vol. 5 (6), 30–33 (1979).
44. R. Kochling and D. Leininger, in Proc. Conf. on Fluidized Bed Combustion, VDI-Berichte No. 322, Nov. 1978, (Dusseldorf, FRG, 6–7, 1978) pp. 131–138.
45. D. Leininger and T. Schider, Glueckauf-Forschungsh 41 (1), 1–6 (1980).
46. D.G. Montgomery, in Fly Ash and Coal Conversion By – Products: Characterization, Utilization and Disposal I, edited by G.J. McCarthy and R.F. Lauf, Mat. Res. Soc. Symp. Proc. Vol. 43 (Materials Research Society, Pittsburgh, 1985), pp. 119–128.
47. C.A. Hamer, Alumina from Fluidized-bed Combustion of Hat Creet B.C. Coaly Waste, Part I: Acid Extraction, Division Report MRP/MSL 79-21(IR) CANMET 1979. Part II: Alumina Recovery and Purification by an HC1 Process, Division Report MRP/MSL 79-144(IR) CANMET 1979. Part III: Alumina Recovery and Purification by an HC1-caustic Process, Division Report MRP/MSL 80-152(IR) CANMET 1980.

48. C.A. Hamer, <u>Acid Extraction of Alumina from Canadian Non-bauxite Sources at CANMET</u>, CANMET Report 81-2E, Feb. 1981.

49. W.R. Livingstone, D.A. Rogers, R.J. Chapman and N.T. Bailey, Hydrometallurgy <u>13</u> (3), 147-50 (1985).

50. M.E.D. Taylor and F.D. Friedrich, <u>The Summerside Project, Canadian State of the Art in AFBC Boilers</u>, Division Report ERP/ERL 82-10(TR) CANMET 1982.

51. D.P. Kalmanovitch, V.V. Razbin, E.J. Anthony, D.L. Desai and F.D. Friederich, in <u>Proc. 8th Int. Conf. Fluidized Bed Combustion, Vol. I</u>, (1985) pp. 53-64.

52. T.W. Constable, Canada Centre for Inland Waters. Personal Communication, 1985.

53. P. Yurkiw and M.T.C. Brown, Hardy and Associates (1978) Ltd. Personal Communication, 1985.

54. V.V. Razbin, Personal Communication, 1986.

ENHANCED RESOURCE RECOVERY BY BENEFICIATION AND DIRECT ACID LEACHING
OF FLY ASH

E.E. BERRY*, R.T. HEMMINGS* and D.M. GOLDEN**
*Ontario Research Foundation, Sheridan Park, Mississauga, Ontario, Canada,
L5K 1B3
** Electric Power Research Institute, P.O. Box 10412, Palo Alto, CA 94303.

Received 30 January, 1987; Communicated by G.J. McCarthy

ABSTRACT

An investigation was conducted into the use of beneficiated ash fractions
as feed materials for direct acid leaching to recover Al and other metals. A
low-Ca (1.4% CaO) ash was selected from which four fractions (classifier
fines, classifier rejects, magnetic concentrate, and non-magnetics) were produced by mineral beneficiation methods. The raw ash and the four ash fractions were leached with 6 M HCl under reflux. Mass balance data were obtained
to determine extraction efficiency for Si, Al, Fe, Ca, Mg, Na, K and Ti. It
was found that size separation of the ash had a marked influence in enhancing
extraction efficiency from the fine fraction for all of the elements studied
except Fe. Extraction of Fe from the magnetic fraction was less efficient
than from the non-magnetic fraction. Examination of the solid residue after
leaching showed that, although most of the acid-soluble components were removed from the glassy phases of the ash, a significant part of the ash comprised non-reactive glass of high-Si content. A relationship between extractable Al in ash and leachable alkali and alkaline earth metals was identified
and discussed in terms of glass modification theory.

INTRODUCTION

Since 1979, EPRI has sponsored research to examine several processes for
the recovery of metal values from coal combustion residues [1-4]. An early
conclusion from these studies was the identification of direct acid leaching
(DAL) with HCl as a promising minimum-treatment method for the recovery of
aluminum and other metals from fly ash. The process requires refluxing fly
ash in HCl, whereby Al, Fe and other metal values are dissolved and a largely
aluminosilicate mass remains (DAL-residue). For economical application of the
DAL-process, it is necessary that: (a) process conditions are optimized,
including pre-leach process operations; and (b) suitable markets are found for
DAL-residues or the products made from them [4].

The work described in this paper comprises part of a study conducted by
ORF for EPRI directed to the following primary objectives:

· Demonstration of the use of pre-treatment (beneficiation) of a selected
  raw fly ash to upgrade it as a feed for the DAL-process;
· Investigation of the extractability of aluminum, iron and other
  components from beneficiated fractions of the selected ash;
· Confirmation of the scale-up of the DAL-process from the laboratory to
  pilot-plant level;
· Demonstration of the production of filler grade products from the DAL-
  residue, evaluation of their performance as fillers in polymer composites
  and evaluation of the potential marketability of DAL-residues in the
  fillers industry.

This paper summarizes the findings of work related to the first two of
these objectives. A full description of the project is presented in the Final
Report [5] and a summary of the investigation of DAL-residues as fillers has
been published elsewhere [6].

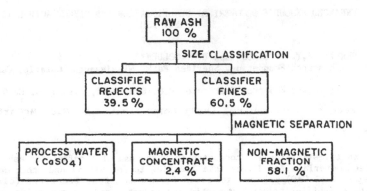

Figure 1. Schematic flowsheet of ash beneficiation protocol.

EXPERIMENTAL

Materials

The ash selected for study was from Plant Bowen, Cartersville, Georgia. Raw fly ash from this plant (~1100 lbs.) was sampled in a dry state over a 5 h period from discharge silos that collect ash from thermal units 1 and 2. The raw ash was subjected to the beneficiation protocol illustrated in the flowsheet (Fig. 1) to produce the following fractions: raw ash, classifier rejects (+35 μm), classifier fines (-35 μm), non-magnetics, and magnetics.

Materials Characterization

The raw ash, and each beneficiated fraction, was characterized with regard to physical properties (specific gravity, surface area), particle size distribution (Coulter counter), chemical analysis (ICAP), crystallinity and phase composition (XRD), and morphology (SEM). Similar properties were determined for selected post-leached solids (DAL-residues). Full details of the methods used are given in the Final Report [5].

Direct Acid Leaching

Following the findings of extensive studies of ash leachability reported by ORNL [1], the laboratory scale DAL-processing of the ash was conducted using refluxing 6 M HCl (105°C) for 2 h at a pulp density of 40% solids. Leaching was conducted in all-glass (Pyrex) apparatus. After refluxing, the residue was filtered and washed with ~100 mL of 1 M HCl, followed by deionized water. All leachate solutions and washings were collected and diluted to 500 mL with 1 M HCl prior to chemical analysis. The leachates were maintained in acidic conditions at all times to prevent precipitation of silica. The leaching procedures were conducted as "mass-balance" experiments in which all solid and liquid fractions were collected and analyzed separately [5].

RESULTS AND DISCUSSION

Fly ash as collected at the power plant is a heterogeneous mixture within which substantial chemical and physical speciation occurs between particles of different types [7]. In particular, chemical differences can be observed between particles of different sizes [8-10], densities [11] and between magnetic

Table I

Physical Properties of Beneficiation Products

Property	Raw Ash	Classifier Rejects	Classifier Fines	Magnetics	Non-Magnetics
Avg. true particle density (g.cm^{-3})	2.36	2.25	2.47	3.67	2.47
Surface area (m^2/g)	0.91	0.66	1.00*	0.68*	2.70**
Particle size characteristics:					
Mean size (μm)	23.4	52.1	10.7	13.3	10.6
Median size (μm)	24.4	51.1	11.3	14.1	11.2
d95 (μm)	90	130	32	28	32

* Krypton Adsorption
** Nitrogen Adsorption

and non-magnetic ash fractions [12]. Therefore, it is reasonable to postulate that particles of such different chemical compositions derived by beneficiation from the same ash source will respond differently to acid leaching.

Beneficiation techniques were used to segregate different types of ash particles, which were then characterized and subjected to direct acid leaching. The leachability of different elements from the various process fractions from raw ash was examined from three perspectives: first, to evaluate the ash as a source of extractable metal values in comparison with other ashes examined by ORNL [1]; second, to evaluate the contribution of pre-leach beneficiation to the extractability of metal values; third, to contribute to an understanding of the fundamental nature of the DAL-process and the relationships between ash properties and metals extractability.

Effects of Pre-Leach Beneficiation on Speciation of Ash Fractions

Air-classification of the raw ash produced 60.5% of a fine ash fraction (95%, -35 μm) and 39.5% of a corresponding reject (2%, -20 μm), as shown in the particle size distributions (Fig. 2). Other physical properties of the ash were largely unaffected by size separation (Table I).

Minor differences in chemical speciation are evident between these two fractions (Table II). Iron and Cr were significantly more concentrated in the reject fraction, whereas potassium and sulphur were more concentrated in the fine fraction, as were most trace elements other than Cr.

X-ray diffraction (XRD) analysis (Fig. 3) indicated little mineralogical difference between either of the ash fractions or raw ash. The principal crystalline phases identified were quartz, mullite, magnetite and hematite, in conjunction with considerable quantities of an amorphous or glassy phase. Overall, the fine fraction was found to contain slightly more glass than either the raw ash or the reject fraction, suggesting a relationship between glass content and particle size.

Representative SEM photomicrographs showing the general morphology of the raw ash, classifier fines and classifier rejects are given in Figure 4 (a-c). The raw ash is seen to be a heterogeneous mixture of particles of various shapes and sizes, with many spheres and numerous irregularly-formed particles.

Table II

Chemical Analyses (wt.%) of Raw Materials and Leached Solids

Constituent	Raw Ash Feed	Raw Ash DAL-res	Classifier-Rejects Feed	Classifier-Rejects DAL-res	Classifier-Fines Feed	Classifier-Fines DAL-res	Magnetic Fraction Feed	Magnetic Fraction DAL-res	Non-Mag. Fraction Feed	Non-Mag. Fraction DAL-res
SiO_2	52.31	62.08	51.75	59.39	52.22	62.27	21.22	32.71	53.47	63.27
Al_2O_3	27.18	24.70	26.23	25.99	28.44	24.24	12.01	15.35	29.30	23.29
Fe_2O_3	10.81	4.26	13.33	4.89	9.09	4.16	63.14	47.63	7.62	3.81
CaO	1.37	0.51	1.57	0.64	1.37	0.53	1.00	0.79	1.18	0.49
MgO	1.04	0.83	0.98	0.91	1.11	0.84	0.65	0.64	1.13	0.79
Na_2O	0.23	0.20	0.21	0.20	0.26	0.20	0.07	0.10	0.26	0.20
K_2O	2.79	2.41	2.65	2.79	3.00	2.70	0.71	1.04	3.06	2.40
TiO_2	1.46	1.50	1.39	1.45	1.51	1.52	0.63	0.83	1.56	1.52
MnO	0.03	0.02	0.03	0.02	0.03	0.02	0.09	0.05	0.03	0.02
P_2O_5	0.25	0.09	0.15	0.05	0.34	0.14	0.37	0.30	0.32	0.12
BaO	0.12	0.11	0.11	0.10	0.12	0.12	0.05	0.06	0.08	0.06
SrO	0.07	0.06	0.07	0.07	0.08	0.06	0.04	0.05	0.08	0.05
SO_3	0.51	0.09	0.29	0.07	0.69	0.22	0.06	0.03	0.06	0.04
C	2.16	2.53	2.56	2.88	1.84	2.05	0.34	0.50	1.96	2.17
LOI	3.06	4.15	2.57	3.21	2.73	3.99	0	0.39	2.67	4.12
Total*	100.33	99.52	101.32	99.44	100.10	98.93	100.38	100.08	100.15	98.22
(Trace Elements , ppm)										
Be	17	13	14	13	20	15	9	10	21	14
Co	48	30	39	29	54	38	99	91	54	33
Cr	145	90	120	90	155	90	460	260	150	95
V	240	150	210	160	270	150	490	400	270	140
Cu	140	90	110	70	180	130	210	150	180	140
Pb	22	2	6	<1	32	2	21	2	36	3
Ni	21	2	14	1	32	4	68	17	36	3
Cd	<1	<1	<1	<1	<1	<1	1	<1	<1	<1
Ag	0.2	<0.2	<0.2	<0.2	<0.2	<0.2	<0.2	<0.2	<0.2	<0.2
Mo	18	2	8	2	28	2	16	8	14	2
Zn	39	1	14	<1	52	4	110	12	64	4
Zr	230	307	258	317	235	304	132	184	221	298

*Note: Total = sum of oxide analyses and C, not including LOI.

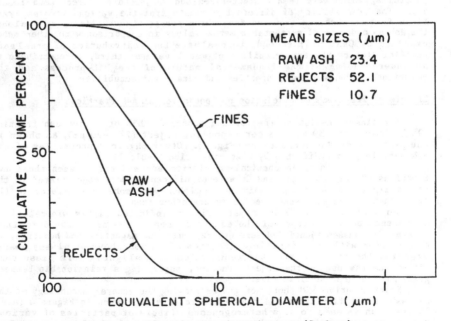

MEAN SIZES (μm)

RAW ASH   23.4

REJECTS   52.1

FINES   10.7

Figure 2. Particle size distributions (Coulter).

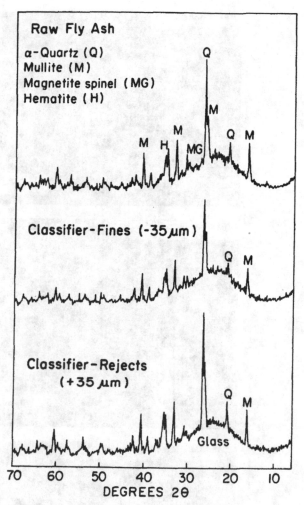

Figure 3. X-ray diffraction patterns (CuK-alpha) from ash size fractions.

The size-classification procedure successfully removed coarse, non-spherical particles (mostly partly combusted coal) together with some spherical fragments and detrital quartz (Fig. 4c). Although some irregularly-shaped particles remained, the classifier fines were composed largely of spherical particles together with some very fine-sized, "lacy" structured carbon that could not be removed by air-classification (Fig. 4b).

Effects of Pre-Leach Magnetic Separation on Speciation of Ash Fractions

As would be anticipated, major chemical speciation (see Table II) was induced by magnetic separation of the classifier fines. The magnetic fraction was rich in Fe (63.1% as Fe_2O_3), identified by XRD as principally magnetic spinel with associated hematite and very little residual glass (Fig. 5). The

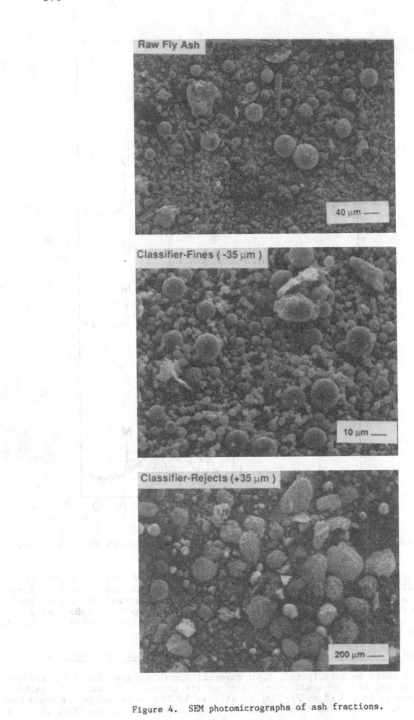

Figure 4. SEM photomicrographs of ash fractions.

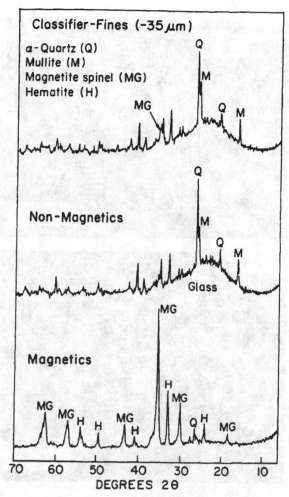

Figure 5. X-ray diffraction patterns from products of magnetic separations of classifier fines.

non-magnetic fraction was correspondingly depleted of iron (7.6% as Fe_2O_3), being largely composed of aluminosilicate glass, quartz and mullite.

Trace element speciation was significant for Mn, Co, Cr, V, Cu, Ni and Zn, all of which were associated with the magnetic fraction (Table II). Notably, Ti, Be, Pb and Zr, were excluded from the magnetic fraction, these elements being presumably associated with glass, mullite or quartz phases, all of which are largely absent from the magnetic fraction. The exclusion of Ti and Pb from the magnetic fraction is unexpected in that both elements can enter Fe-spinel structures in the same manner as Mn, Co, Cr, V, Cu, Ni and Zn. It could be postulated that this latter group of elements is in some way associated with the iron sources in the original coal (perhaps as sulphides)

372

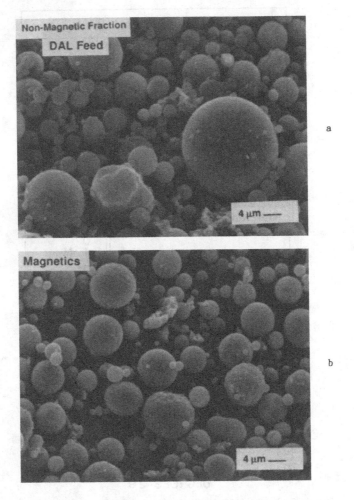

Figure 6. SEM photomicrographs of non-magnetic and magnetic ash fractions.

that form spinel compounds in ash. Alternatively, Ti and Pb may be associated with clay-like coal minerals and thus enter into the glass or mullite phases.

The increase in the average density of the magnetic concentrate over the non-magnetic material (Table I) is expected and is consistent with the enrichment of Fe-containing phases in the magnetic product. The mean particle size of the magnetic fraction is also slightly larger than the non-magnetic fraction.

Scanning electron micrographs of particles from the magnetic fraction (Fig. 6b) showed them to be typically spherical in form with some indication of surface crystallinity which was not apparent among the non-magnetic materials. The non-magnetic fraction was predominantly spherical (Figure 6a); however, the presence of ultra-fine carbon is indicated, along with other irregularly-shaped particles.

Direct Acid Leaching of Ash Fractions

On leaching with 6 M HCl some differences in the mass of dissolved material were observed between the various ash fractions (Table III), the main distinction being the substantially higher solubility of the magnetic fraction.

The principal major elements that dissolved from all of the fractions (Table IV) were Al, Fe, Ca, Mg, Na, K, Ti and P. Silicon was largely insoluble, although there was some indication that it was more mobile from the magnetic fraction. This may indicate either the presence of a relatively soluble high-Fe glass component or the partial inclusion of Si^{IV} in acid soluble spinels in the magnetic concentrate.

Minor and trace element leaching varied considerably; Mn, P, Pb, Ni, Mo and Zn were all acid soluble from all particle classes, although P, Ni, Mo and Zn were least soluble from the magnetic fraction. With the exception of Zr, this element being largely insoluble from all fractions, the trace elements were slightly more soluble from the fine and non-magnetic fractions than from the raw ash.

A principal concern of this project was the influence of beneficiation on the acid solubility of Al, Fe, and the alkali and alkaline earth elements that would consume HCl and, hence, adversely influence the economic viability of acid extraction.

Previous work at ORF [11] has shown that size classification of ash can lead not only to reduced particle size and carbon content but also to a degree of mineralogical and chemical speciation. One benefit of size classification was shown to be the increased concentration of particles of higher glass content in the finest size fractions. It is expected from the work of Hulett et al. [2] that ash fractions of high glass content are desirable in the DAL-process due to the greater reactivity of aluminosilicate glass toward HCl compared with chemically less reactive phases such as mullite, quartz, or ferrite-spinel. Similarly, it can be expected that the distribution of recoverable Al and Fe would be affected by magnetic separation. It might be anticipated that the recovery of an iron-rich magnetic concentrate would enhance the recovery of Fe and other strategic metal values associated with the magnetic phases. At the same time, the production of an iron-depleted, non-magnetic fraction could be expected to reduce the load on "downstream" ion-exchange removal of Fe during the processing required for commercial recovery of Al.

To compare the ease of extraction of elements from the different particle classes, "extraction efficiency" (E%) was defined in this study:

$$E(\%) = \frac{X_1(W_1)}{X_f(W_f)} \times 100$$

where, X_1, X_f = weight fraction of element i in leachate and feed, respectively, and W_1, W_f = weight of leachate and feed, respectively.

Differences in elemental extractability between ash sources or process fractions arise principally from the varying acid solubilities of the component phases in ash. To a lesser extent, extraction efficiency also incorporates the effects of physical factors such as particle size and surface area of the ash, both of which may modify the rate of reaction. However, in the absence of more extensive kinetic data, it is assumed that the process conditions used in this study lead to essentially complete dissolution of leachable metal values such that physical differences between ashes are less important than differences in chemical speciation.

Extraction efficiencies, calculated from the laboratory acid leaching experiments, are summarized for the major elements in Figure 7. Aluminum extraction efficiency from the raw Bowen ash was 27%. Whereas this is low in comparison with the extraction from some ashes, reported in the ORNL study [1] it is typical for low-Ca ashes. It is clear, however, from consideration of the extraction data from other fractions (Table IV), that pre-DAL beneficiation had a marked and positive effect on Al extraction efficiency which rose

Table III

Acid Solubility and Al Extraction of Ash Fractions

Fraction	Acid Soluble Material (wt.%)	Al Extraction (%)
Raw Ash	16.3	27
Classifier Rejects	15.7	15
Classifier Fines	18.0	39
Magnetics	36.3	17
Non-Magnetics	15.0	37

Table IV

Chemical Analyses of Leachate Solutions

Element	Leachate Concentration (ppm)*				
	Raw ASH	Classifier Rejects	Classifier Fines	Magnetic Fraction	Non-Magnetic Fraction
Sl	8	6	6	20	3
Al	3100	1600	4700	750	4700
Fe	4600	5700	4000	19000	2400
Ca	580	640	650	260	540
Mg	190	100	270	100	280
Na	60	30	90	10	80
K	50	200	850	10	850
Ti	100	64	140	37	140
Mn	10	9.6	11	33	9.2
P	56	34	77	44	73
Ba	4.3	11	3.4	8.3	20
Sr	29	19	37	6.2	39
Zr	0.6	0.6	0.6	1.6	0.4
Be	0.7	0.4	1	1.1	0.9
Co	2.6	1.8	3.1	5.9	2.7
Cr	6.7	0.6	9	25	7
V	12	6.3	17	23	15
Cu	7.7	4.1	12	10.1	11
Pb	5.6	2.9	7.6	5.9	6.9
Ni	1.5	1.3	1.1	8.7	<0.1
Cd	0.8	<0.1	0.6	<0.1	<0.3
Ag	<1	<1	<1	1	1
Mo	2.4	0.1	3.3	2	2.2
Zn	2	0.1	0.2	13	2
S	163	67	157	-	-

*NOTE: All leach experiments were conducted using 40g of feed ash fraction, all leachate solutions were made up to a volume of 500 ml.

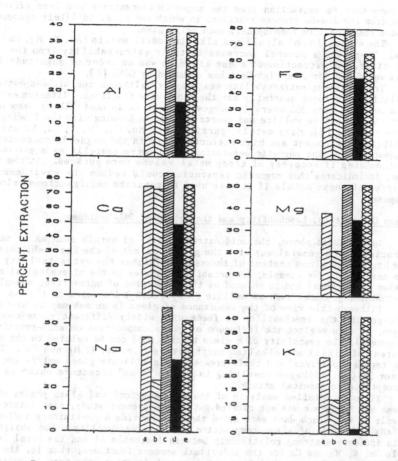

Figure 7. Comparison of major element extraction efficiency (E%) for ash fractions: (a) raw ash; (b) classifier rejects; (c) classifier fines; (d) magnetic concentrate; (e) non-magnetic fraction.

to approx. 40% for the classifier fines and non-magnetic fractions. By comparison, extraction of Al from the classifier rejects and magnetic concentrate was only 15 and 17%, respectively, indicating low recovery potential.

The greatest Al-extractability was found for the fractions that may be expected to contain the largest glass content (classifier fines and non-magnetics), consistent with the findings of Hulett et al. [2]. The low Al-extractability of the classifier rejects is attributed to aluminum being present in an unreactive mullite phase. The low Al-extractability of the magnetic concentrate may arise from Al substitution in comparatively unreactive ferrite spinel phases. It is probably significant in this regard that little alumino-silicate glass is evident from XRD examination of this fraction (Fig. 5).

Pre-DAL size classification had little effect on Fe extractability, whereas magnetic separation produced a high-Fe magnetic concentrate from which high Fe recoveries might be obtained if leached separately. It is interesting to note that Fe extraction from the magnetic concentrate was less efficient than from the low-Fe process fractions in which the metal is likely present in a more reactive form than spinel, such as in glass.

The extraction of alkali and alkaline earth metals (Na, K, Mg, Ca) was low. There was a general increase in their extractability from the fine fractions, but extraction of these elements was an order of magnitude less than was found for some high-Ca ashes examined by ORNL [1].

Trace element extractability was complex, although the highest extractabilities were found generally for the finer size fractions. Titanium extraction was less than 20% overall, suggesting that it is most likely associated with the unreactive mullite and quartz phases, in keeping with the findings of Hulett et al. [2]. Many metals, particularly Mn, Co, Cr, V, Cu, Ni and Zn, exhibited enrichment and high extractability in the magnetic concentrate, suggesting that the magnetic separation would serve usefully as a preconcentration step if recovery of trace metal values were pursued. At the same time, it indicates that magnetic separation could reduce the environmental loading of these metals if it were used to separate ash fractions prior to disposal.

Glass Composition, Leachability and the Nature of DAL-Residues

As discussed above, the acid extractability of metals such as Al may be associated in a general way with the glass content of the feed ash; that is, the higher the glass content of the ash, the higher the extractability. In the case of Al, for example, extractability relates to the mineralogical speciation of the metal and is reduced as the proportion of unreactive, non-glassy Al-containing phases such as mullite ($3Al_2O_3 \cdot 2SiO_2$) increases in the ash.

Although this view of the importance of glass in an ash may be useful in interpreting the leachability of feeds with widely different phase compositions, it does neglect the influence of glass composition on acid-reactivity. In general, the reactivity of a glass towards acid can be related to the total content of alkali and alkaline earth metals (i.e. Na, K, Mg and Ca). These are termed "modifiers" and their presence in a silicate glass matrix tends to sever Si-O-Si linkages resulting in a more "open" structure which is more susceptible to chemical attack.

Although detailed analysis of the mineralogical and glass phases of the Bowen ash fractions was not carried out in the present study, some interesting trends in the leach data were noted that may provide a preliminary indication of the influence of glass composition on acid leachability. The analytical data indicate a strong relationship between leachable Al and the total leachable Na, K, Mg and Ca for the individual process fractions; that is, the more Na, K, Mg and Ca extracted from an ash or ash fraction, the more Al extracted (Fig. 8). Thus, it would appear that the individual process fractions have quite different glass compositions which result in varying levels of reactivity towards acid. Furthermore, it can be concluded that the greatest reactivity is found for those fractions in which the glasses are most heavily modified with alkali and alkaline earth metals. With this view, it follows that the total glass content of an ash per se is related only indirectly to acid leachability and that glass chemistry is probably of greater importance.

Further support for this contention comes from XRD analysis of the DAL-residue from the non-magnetic fraction, which indicates that a considerable proportion of the glass in this material remained after acid leaching (Fig. 9). Whether the residual glass in the spent ash was an "acid-unreactive" component or was altered by the leaching process is not clear from the present data. Some indication that the former mechanism might be applicable has been obtained in related work at ORF [11] in which highly modified glasses (CaO >25%) have been shown to co-exist with glasses of relatively low modifier content (CaO <3%) in the same ash sample. Clearly, further investigation of the glass chemistry of candidate ashes will be of significant value in the selection of materials for DAL processing.

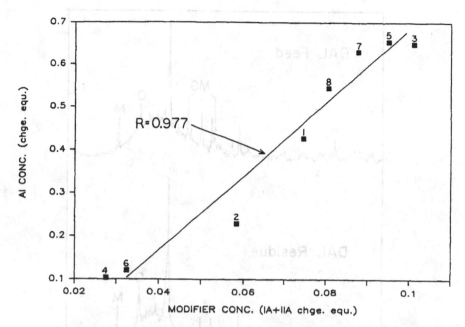

Figure 8. Variation of leachable Al with leachable Na, K, Mg and Ca contents of ash process fractions and linear regression coefficient. 1. raw ash; 2. classifier rejects; 3. classifier fines; 4. magnetic concentrate (Davis tube); 5. non-magnetic product (Davis tube); 6. magnetic concentrate (pilot plant); 7. non-magnetic product (pilot plant); 8. pilot scale leach of non-magnetic product (pilot plant).

Rather unexpectedly, there was no direct evidence that ash particles of fine size were selectively dissolved. The particle size distributions of the pre-leached and post-leached materials were found to be very similar (Fig. 10). This is consistent with the observation that the mass of material dissolved from the fine sized non-magnetic fraction (15.0%) was similar to the mass dissolved from the coarse fraction (15.7%), although much more of the Al present was extracted from the fine particles.

The fine particles contained more Al in glass phases rather than mullite and, thus, it may be concluded that selective dissolution of glassy material from the ash is a probable mechanism for Al-leaching by HCl. In contrast to what is typically found with ash particles etched with HF [2], examination of the DAL-residue by SEM did not reveal any significant etching of the surface structure (Fig. 11). Further work is required to understand the microstructural consequences of leaching with HCl and the influence that ash-glass composition may have on ash reactivity.

CONCLUSIONS

1. Overall, the extractability of Al from the raw Plant Bowen ash was similar to that found by ORNL for other low-Ca ashes. However, relative to the high-Ca ashes reported by ORNL, the extractability of Al from Bowen ash was poor, as with other low-Ca sources.

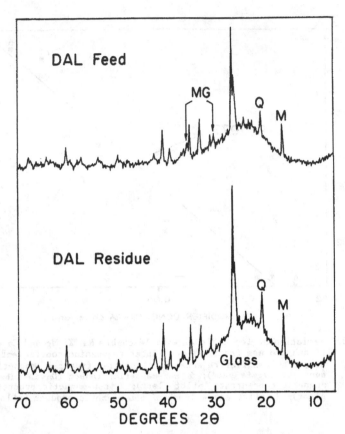

Figure 9. X-ray diffraction patterns from non-magnetic classifier fines
(DAL-feed) and DAL-residue. [Principal phases as in Fig. 3.]

2. Size classification resulted in only minor differences in the overall
chemical composition of the ash fractions, the main effect being seen on some
concentration of Fe-rich particles in the classifier reject fraction. Some
evidence was obtained by XRD that the classifier reject fraction contained
less glassy material and more quartz and mullite than did the fine fraction.
Although the chemical composition of the size-classified products was similar,
their leachability was markedly different. It was found that 39% of the aluminum present in the fine fraction was leachable in comparison with 15% from
the classifier rejects. Leachability of iron from the two size fractions was
of similar magnitude (77%, coarse fraction; 79%, fine fraction).

3. As expected, magnetic separation of the classifier fines fraction
resulted in a major concentration of iron-rich particles in a magnetic
fraction. Acid leaching of this fraction yielded little Al (17% of the total
Al); only 51% of the total Fe in the magnetic concentrate was acid-leached,
the remainder being present in acid insoluble components. Aluminum extractability from the non-magnetic fraction was similar (37%) to that from the fine
fraction as was the leachability of Fe (71%). The major benefit resulting
from magnetic separation lies in the segregation of an iron-rich stream, prior

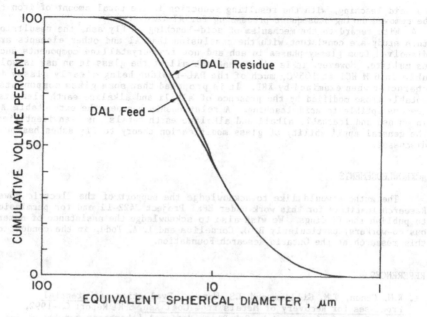

Figure 10. Particle size distributions (Coulter) of non-magnetic classifier fines (DAL feed) and DAL-residue.

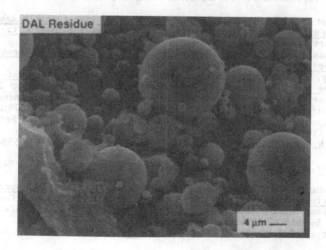

Figure 11. SEM photomicrograph of DAL-residue from leaching of the non-magnetic fraction.

to acid leaching, with the resulting reduction in the total amount of iron to be removed during subsequent processing for Al recovery.

4. With regard to the mechanism of acid-leaching of fly ash, the results of this study are consistent with the conclusion that Al and other elements are dissolved from glassy phases in ash and not from crystalline components such as mullite. However, it is clear that not all of the glass in an ash is soluble in 6 M HCl at 105°C, much of the DAL-residue being clearly glassy in character when examined by XRD. It is proposed that some glass components, notably those modified by the presence of alkali and alkaline earth elements, are susceptible to acid leaching. A relationship between the extractable Al in an ash and leachable alkali and alkaline earth metals has been identified. The general applicability of glass modification theory to fly ashes has been discussed.

ACKNOWLEDGEMENTS

The authors would like to acknowledge the support of the Electric Power Research Institute for this work under EPRI Project 2422-11 and for permission to publish the findings. We wish also to acknowledge the assistance of numerous co-workers, particularly B. J. Cornelius and I. A. Todd, in the conduct of this research at the Ontario Research Foundation.

REFERENCES

1. R.M. Canon, T.M. Gilliam and J.S. Watson, Evaluation of Potential Processes for Recovery of Metals from Coal Ash, EPRI Report CS-1992, Vols. 1 and 2, (prepared by Oak Ridge National Laboratory for the Electric Power Research Institute, 1981).
2. L.D. Hulett, A.J. Weinberger, N.M. Ferguson, K.J. Northcutt and W.S. Lyon, Trace Elements and Phase Relations in Fly Ash, EPRI Report EA-1822, May, 1981.
3. R.F. Wilder, P.J. Barrett, L.W. Henslee and D. Arpi, Recovery of Metal Oxides from Fly Ash, EPRI Report CS-3544, 3 Volumes, (prepared by Kaiser Engineers California for the Electric Power Research Institute, June, 1984).
4. R.F. Wilder and D.L. Markman, Recovery of Metal Oxides from Fly Ash, Including Ash Beneficiation Product: Product Market Survey Report, EPRI Report RP 2422-8, May, 1985.
5. R.T. Hemmings and E.E. Berry, Evaluation of Plastic Filler Applications for Leached Fly Ash, EPRI Report CS-4765, Sept., 1986.
6. E.E. Berry, R.T. Hemmings and D.M. Golden, in Proc. 42nd Ann. Conf. SPI Composites Inst. (Soc. Plastics Ind., New York, N.Y., 1987) paper 25E.
7. G.L. Fisher, B.A. Prentice, D. Silberman, J.M. Ondov, A.H. Biermann, R.C. Ragaini and A. R. McFarland, Env. Sci. Technol. 12, 447-451 (1978).
8. R.L. Davison, D.F.S. Natusch, J.R. Wallace and C.A. Evans, Env. Sci. Tech. 8, 1107-1113 (1974).
9. R.D. Smith, J.A. Campbell and K.K. Nielson, Env. Sci. Tech. 13, 553-558 (1979).
10. J.M. Ondov, R.C. Ragaini, and A.H. Biermann, Env. Sci. Tech. 13, 946-953 (1979).
11. R.T. Hemmings and E.E. Berry, in Fly Ash and Coal Conversion By-products: Characterization, Utilization and Disposal II, edited by G.J. McCarthy, F.P. Glasser and D.M. Roy, Mat. Res. Soc. Symp. Proc. Vol. 65 (Materials Research Society, Pittsburgh, 1986) 91-104.
12. J.D. Watt and D.J. Thorne, J. Appl. Chem. 15, 585-594 (1965).

CARBOCHLORINATION OF FLY ASH IN A FUSED SALT SLURRY REACTOR

M.S. DOBBINS* and G. BURNET**
*Chemical Engineering, Corning Glass Works, Corning, NY 14831.
**Ames Laboratory, U.S.D.O.E. and Department of Chemical
 Engineering, Iowa State University, Ames, IA 50011.

Received 17 October, 1986; refereed

ABSTRACT

Carbochlorination of the metal oxides in fly ash by suspending the solid reactants in a $NaCl-AlCl_3$ melt at 530-850°C and then sparging chlorine into the melt has been investigated. A mechanically agitated, semi-batch reactor was used to test the effects of temperature, oxide and carbon loading, salt composition and gas flow on the reaction rate. The process was modeled using the carbochlorination of pure alumina, the rate of which was found to be chemical reaction controlled at temperatures below about 650°C and gas-liquid mass transfer controlled at higher temperature. The carbochlorination rate of the mixed oxides in coal fly ash was also mass transfer controlled at higher temperatures when aluminum recoveries were less than about 50%. At higher aluminum recoveries, the overall rate was limited by the rate of ash dissolution into the melt.

INTRODUCTION

Carbochlorination has been demonstrated as a method for recovering metals from coal combustion wastes and other polymetallic materials [1]. In this method, Cl_2 reacts with metal oxides in the presence of C to produce volatile metal chlorides which are recovered and separated. Prior work on the recovery of metals from coal fly ash by carbochlorination has been limited to gas-solid reaction systems [2,3].

Fly ash is a mixture of metal oxides, the composition of which depends on the coal source and method of combustion. The ash is being generated at an annual rate which exceeds 52 MM Mg/yr [4]. Recovery of 80% of the aluminum, titanium and iron present in this amount of fly ash could provide 80% of the United States demand for primary aluminum, and 47% and 5% of the demand for titanium and iron respectively. At the same time, utilization would substantially reduce the problems of disposal.

The overall goal of the chlorination research has been a process for recovery of aluminum. Titanium and iron are secondary products that are obtained during the separation and purification of the $AlCl_3$. Germanium, gallium, uranium, molybdenum and other trace elements might also be recovered if a particular ash were rich enough in them. Previous work with Al_2O_3 indicates that slurrying the ash and the carbon reductant in molten $NaCl-AlCl_3$ could result in a high level of reactivity and good yields [5].

REACTIONS

As normally carried out, carbochlorination is a reaction between two solids and a gas when solid carbon, the preferred reductant, is used. Such reactions tend to be rate limited by interaction between the two solids. The reactions can be changed if the reactants are slurried in a molten salt medium that also acts as a solvent. Dissolved oxide species react with the molten salt or with the carbon and chlorine to produce volatile metal chlorides.

When molten $NaCl-AlCl_3$ is used, the $AlCl_3$ can convert the oxides other than Al_2O_3 commonly found in fly ash (e.g., Na, K, Ca, Mg, Fe, Si and Ti) to their respective chlorides at the expense of forming alumina. For example,

$AlCl_3$ will chlorinate Fe_2O_3 directly:

$$Fe_2O_3 + 2AlCl_3 \longrightarrow 2FeCl_3 + Al_2O_3 \qquad (1)$$

A second and simultaneous reaction mechanism involves the stepwise carbochlorination of Al_2O_3 [6]:

$$Al_2O_3 + AlCl_3 \longrightarrow 3AlOCl \qquad (2)$$

$$AlOCl + Cl_2 + C \longrightarrow AlCl_3 + CO \qquad (3)$$

$$AlOCl + Cl_2 + CO \longrightarrow AlCl_3 + CO_2 \qquad (4)$$

Reactions 2-4 were investigated separately by carbochlorinating pure gamma-alumina in molten $NaCl-AlCl_3$. The carbochlorination portion of the reaction process could then be studied without the complicating influence of the melt-oxide reactions.

EXPERIMENTAL

The reactor, support equipment and procedures used in this study have been described in detail elsewhere [7,8]. The reactor was a mechanically stirred slurry reactor of quartz construction, capable of handling up to 85g of oxide in 375g of $NaCl-AlCl_3$ melt at temperatures as high as 800°C. The reactor was operated in a semibatch mode with Cl_2 bubbled continuously into the shear zone of the stirrer, which was placed off-center to eliminate vortexing in the unbaffled reactor vessel.

The oxide reaction rates were monitored by measuring the consumption of Cl_2, the production of CO and CO_2, and the composition of slurry samples taken during reaction. Gas and slurry samples were taken at 5 and 15 minute intervals respectively. The duration of the experiments ranged from 1 to 3 hours.

CARBOCHLORINATION OF ALUMINA

The effects of temperature, melt composition, melt:alumina ratio, melt:carbon ratio and Cl_2 flowrate on the carbochlorination of Al_2O_3 were determined from experiments selected using a Plankett-Burman statistical design. The range for each of the experimental parameters used is shown in Table I.

Of the five parameters tested, only temperature and carbon loading had an effect on reaction rate in the range of variables investigated. In the case of salt:alumina ratio, up to 95 percent of the alumina was chlorinated with no significant reduction in reaction rate as the ratio was increased. The lack of rate dependence for such high alumina conversions confirmed that a dissolved alumina specie was being chlorinated and that the dissolution rate of alumina was faster than the chlorination rate.

Melt composition, which influences the solubility of both Al_2O_3 and Cl_2, also had no observable effect on the rate of Al_2O_3 chlorination. Alumina solubility is a known function of the $NaCl:AlCl_3$ ratio [6] but this did not influence the rate. Neither was there any effect due to possible changes in the solubility and transport of the Cl_2. Chlorine flowrate as a variable also had a negligible influence.

Table I

Variable levels for the Plankett-Burman designed experiments

Variable	High Level	Low Level
Temperature, °C	800	600
NaCl:AlCl₃ molar ratio	48:52	40:60
Salt:Alumina wt. ratio	7.5:1	15:1
Salt:Carbon wt. ratio	21:1	42:1
Chlorine flowrate, cm³ min⁻¹	255	90

Carbon loading, on the other hand, did affect the chlorination rate even though found to be statistically insignificant. At 600°C the reaction rate was proportional to the carbon loading while at 800°C the rate increased with carbon loading, although not proportionately.

Temperature was the only process variable that showed statistical significance in the Plankett-Burman study and therefore, was investigated in greater detail [7,8]. A significant temperature dependence was found to exist between 530 and 600°C, while from 650 to 800°C the reaction rate was much less sensitive to temperature. In the low temperature region, the activation energy was 29.7 Kcal/mole Al_2O_3, which is consistent with chemical reaction (kinetic) control, while the activation energy for the 650 to 800°C region was 7.0 Kcal/mole Al_2O_3 which is indicative of mass transfer control.

Based on the overall investigation using Al_2O_3, it was concluded that the carbochlorination reaction occurs at the carbon surface, is reaction rate controlled at temperatures below 650°C and is gas-liquid mass transfer controlled at high temperatures. Further evidence for the latter was developed from an experiment on the effect of stirring speed versus reaction rate at 750°C, which showed that the rate was proportional to changes in gas hold-up in the slurry [7].

CARBOCHLORINATION OF FLY ASH

In the case of fly ash, the carbochlorination reaction was expected to be affected by the melt composition. Work at 600°C with fly ash-melt mixtures without carbon or Cl_2 present showed that increasing the $AlCl_3$ content of the melt caused an increase in the dissolution of the ash. To see what effect this had on the carbochlorination reaction, two experiments using melt compositions of 40% $AlCl_3$ - 60% NaCl and 48% $AlCl_3$ - 52% NaCl were performed. These experiments were designated as 40A and 48A, respectively. The other experimental conditions were selected as follows to be most favorable to the carbochlorination reaction:

Temperature	750°C
Melt:Carbon ratio	21:1 (wt.)
Melt:Fly Ash ratio	42:1 (wt.)
Cl_2 flowrate	245 cm³ min⁻¹

Table II

Partial chemical composition of the −325 mesh
Halomet nonmagnetic fly ash fraction

Component	Weight Percent	Component	Weight Percent
SiO_2	47.66	CaO	3.66
Al_2O_3	32.26	TiO_2	1.09
Fe_2O_3	10.41	Na_2O^a	1.09
MgO	0.92	K_2O^a	1.16
		L.O.I.	1.87

[a]Unsieved ash composition

The 40A experiment was run for a total of 62.5 minutes and the 48A experiment for 180 minutes. The ash used was derived from a Pennsylvania #6 coal, and had been commercially processed by the Halomet Corporation to remove the magnetic fraction of the ash for use as a substitute magnetite. Use of the nonmagnetic fraction is consistent with the overall carbochlorination ash utilization plan.

The residual material from the Halomet process was used in this work after being sieved to remove a small +325 mesh fraction that otherwise reduced accuracy when slurry samples were taken from the reactor. A partial chemical analysis of the nonmagnetic ash fraction is given in Table II. The elements present are reported as oxides.

Silicon, aluminum and iron oxides make up over 90% of the ash and determine its overall chlorination behavior. Figure 1 shows the conversion of these three components as a function of time for both the 40A and 48A experiments.

Iron oxide conversion was virtually identical for the two experiments which would be expected since Fe_2O_3 reacts readily with $AlCl_3$ by reaction 1. This similarity indicates that differences observed in the SiO_2 and Al_2O_3 conversion rates between the two experiments are real and not artifacts caused by experimental variations such as in slurry sampling.

In contrast to the Fe_2O_3, the SiO_2 and Al_2O_3 conversions are significantly different. In 40A there was a rapid increase in the conversion of both oxides between 7.5 and 22.5 minutes, while the rates were virtually constant throughout 48A. After 30 minutes, the SiO_2 and Al_2O_3 reaction rates in 48A surpassed those of 40A becoming 3.9 and 1.3 times greater respectively. As a result, the total ash conversion rate for 48A was much faster but was much less selective between SiO_2 and Al_2O_3.

The oxide conversions shown in Figure 1 were supported by the CO_2 production rates for each experiment as shown in Figure 2. The 40A experiment showed a greater CO_2 rate at the start of the experiment than did 48A and a lower rate at the end, which agrees with the trends observed in the oxide conversion rates. The total amount of CO_2 produced by the end of each chlorination was within ±3% of the amount predicted from the oxide conversion data. Carbon monoxide production is not included since in all runs it accounted for less than 2% of the carbon oxides formed.

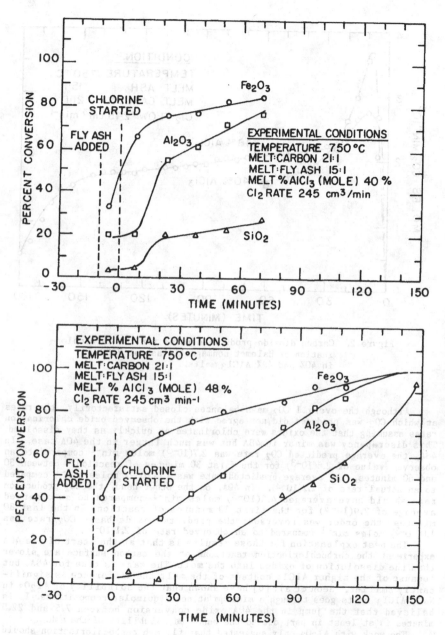

Figure 1. Conversion of iron, aluminum and silicon oxides for
the carbochlorination of Halomet nonmagnetic ash at
750°C in a 40% AlCl₃ melt (upper) and a 48% AlCl₃
melt (lower).

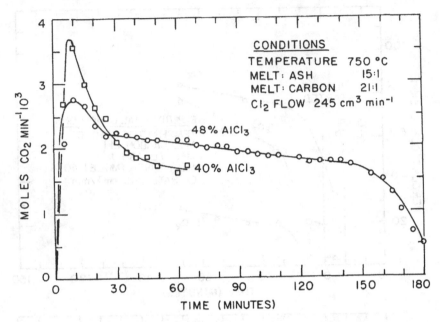

Figure 2. Carbon dioxide production rates for the carbochlorination of Halomet nonmagnetic ash at 750°C
in 40% and 48% AlCl₃ melts.

Although the overall CO_2 mass balances closed satisfactorily, the rates
at which CO_2 was produced did not agree with the observed oxide chlorination
rates assuming that the oxides were chlorinated as quickly as they dissolved.
This discrepancy was minor in 48A but was much larger in the 40A case. In
48A, the average predicted CO_2 rate was $2.7(10^{-3})$ moles min^{-1} compared to an
observed value of $2.6(10^{-3})$ for the first 30 minutes of reaction. Between 30
and 60 minutes, the average predicted rate was $1.9(10^{-3})$ moles min^{-1} compared
to an actual rate of $2.2(10^{-3})$. In 40A, on the other hand, the CO_2 production
rate should have averaged $4.2(10^{-3})$ moles min^{-1} compared to an observed
average of $2.9(10^{-3})$ for the first 30 minutes of reaction. In the last 30
minutes, the order was reversed; the predicted oxide-based CO_2 rate was
$1.0(10^{-3})$ moles min^{-1} compared to an observed rate of $2.1(10^{-3})$.
The best explanation for these results is that at the start of the 40A
experiment the carbochlorination reactions at the carbon surface are slower
than the dissolution of oxides into the melt. The same is true for 48A, but
because of the higher AlCl₃ content of the melt the difference is significantly smaller. Seon et al. [6] have shown that the solubility of Al₂O₃ in
NaCl–AlCl₃ melts goes through a minimum at near equimolar compositions. It is
believed that the jump in the 40A oxide conversion between 7.5 and 22.5
minutes is at least in part due to the enhanced solubility of the ash.
The work with Al₂O₃ only suggested that fly ash carbochlorination should
be gas-liquid mass transfer controlled and that the CO_2 production rate should
be about $2.5(10^{-3})$ g-moles min^{-1}. Both 40A and 48A, however, exceeded this
value for the first 30 minutes of reaction, 40A by over 50%. This apparent
violation of the chlorine mass-transfer limitation is attributed to FeCl₃ in

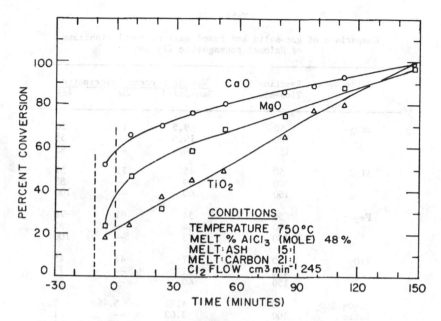

Figure 3. Conversion of calcium, magnesium and titanium oxides
for the carbochlorination of Halomet nonmagnetic ash
at 750°C in a 48% AlCl₃ melt.

the melt which enhances the absorption and transport of chlorine by the
following reaction [9]:

$$FeCl_3 \longrightarrow 0.5Cl_2 + FeCl_2 \tag{5}$$

In addition, fly ash carbochlorination appears to be rate limited by
dissolution of the ash. The 40A carbochlorination was obviously limited by
the ash dissolution rate after 30 minutes of reaction since the oxide conversion rate was approximately half that predicted from the CO_2 content of the
product gas. In 48A, the CO_2 production rates matched the oxide conversion
data so the carbochlorination and ash dissolution rates were essentially the
same. The decline in CO_2 production as the experiment progressed, in contrast
to a constant rate of production as was the case for pure Al_2O_3 carbochlorination, is additional evidence that the reaction is limited by dissolution of
the ash.

The conversions of calcium, magnesium, and titanium oxides in the ash
were also measured. These three oxides made up a total of 5.7% of the Halomet
nonmagnetic ash, which is typical of bituminous coal ashes [3].

Conversion rates from 48A for the oxides are shown in Figure 3. The 40A
conversions were similar. The high initial CaO conversion was observed in
both experiments and is attributed to the removal of calcium anhydride from
the ash.

Both MgO and TiO_2 tend to be concentrated in the glass phase of the ash
with MgO substituting slightly into the crystalline phases [10]. The low
conversions of MgO and TiO_2 relative to the other oxides indicate that the

Table III

Comparison of gas-solid and fused salt carbochlorinations
of Halomet nonmagnetic fly ash

Oxide	Reaction Time-min	Weight Percent Chlorinated		
		Gas-Solid	40A	48A
SiO_2	50	9.5	13	20
	100	19.5	--	51
	150	20	--	95
Al_2O_3	50	45	71	51
	100	59	--	83
	150	60	--	100
Fe_2O_3	50	54	81	78
	100	64	--	91
	150	73	--	100
TiO_2	50	27	47	45
	100	31	--	77
	150	40	--	98
$Al_2O_3:SiO_2$ Ratio	50	4.74	5.46	2.55
	100	3.03	--	1.63
	150	3.00	--	1.05

glass phase of the ash is not as readily attacked by $NaCl-AlCl_3$ melts as are the crystalline phases.

COMPARISON WITH GAS-SOLID CARBOCHLORINATION OF ASH

To further evaluate the results of this investigation, a comparison was made between fused salt and gas-solid carbochlorination of the same fly ash. Table III contains a summary of conversion data for both carbochlorinations. The gas-solid runs were carried out at 850°C using a 10 gram loose powder sample, a chlorine flow rate of 44 cm^3 min^{-1} and an ash-carbon weight ratio of 10:3.

The gas-solid temperature was 100°C higher than the fused salt temperature which should have improved the gas-solid conversion ratios. The gas-solid chlorine:ash ratio was only $1.8(10^{-4})$ g-mole $min^{-1}g^{-1}$ compared to 4.17 g-mole $min^{-1}g^{-1}$ for the fused salt work but this difference was probably not important since chlorine was not a limiting reagent in any of the runs made.

Table III shows that the extent of reaction is significantly greater for fused salt carbochlorination for the same duration of reaction. The 40A conditions are the best for selectively removing alumina. The 48A run becomes more attractive if the primary process goal is maximizing the removal of titanium and alumina.

The 100% recovery of Al_2O_3 accomplished in 48A in 150 minutes has not been reported previously for any carbochlorination method. Essentially all of the ash was consumed. From a practical standpoint, such complete conversion may be of little value since about 50% of the chlorine used would be consumed in the production of $SiCl_4$, a product of doubtful market potential in the volume that would be produced. In general, the carbochlorination becomes more attractive the higher the $Al_2O_3:SiO_2$ conversion ratios obtained.

The choice of reaction system and operating conditions requires an economic analysis that is beyond the scope of this paper. There are fundamental differences in the two reaction systems which would influence selection of conversions and of product recovery and separation methods.

One possible process scheme yet to be fully researched but incorporating the fused salt reactor concept is shown in Figure 4. Fly ash is first passed through a magnetic separator to remove a marketable iron-rich fraction [3]. The nonmagnetic portion of the ash then becomes the feed to a carbochlorination reactor along with powdered coke and make-up NaCl. A bubble column reactor would show promise because of its mechanical simplicity.

The metal chlorides exiting in the vapor phase would be $SiCl_4$, $TiCl_4$, $AlCl_3$, $FeCl_3$ and $NaAlCl_4$. The last three would be condensed and separated from the remaining gas mixture which would enter an absorber and distillation train for recovery and purification of the $SiCl_4$ and $TiCl_4$. The physical properties of the latter two chlorides indicates that the distillation/absorption scheme shown in Figure 4 should work. Chlorine and $SiCl_4$ as overheads from the distillation towers would be sent to a recovery/recycle operation while the $TiCl_4$ in the final bottoms would be a product.

The condensed $AlCl_3$, $FeCl_3$ and $NaAlCl_4$ would either be sent to the product purification system, or refluxed back into the reactor. The reflux ratio would depend on the amount of $AlCl_3$ removed with the ash residue, which would be a function of the melt and slurry compositions.

Yet another portion of the process would handle removal of unreacted solids from the reactor assuming only partial conversion of the ash. The solids would be removed from the reactor as a melt slurry. The solid:liquid ratio would be controlled so that only as much $AlCl_3$ would be removed as was produced by carbochlorination. After exiting the reactor, the slurry would be cooled and solidified into prills in a shot tower. This would be followed by dissolution of the prills in a dilute acid solution and separation of the ash residue by filtration.

The filtrate containing the metal chlorides would go to a solvent extraction section where $FeCl_3$ would be removed using tertiary amines in an organic solvent as demonstrated by Sheng, et al. [11]. No process details are shown for the final purification of $AlCl_3$. In the Sheng work, $AlCl_3$ was recovered by sparge crystallization using HCl. The NaCl that would be in the slurry would be insoluble in the presence of HCl so the one-step crystallization process would produce a mixed $NaCl-AlCl_3$ $(6H_2O)$ product and thus would be unacceptable.

CONCLUSIONS

The carbochlorination of gamma-alumina is independent of slurry loading, melt composition and Cl_2 flow rate in the range of conditions investigated. The carbochlorination reaction takes place at the carbon surface and is reaction rate controlled at temperatures below 650°C, and gas-liquid mass transfer controlled at higher temperatures.

The carbochlorination of fly ash is strongly affected by melt composition. The carbochlorination of the SiO_2 is greater in a 48% $AlCl_3$ melt than in a 40% $AlCl_3$ melt under identical conditions even though the overall solubility of the fly ash oxides is greater for 40% $AlCl_3$. Carbochlorination in a melt containing 40% $AlCl_3$ becomes dissolution rate limited much more quickly than in a 48% $AlCl_3$ melt. The carbochlorination of Fe_2O_3 and of the minor oxides present is not appreciably affected by melt composition. Titania and MgO conversions were low indicating that the glassy fraction of the ash where these oxides concentrated was not strongly attacked.

A comparison of fused salt and gas-solid carbochlorination of a Halomet nonmagnetic ash fraction under similar reaction conditions shows that a larger fraction of the ash is converted to chlorides using the fused salt system.

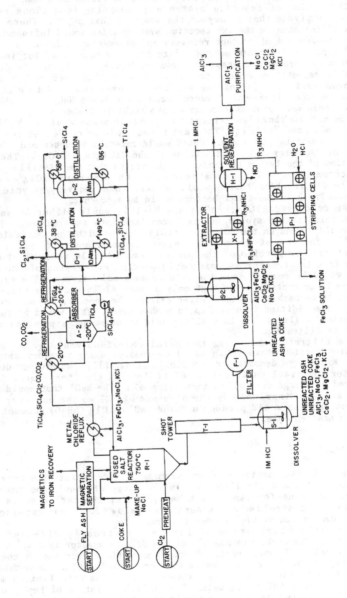

Figure 4. Process schematic for a HiChlor process using a fused salt slurry reactor.

ACKNOWLEDGEMENT

This research was supported by the U.S. Department of Energy, contract No. W-7405-Eng-82 and by the Iowa State Mining and Mineral Resources and Research Institute under U.S. Bureau of Mines Allotment Grants G1144119 and G1154119.

REFERENCES

1. G. Bombara and R. Tanzi, J. Metals, 36, 74 (1984).
2. D.J. Adelman, unpublished Ph.D. dissertation, Iowa State Univ., Ames IA, available from University Microfilms, (1985).
3. G. Burnet, M.J. Murtha and J.W. Dunker, Recovery of Metals from Coal Ash--An Annotated Bibliography (Rev), USDOE IS-4833 (rev), UC-90, Ames Laboratory, Ames, IA (1984).
4. Amer. Coal Ash Assoc., Washington, DC, Ash Production - Ash Utilization, (1984).
5. J. Hills and W. Durrwachter, Angew. Chem. 72, 850 (1960).
6. F.G. Seon, G. Picard and B. Tremillion, Electrochimica Acta. 28, 209 (1983).
7. M.S. Dobbins, unpublished Ph.D. dissertation, Iowa State Univ., Ames, IA, available from University Microfilms, (1986).
8. M.S. Dobbins, Kinetic Studies of the Carbochlorination of Dispersed Oxides in a Molten Salt Reactor, unpublished paper presented at the AIChE Nat'l. Mtg., Miami, FL (1984).
9. A.B. Bezukladnikov, E. Ya Tarat and D.P. Baibakov, Z. Prikl. Khim. 47, 1722 (1974).
10. L.D. Hulett, A.J. Weinburger, K.J. Northcutt and J. Ferguson, Science 210 , 1356 (1981).
11. Z. Sheng, M.J. Murtha and G. Burnet, Sep. Sci. and Tech., 18, 1647 (1983).

Glossary of Terms

Cement Chemist's Shorthand Notation

AFm	monosulfate phase ($3CaO \cdot Al_2O_3 \cdot CaSO_4 \cdot 13H_2O$), a hydration product in cement mixtures.
AFt	ettringite phase ($3CaO \cdot Al_2O_3 \cdot 3CaSO_4 \cdot 32H_2O$), a hydration product in cement mixtures.
C_3A	tricalcium aluminate ($3CaO \cdot Al_2O_3$).
CH	calcium hydroxide (portlandite) ($Ca(OH)_2$).
C_3S (alite)	tricalcium silicate ($3CaO \cdot SiO_2$).
C_2S (belite)	dicalcium silicate ($2CaO \cdot SiO_2$).
C–S–H	amorphous or semi-crystalline calcium silicate hydrates (may be of varying stoichiometries)

Common Abbreviations

AFBC (FBC)	Atmospheric pressure Fluidized Bed Combustion
EP	Extraction Procedure
FA	Fly Ash
FGD	Flue Gas Desulfurization
OPC (PC)	(Ordinary) Portland Cement
PFA	Pulverized Fuel Ash
LOI	Loss On Ignition
RH	Relative Humidity
SRM	Standard Reference Material
w:c, w:s	water to cement or water to solid ratio (also w/c, w/s and L/S)

Abbreviations for Analytical Methods

AAS (AA)	Atomic Absorption Spectroscopy
DCP	Direct Current Plasma Spectrometry
DSC	Differential Scanning Calorimetry
DTG	Derivative Thermogravimetric Analysis
EDX (EDXA, EDS)	Energy Dispersive X-ray Analysis
EMPA	Electron Microprobe Analysis
ESCA	Electron Spectroscopy for Chemical Analysis
FTIR	Fourier Transform Infrared Spectroscopy
ICP (ICAP)	Inductively Coupled (Argon) Plasma Spectrometry
SEM	Scanning Electron Microscopy
TGA	Thermogravimetric Analysis
XPS	X-ray Photoelectron Spectroscopy
XRD	X-ray Diffraction

Organizations and Institutions

AASHTO	Am. Assoc. of State Highway and Transportation Officials
ASTM	American Society for Testing and Materials
EPA	Environmental Protection Agency
EPRI	Electric Power Research Institute
GRI	Gas Research Institute
MRL	Materials Research Laboratory

NBS   National Bureau of Standards
ORF   Ontario Research Foundation
ORNL   Oak Ridge National Laboratories
PSU   Pennsylvania State University
UND   University of North Dakota

Selected Mineral Names and Formulae

anhydrite	$CaSO_4$
bassanite	$CaSO_4 \cdot 0.5H_2O$
calcite	$CaCO_3$
ettringite	$Ca_6Al_2(SO_4)_3(OH)_2 \cdot 25H_2O$
ferrite spinel	$(Mg,Fe)(Fe,Al)_2O_4$
gaylussite	$Na_2Ca(CO_3)_2 \cdot 5H_2O$
gypsum	$CaSO_4 \cdot 2H_2O$
hematite	Fe_2O_3
hydrotalcite	$[Mg_{0.75}Al_{0.25}(OH)_2](CO_3)_{0.125} \cdot 0.5H_2O$ or $Mg_6Al_2CO_3(OH)_{16} \cdot 4H_2O$
lime	CaO
magnetite	Fe_3O_4
melilite	$Ca_2(Mg,Al)(Si,Al)_2O_7$
merwinite	$Ca_3Mg(SiO_4)_2$
mullite	$Al_6Si_2O_{13}$
periclase	MgO
portlandite	$Ca(OH)_2$
quartz	SiO_2
spinel	$MgAl_2O_4$ (see also ferrite spinel)
stratlingite	$Ca_2Al_2SiO_7 \cdot 8H_2O$

Author Index

Subject Index

Printed in the United States
By Bookmasters

Printed in the United States
By Bookmasters